Nervous Acts

Nervous Acts

Essays on Literature, Culture and Sensibility

George S. Rousseau

First published 2004 by
PALGRAVE MACMILLAN
Houndmills, Basingstoke, Hampshire RG21 6XS and
175 Fifth Avenue, New York, N.Y. 10010
Companies and representatives throughout the world

PALGRAVE MACMILLAN is the global academic imprint of the Palgrave Macmillan division of St. Martin's Press, LLC and of Palgrave Macmillan Ltd. Macmillan® is a registered trademark in the United States, United Kingdom and other countries. Palgrave is a registered trademark in the European Union and other countries.

ISBN 1–4039–3453–3 hardback
ISBN 1–4039–3454–1 paperback

This book is printed on paper suitable for recycling and made from fully managed and sustained forest sources.

A catalogue record for this book is available from the British Library.

Library of Congress Cataloging-in-Publication Data
Rousseau, G. S. (George Sebastian)
    Nervous acts : essays on literature, culture, and sensibility / by George Rousseau.
        p.   cm.
    Essays previously published in various sources, 1965–2000.
    Includes bibliographical references and index.
    ISBN 1–4039–3453–3 — ISBN 1–4039–3454–1 (pbk.)
    1. Literature and medicine.   2. Nervous system.   3. Literature, Modern— History and criticism.   4. Enlightenment—Europe.   I. Title.

PN56.M38R68 2004
809'.933561—dc22

                                                                2004051264

10   9   8   7   6   5   4   3   2   1
13 12 11 10 09 08 07 06 05 04

Printed and bound in Great Britain by
Antony Rowe Ltd, Chippenham and Eastbourne

To the memories of my first two teachers
Henry Steele Commager (1902–1998)
and
Marjorie Hope Nicolson (1894–1979)
who awakened me

Contents

Acknowledgements

The essays that follow were published between 1969 and 1993, each having been conceived for a special occasion and, once born, having taken on a life of its own, even an afterlife: written when I was asked to present something at a conference, contribute a chapter to a book, or – as in the case of chapter 6 – when, having immersed myself in the writings of Foucault on the history of sexuality before he was known in America, I was determined to learn how a discourse of nymphomania had actually been born. It never occurred to me when writing each essay, and engaging in the necessary research, that they would ever be reprinted; or, if reprinted, that some day I would select them from a large number of other essays for the constant thread that runs through all: the huge, cumulative effect of anatomy and physiology in the Enlightenment 'Republic of Letters' and, more specifically, to tease out those influences in the rise of the pan-European Sensibility movement.

The latter – Sensibility – constitutes a significant chapter in Western civilization that still leaves many questions unanswered despite a massive amount of writing in recent decades. Sensibility has been well scrutinized, yet a lingering sense continues that our knowledge remains incomplete. One gap is the broad scientific underbelly of the Enlightenment and it assumes, in my making a case for it, that cultures, like grand mosaics, are organically whole designs containing the entirety of their tiles, not merely a few tiles scattered here and there. Each of my essays deals with early modern neuroanatomy – my old trope of 'nerves, spirits and fibers' – by starting 'inside' the discourse itself, and then opening up the implied model and its discursive forms to larger problems associated with its conception and incorporation into the social fabric in which it originally fit. And all the essays treat of neuroanatomy primarily as textual embodiments, some in didactic treatises, others in fictions, like those of Tobias Smollett and Laurence Sterne, seemingly so far removed from the concerns of medical writers that one wonders if they can have shared anything. Yet it was this 'commonality' that I originally sought to explore and whose limits I continued to test when writing each. Now that they are practically bound in one spine, spanning the work of more than thirty years, I can see how I was groping – intuitively and often unknowingly – for a grand hypothesis within the cupola of literature and science that I predictably never found.

All the essays appear here without alteration as they originally appeared. The introduction (chapter 1) is new: written last, in 2003, the reader may wish to read it last. It aims to explore two contradictory impulses that continue to exist in unabated tension 35 years after the publication of the earliest of the essays and seemingly incapable of resolution: first, the need to describe, and then to reflect upon, aspects of the intellectual journey this work has entailed for the way it shed lights on the triumphs and pitfalls of interdisciplinary scholarship in our time; secondly, the provision of an adequate intellectual context for the rise of European Sensibility and Romanticism that includes the broad arts of medicine and science, especially anatomy and physiology. The short epilogue, written just before this book went into print, distills aspects of the first – the intellectual journey. It glances backward to the time of origins and remembers what it was like to problematize 'histories of the body' before that became a fashionable pursuit, and closes by raising a few of the many current moral and ethical challenges to neuro-anatomy. The epilogue also dares to gaze, again briefly, into a future twenty-first-century environment in which the quantum of our daily stress seems, at least for many of us, to have reached breaking limits. The 'nerves' – literal, metaphoric, symbolic, as cultural systems of meaning – are implicated in all these endeavors – hence my tremulous 'nervous acts.'

I must express my gratitude to the journals and collections in which these essays first appeared: chapter 2 in *Eighteenth-Century Studies*; chapter 3 in *Tobias Smollett: Bicentennial Essays Presented to Lewis M. Knapp* edited by G. S. Rousseau and P. G. Boucé (New York: Oxford University Press, 1971); chapter 4 in Harold Pagliaro (ed.), *Racism in the Eighteenth Century* (Cleveland: The Case Western University Press, 1973); chapter 5 in R. F. Brissenden (ed.), *Studies in the Eighteenth Century: Proceedings of the Fifth David Nicol Smith Seminar at the Australian National University* (Canberra: The Australian National University Press, 1975); chapter 6 in P. G. Boucé (ed.), *Sexuality in Eighteenth-Century Britain* (Manchester: Manchester University Press, 1982); chapter 7 in Joan Pittock and Andrew Wear (eds), *Studies in Cultural History* (London: Macmillan, 1989); chapter 8 in Frederick Amrine (ed.), *Literature and Science as Modes of Expression* (Dordrecht: Kluwer Academic Publishers, 1991); chapter 9 in Sander Gilman et al. (eds), *Hysteria Beyond Freud* (Los Angeles and Berkeley: University of California Press, 1993).

The original essays would never have been written were it not for the encouragement of colleagues at Princeton University while I was a graduate student there – where the first seeds were planted of the idea that a 'culture of neurology' had existed through much of Western civilization from the time of Galen forward – and later at Harvard University and the University of California at Los Angeles, especially at the UCLA Brain Research Institute where brain theorists Professors Saul Zamenehoff and Dr. John Campbell, now Professor of Anatomy, were avatars of knowledge expounding to their enraptured listener. Over the years others have also been willing to expatiate over, and listen to, segments of this neurological heritage. I learned a great deal from German Berrios, Gert Brieger, Julia Briggs, Laurence Brockliss, Laurel Brodsley, Theodore K. Browne, the late Leila Brownfield, Adam Budd, John Carey, Arthur Donovan, Bernardino Fantini, Robert G. Frank, Sander Gilman, David Haycock, Clark Lawlor, Mark Micale, John Neubauer, Michael Neve, Maximillian E. Novak, Tony Nuttall, Vivian Nutton, Stuart Peterfreund, Richard Popkin, David Raubenheimer, Andrew Rees, Betty Rizzo, William Schupbach, Sonu Shamdasani, Elaine Showalter, Geoff Sill, Ray Stephanson, Fernando Vidal, Caroline Warman and the late Roy Porter, who lived long enough to read each of the essays and even profess that they had been beneficial in his own pronouncements on 'modern nervousness'.[1]

The reception of the essays over four decades has been gratifying. But it has also been something of a trumpet call to the lingering artificial academic division of arts (humanities) and sciences (especially medicine), and to glimpses of the precarious ways in which discourses get inscribed in history and then have equally unpredictable after-lives. I still remember the thrill I derived of possibly having hit on something important when Raymond Stephanson, an authority on the eighteenth century, singled out 'Nerves, Spirits and Fibres' as the *summum bonum* of my work. Too generously he wrote in the *University of Toronto Quarterly* that 'When historians of a later age assess twentieth-century Anglo-American critics and their work, I suspect that this thesis [i.e., the paradigm of nerves-spirits-fibers as intrinsic to the rise of sensibility and the cult of sentiment in the late Enlightenment and during Romanticism] will be rated as Rousseau's most significant contribution.'[2] Only time will tell whether they remain enduringly 'intrinsic' but the struggle to create adequate and broad historical contexts that incorporate science and medicine is far from over, and its legacy in the rise of European Sensibility is not yet guaranteed. During that same year – 1993 – Simon Richter, now a distinguished Professor of German at the University of Pennsylvania, commented in *Theory and Interpretation: the Eighteenth Century* that these essays demonstrate that literature and science can no longer be construed apart from each other and that 'the literature of sentiment and physiology of sense must be viewed, in parallel, together.'[3] These, and other similar, judgments aggrandized the small part I had played in the retrieval of a labyrinth of scientific and literary discourses in history that has continued into the twenty-first century. Nevertheless, they also awakened me to the necessity of pursuing *techne* – the wise old Platonic craft – in the arts of persuasion in transdiciplinary scholarship; for even in the heyday of interdisciplinarity before the 1980s it was never easy to persuade scholar-critics to furrow further afield than in their monodisciplinary domains.

Few essayists can have been so lucky in their editors. In Basingstoke I found a remarkable team – especially Luciana O'Flaherty, Daniel Bunyard, Tim Kapp, and Nick Brock – who did everything humanly possible to ensure that this book would be published in 2004.

George Rousseau
Oxford England
Easter 2004

Notes

1. See Roy Porter (1992); Roy Porter (2001).
2. See Raymond Stephanson (1993), p. 393.
3. See Simon Richter (1993), p. 92.

Part I
Introduction

1
'Originated Neurology': Nerves, Spirits and Fibers, 1969–2004

1 Nerves and neurons in human evolution

Early in my graduate school years, when I was reading indiscriminately in Princeton University's great Firestone Library, I came upon a passage in John Evelyn's *History of Religion* that set me wondering: 'The soul, as seated more conspicuously in the brain, does, by the originated neurology, give intercourse to the animal spirits.'[1] The complex parts of this seemingly inconspicuous sentence were remarkable and required unpacking. Besides, at that time in the 1960s Evelyn's profile was hugely different from what it has become today. Although he was one of the founders of the Royal Society and an engineer of the Restoration of Charles II, he was then conceived as a purely antiquarian mentality; looking backward and curiously eclectic in his choice of recondite subjects (genealogy, forestry, gardens, numismatics, religion, town planning); a figure in the mainstream of neither the humanities nor the sciences but stranded on an isthmus somewhere in between.[2] His histories, personal life and didactic literature had not yet become intriguing in the way we construe them now, and his biography – even his recently discovered amorous life – was virtually non-existent.[3]

Nothing *biographical* about the writer struck me: the source could as well have been Pepys or another Restoration wit or philosopher. The vocabulary was odd, or at least unusual, for the 1670s, as well as the organic relation of the 'body parts' within the passage. But the writer seems to be up to the minute in his anatomy. He knows that the soul has relocated to the brain, which not everyone did in the mid-seventeenth century, not even all persons educated in universities. He confidently claims that an anatomico-physiological process – 'originated neurology' – gives rise to animal heat and motion[4] and by this he appears to have something clearly in mind. 'Originated neurology' is more inclusive than it suggests: the phrase resounds in the ear, auguring a type of 'origins' for nerves, or – at least – a chamber of neurology that controls human beings beyond their mere animal spirits. The two words, yoked in conjunction, are noteworthy: by 'originated' he implies that the brain is the source of everything occurring in the body: *all* physiological processes – an idea still rather bold in the 1670s.[5]

Evelyn's contemporary, the great English anatomist and Oxford physician Thomas Willis (1621–75), had Anglicized and applied the word 'neurology' in the 1660s, although an older Latin form as 'neurologia' – writing about the nerves – antedated his uses. Willis invoked the term neurology in his various Restoration treatises on the brain, aware that he was charting a new province of scientific knowledge.[6] This was one that included, rather than excluded, figurative language and encouraged metaphor: there being no other way, Willis thought, to understand the brain's iatrochemical 'eruptions and fermentations... flows and sublimations.' His chemical conception of the brain as a body system (not organ or part) in continual flux is anything but predictable: a 'nervous stock' so difficult to comprehend that only figurative language of the type used by poets and myth-makers prone to analogy, can begin to describe it. More locally within the confined organ itself (the brain) he changed the course of thinking and – as I argue in several chapters of this book – gradually revolutionized what would become European (and further afield American) neuroscience by expanding its horizon. In all this novelty he still remains an unsung hero in the history of science and medicine, but his new ideas were not strangers in Evelyn's close-knit Restoration milieu. That Evelyn selectively conjoins aspects of their anatomy and physiology to generate his memorable 'originated neurology' was also daring: not for any sexual connotation ('intercourse' was perfectly legitimate in Restoration parlance as a neutral word for intertraffic, entirely lacking our biological conjugation), but for anticipating the existence of a modern nervous system; as had Shakespeare before him, one where the parts depend upon each other in a continuous loop and, more firmly, on a commanding brain. The passage further startles by appearing in the middle of a putative history of religion.

Evelyn's coinage reverberated in my head over many months in 1964, especially as I became deflected by other work, as graduate students inevitably are. Eventually I saw its relevance: to the history of British literature of the classical period, to the relations of literature and science, and literature and medicine, during the Enlightenment and especially as a pan-European eighteenth-century Sensibility movement began to assert itself.[7] Could it be that Evelyn's near-contemporaries – Thomas Browne, Willis, Pepys, John Locke, Swift and Pope, Addison and Steele, others Wits and Scriblerians, and Evelyn himself – were awakening to the utility, as well as the bald fact, of man's *nervous system*? Mankind's ownership of an anatomical system seems to have struck them less than the notion that this nervous apparatus (if this metaphor captures what it was for them) embedded a gift: as words had been historically since time immemorial. The dawning augured the possibility of nuanced states of consciousness, a new emotional range promising succor in a dangerous world fraught with fear and terror.

This reception, rather than rise, of the nervous system was what I sought, especially its embrace as a great gift rather than a menace or demon to be reckoned with. It was the theology – the body cosmology – of early modern nerves I was chasing. But why, I wondered, had its reception been culturally loaded in this Evelynesque way? Many years later I found another passage written in rhapsodic tones of similar awe about the nervous system by the American neuroscientist Gerald M. Edelman, our contemporary author of many books about the Darwinian evolution of nervous systems.[8] I stumbled upon his pronouncement before I had even heard of scientist J. Z. Young's 'mnemon hypothesis' pertaining to the development of memory in relation to the nervous system.[9] Still I fumbled: much of the erratic course of my peregrinations was owing to insufficient knowledge of the history of physiology since the 1950s. I was no professional anatomist and had to work intuitively, albeit with guidance, especially for the cutting-edge profile of this swiftly changing field. Edelman's passage noted that 'whether richly structured or simple, nervous systems evolved to generate individual behaviour that is adaptive within a specific econiche in relatively short periods of time.'[10] Here then was the 'gift' historicized and Darwinized. These 'evolving' nervous systems appeared not so essentially different from Evelyn's 'originating neurology' – despite the absence, in Edelman, of a soul loosely lodged in the brain.

Those months in 1964–65 sped by without time to reflect further on the implications of Evelyn's passage. I remember how mystified I was to discover it concealed within a discourse on religion, a textual location which today would give me positively no pause. Besides, in the 1960s the structured academic disciplines within universities were fiercely territorial, fencing subjects off from other fields with the zeal of barking dogs. No wonder that a notorious 1959 'two-cultures controversy' fueled by C. P. Snow and F. R. Leavis on the 'other side of the ocean' (as my Princeton professors euphemistically referred to Great Britain) had only recently erupted.[11] The fact that science was then clearly in the ascendancy, as a consequence of the Cold War and the race to reach the moon by accelerated technology, intensified the tensions between the camps. Furthermore, anyone who was then a graduate student at Princeton, or any other ivy-league university, knew that academically the first allegiance was to your subject. Yet Evelyn's 'originated neurology' possessed no unambiguous allegiance: here was an Oxford (Balliol) educated antiquarian, a gentleman-scholar happening upon the 'gift' of the nervous system through the glass-piece of the history of religion. To which subject did Evelyn think his 'originated neurology' belonged? Perhaps he intended the nerves to be theological after all, I speculated. Finally, the post-Evelynesque edifice of knowledge after his death in 1706 seemed to me to require rearrangement.[12] The historical rise of nerves could not be settled: it was chaotic and unruly, fundamentally messy.

I reached no conclusions, nor did anyone in my literary milieu then want to hear about 'nerves in history,' yet it was ironic that I was surrounded – at Princeton and elsewhere then in vibrant American academia – by young scholars asking hard questions about context and method. The word *interdisciplinary* was then new; still sufficiently novel to be hard to pronounce and spell, but it was almost never invoked in practice (that came later, in the 1970s) and had made few of the advances it would later maken within the academy.[13] Our intellectual excitement as research students was fueled by an inner drama playing itself out over the fate of the humanities and the sciences. Which would win out? Which was the stronger? Which older and therefore the more durable and consequential for knowledge? In all these questions and allied topics the Darwinian metaphor about evolving anatomies, especially evolving nervous systems, kept haunting me. My memory is that the copula of science and the humanities constantly loomed in our daily conversations, even if we were unaware of its contexts.[14]

Although mesmerized by the sciences, especially anatomy and astronomy, I was of the party of the humanists. I had studied Classics and Comparative Literature and was writing a doctoral dissertation about doctors and medicine in eighteenth-century literature.[15] No matter how curious about practices I was entrenched in *words*: word before things, and words as the route to the most ordinary understanding of things, almost the opposite of the early Royal Society's edict *nullius in verba* ('on the word of no one'). I had briefly dipped into medicine, especially philosophical writing about the body, healing and suffering, and even contemplated defecting to medical school and becoming a brain surgeon. To compound matters, I had spent my pre-university years – if I may speak autobiographically – in conservatories of music training to become a concert pianist. The piano instilled all sorts of habits of discipline and care for the bodily self too often overlooked because not articulated by performers. More fundamentally, serious application to the tune of five or six hours of practice each day made me aware at that young age that instrumental virtuosity depended on the muscles, ligaments, tendons, arms, shoulders, neck – the whole anatomical maze of the upper torso. The opposable thumb was particularly crucial: it permitted the virtuosity of keyboard performers, to a degree many often overlooked.[16]

Throughout my teenage years I had attended dozens, perhaps hundreds, of concert recitals and believed, before the age of twenty, that concert performance was – more than anything else – the perfect balance of the whole human nervous system: especially the fingers and synapses from fleet digital fingers to the brain and vice-versa. Hence the commonplace phrase that pianist A or B 'has nerves of steel.'[17]

My years in graduate school superimposed the paradoxical idea, buttressed by the context of 'two cultures' (arts and sciences), that all anatomies and physiologies were *both* evolving and *not* evolving. That is, they proceeded forward and backward at the same time, yet sometimes stood still. It may have been the Darwinian strain of our discussions that infused the former strand, aided by a dominant Whiggish sense in the America of the 1960s that *everything* was getting better. Whichever pulled hardest the notion of evolving bodies – whether literal human bodies or bodies of knowledge in the arts and sciences – imbued us. The record of great pianists proved it: Chopin revolutionized the piano technique by building on Bach and Beethoven; Liszt perfected Chopin; the Russians improved on Liszt, Horowitz and Rubenstein – then our pianistic demigods – flawlessly playing the most difficult technical compositions – and so it proceeded down to our own late twentieth century when compositions were being written for the piano that would have been considered unplayable a century earlier. Concertos by Max Reger, Prokofiev and Shostakovich proved the point: muscles and tendons stretched as never before.

The domain of the piano, which could seem to be a microcosm of the larger world for the ways in which it enveloped physical activity, fame, fortune, and cultural history, paralleled that of other activities, especially the evolution of knowledge. All knowledge – some of us then believed – progressed by virtue of the scrupulosity with which thought was refined, revised and perfected. Even so, the sciences seemed to be advancing far more quickly than the arts. Why was this so – and what would be the consequences for nervous systems? During the 1970s, when many of the essays collected in this volume were conceived and composed, a moment of scientific optimism crescendoed before it waned – it has been receding discontinuously ever since.[18] In this welter of ideas and trends, pasts and presents, which provide a context for nerves, the durability of imagination and memory lurked.

These two categories had comprised a large part of humanity's history, yet sooner or later everyone immersed in the history of ideas recognizes that memory depends upon imagination, and imagination upon the nervous system and brain, returning us to the point of departure. The paradigm appears plain enough, if profoundly in need of unpacking. But I did not know circa 1964–1970 that I would spend a large chunk of the next two decades doing just that.

By the *interdependence* of the above functions I sought to understand these concepts – memory, imagination, nerves – literally rather than metaphorically or symbolically; in addition, I aimed to comprehend them in their necessary and sufficient conditions, that is, in terms of what was *necessary* for what. You cannot remember something without imagining it first in consciousness: you cannot 'remember' the sun or moon if you have never been able to imagine it. And you cannot imagine anything without a brain large enough to enable you to make the necessary associations for the imaginative act to occur in the first place – a brain commensurate with its apparatus of nerves, synapses and organic tissues. Observation confirms that this is true throughout the animal kingdom, from human forms throughout the mammalian world. What, I began to ask myself in the 1960s, had been the archeology of conceptions of memory and its attendant imagination? Not the record of memory itself but configurations of how it had been imagined.[19]

In those pre-Foucaldian years, when the history of science and medicine had more forest to clear than it does now, and when the discourses of the body had not yet acquired high profile and professional status, replies were not to be found. Set the dials to approximately 1965 or 1970 in the Anglo-Saxon world, and the picture was unclear: a blank slate waiting to be framed. I searched in vain for intellectual histories of imagination (with or without *the* article) and found nothing substantial before the 'instauration' of the European Romantics: especially the systems of Wordsworth and Coleridge in their various 'biographia literaria' (Wordsworth in *The Prelude*, Coleridge in the *Biographia Literaria*).[20] On what foundations had these revolutions of nervous thought built?

If imagination was intrinsic to Romanticism (in whichever of its national appearances and despite its diversity), and if the histories of imagination remained unwritten, then the lacuna – the big black hole – must lie somewhere in the chronological interstice between the late Renaissance and high Enlightenment.[21] These notions were vague, exceeding the brief of academic philosophy and spanning chronological decades, even generations: questions without answers, discourses remaining to be filled.

Archeologies of memory were even scarcer then. When located, they appeared as fragments of the history of psychology, which itself had no coherent or professional profile before the nineteenth century. Renaissance 'theatres of memory' had been studied, as had pneumonic devices and marvelous instruments as *aides mémoires*, but the period I describe here (i.e., the 1960s and 1970s) occurred before the history of science had meticulously studied cabinets of curiosity and stretched its tentacles to prodigies and wonders. Yet if both memory and imagination depended in fundamental ways on the nerves, then the histories of anatomy and physiology – I imagined – existed to pave the way to understanding Romantic innovations in these realms. And if these discourses of the body could be retrieved, the missing links to memory and imagination might be made.

Clearly the connections were then absent: I found internal, localized histories of anatomy and physiology that made no attempt whatever to relate to other domains, yet I wanted to believe they constituted a type of discursive connective tissue. Having been educated in an American graduate school where these links were viewed as legitimate rather than as the wanton signs of Icarean overreaching, I was disappointed. I should have suspected that my academic mindset based on a natural – almost ingrained – interdisciplinarity might not be universal. Or, if sanctioned in America, that such methods might be unacceptable in Britain or Europe. Nevertheless, the evolutionary substratum of my views remained: the post-Darwinian notion (already described above) that everything – including ideas and beliefs – was evolving in predictive ways that must not be confused with advancement.[22] All the more reason, I came to suspect, that even in the early modern world the secrets of imagination lay hidden in the recesses of anatomy and physiology as regulated by the brain. Briefly put, the new anatomy of the seventeenth century problematized the human being in ways adaptive to an evolving culture, but where had the leaps been made? The more I looked backward in time for origins, the murkier they seemed to be.[23]

2   Nerves and selves in the ancient world

Any medical historian will tell you that the modern view of ancient nerves differs enormously from the one the Ancients held about themselves even if their views were disjointed and contradictory.[24] This is an attitude still not appreciated by most intellectual historians, even those who study 'body parts'.[25] Furthermore, until recently nerves in history have proved to be a murderously difficult 'body part' (which it was strictly not of course) to categorize, not least for the way in which its fortunes have depended upon the developing microscope, simple and compound. Nervous anatomy had always been inherently ambiguous and approximate in its claims.[26] The origins of the word 'nerve' – in Greek meaning tendon or sinew – embedded confusion between connective tissues and other, subtler, types of physical continuity within the body. Comb the anatomical systems of the Ancients and there is no agreement about the extent of these physical connective systems. The only concord was that nervous anatomy requires expertise. As late as the twelfth century the Jewish philosopher Moses Maimonides observed: 'One who is not knowledgeable in anatomy may mistake ligaments, tendons and chords for nerves.' However inexact this 'science' then was, it did not diminish the function of the nervous system. Ancient medical practitioners understood that nerves served two proximate functions in the animal body: movement and sensation. The debate focused on how the two functions were interconnected and which principal organ or organs directed the processes. The brain was merely one candidate, not yet even delimited to the human head.[27] Moreover, localization of the nervous system did not extend far enough to explain the operation of the imagination. Cognition and higher associated thought entailed the reciprocity of mind–body: a mystery only illuminated by knowledge of the tortured history of their philosophical relationship.[28]

Even so I was unprepared to abandon the Ancients in this quest for the origins of Romantic 'instauration' and innovation. What puzzled me in large part was the Ancient search itself for the origins of nervous activity, as if the anatomical facts spoke sufficiently, and as if localization would solve the mysteries of motion, sensation and thought. Despite pursuing imagination and memory, I found myself continually bogged down in pinpointing anatomical structures among Aristotle and Galen, the Alexandrian and Roman physicians, Soranus of Ephesus; and this without taking into account the revivals and translations of these Ancient thinkers among the Renaissance anatomists, especially Vesalius (whose brain anatomy was far in advance of anything before him) and Eustachius (who described nerves accurately and gave the first modern description of the tube named after him).

I was impressed by their firm belief that they could unravel the mysteries of nervous physiology if they could peg the nerves anatomically to one – and *only* one – part of the body. This was a collective mission extending over many centuries. Aristotle judged the nerves to be controlled by, and originated in, the heart (here, perhaps, was one source of Evelyn's contention), because in his interpretation the heart was the primary organ of the body and the seat of motion and sensation.[29] But even Aristotle was misled by confusion between ligaments and nerves in drawing this conclusion. Six centuries later, Galen contradicted him, disparaging those 'who know nothing of what is to be seen in dissection.'[30] Instead Galen concluded that the brain was the most important organ of the body, with the nerves emanating from it: 'I have shown in my book *On the Teachings of Hippocrates and Plato* that the source of the nerves, of all sensation, and of voluntary motion is the encephalon [the brain] and that the source of the arteries and of the innate heat is the heart.'[31]

This Galenic leap was monumental. Medical historian Vivian Nutton had demonstrated it in diverse contexts, especially in the legacy it reaped in the anatomical schools of the Renaissance.[32] But sooner or later it was crucial to learn what became of Galen's views in 1600, 1700, even 1800. He had not linked memory to imagination in the associative sense I sought. But he had demonstrated that the spinal cord was an extension of the brain, which carried sensation to the limbs, and he posited that the nerves controlled the actions of muscles in the limbs. Hence two different types of nerves governed the two principal functions of the nervous system – sensation and motion – and these were respectively soft and hard. Galen further promoted an ingenious, if bold, anatomical feature of the nerves when imagining them as hollow tubes.[33] Quite logically he reasoned that the animal spirits – the 'pneuma psychikon' or body's principal source of vitality in his anatomical system – must circulate throughout the body.

Galen's leap to the hollow tube and – even more so – his ubiquitous animal spirits, were consequential for anatomy and physiology long after the Renaissance. It may be that no genuine advance in neuroanatomy occurred afterwards until the sixteenth century. Yet nerves were again transformed after Leonardo and Vesalius, down through Reformation and counter-reformation nerves. Galen was not the first to posit animal spirits or an equivalent doctrine, yet they featured as the centerpiece of his nervous anatomy. His methodology was based on observation and reason; but his enlargement of animal spirits was grounded in an act of faith carrying with it a tidal wave of older (i. e., pre-Galenic) anatomy.[34] No one in the ancient world had seen these 'spirits,' which required at least the simple microscope for evidence of their existence. Requests for proof as evidence were misguided. Yet the more I tried to understand the Galenic doctrine of spirits, the less I comprehended his vast anatomical-medical system.[35] The animal spirits clearly constituted the specific site where the interconnection between memory and imagination, mind and body, would be made – even fought out later on; but I had no armaments to proceed from Galenic spirits to Renaissance spirits. I needed to understand the interstice between Roman 'spirits' and, later, Catholic and Reformation 'spirits' if I were to understand the Enlightenment legacy of Romantic theories of memory and imagination. It was a tall order circa 1970, especially if you allowed the supra-anatomical domain of 'spirits' to filter in.[36]

I continued my backward gaze in the 1970s, locating medieval physicians, in agreement with Galen, who believed nerves to be offshoots of, and systems controlled by, the brain. The Islamic medical philosopher Avicenna (980–1037) wrote in the early eleventh century that 'nerves are one of the simple members: homogeneous, indivisible, elementary tissues (others include the bone, cartilage, tendons, ligaments, arteries, veins, membranes, and flesh).'[37] He offered a precise physical description of them as 'white, soft, pliant, difficult to tear.' Avicenna and his contemporaries described the complex arrangements of nerves in the body, attempting to differentiate their functions. In the *Canon of Medicine*, he noted that 'dryness in the nerves is the state which follows anger,' anticipating the humoral theory that would endure for centuries. Such views suggested that Avicenna thought nerves to be entangled with, and responsive to, the passions, yet another sign of their strong connections to the brain. But nowhere does Avicenna make any such explicit assertion.

More than a century later, 'Master Nicolaus,' as Cusanus was known, offered a more precise vocabulary to express the new complexity of the nervous system, discarding the terms 'soft' and 'hard' for the more familiar notion of sensory and motor nerves.[38] He considered nerves as subservient members of the brain that carried the animal spirits to all the organic members, endowing them with sensation and motion. And he differentiated their points of origin and termination: 'According to some authorities, all the sensory nerves originate from the cellula phantastica, the motor from the cellula memorialis.' Cusanus posited five kinds of sensory nerves classified according to the operations of the five senses (sight, hearing, smell, taste, and touch). Two nerves arise from the 'cellula phantastica' and cross in the middle of the forehead, one of them passing to the pupil of the right eye, the other to the left. Through these nerves, visual spirits are conveyed to the pupils of the eye. Here, then, long before the Renaissance anatomists revived the *corpora fabrica*,[39] was a formed theory of animal spirits embedded in a system of physiology, often forgotten by later mechanists when claiming they were the first to elevate nerves.[40] Yet it was the long sweep of nerves, rather than narrow gaze, I was chasing, particularly for the murmurs and antecedents of synaptic plasticity.

And the brain? Leonardo's famous image of the human head renders brain lobes apparent in a drawing that is clearly based on medieval theories of the body rather than actual dissection. The Renaissance anatomists devised elaborate theories of how the nerves created passions and sensations in different parts of the body through their attachment to principal organs and sexual members.[41] 'By means of nerves the pathways of the senses are distributed like the roots and fibers of a tree,' claimed the Paduan Alessandro Benedetti (1450–1512) in 1497.[42] Brain and nerves were thus sealed into a system of reciprocity long before the seventeenth century, even if the direction of influences remained murky. This concern for the pathways was an old one centering, after Galen, primarily on the effect of the brain on the body.[43] In the ana-tomical worlds of Vesalius and Valli there were many theories but no agreement. But it would be wrong to assume no new perspective to the fabric-of-the-body nerve constellation was superadded: neuroanatomy was a growing concern and continued to be among the Enlightenment anatomists, Galvani, Jacob Winslow and Thomas Soemmering.[44] Decade after decade, in almost every country but especially in Italy, neuroanat-omists refined some portion of the map of the brain and its slavish vassals. The dilemma for someone like me, outside the field, was the assignment of niche and rank, point and emphasis. Hence when I became immersed in an eighteenth-century neuroanatomist called Domenico Cotugno (1736–1822), who seems to have first discovered the aural aqueducts, as well as described cerebro-spinal fluid and elaborated its pathways,[45] I had to discern whether he had made any fundamental change to animal spirits in the historical maps I was drawing up. Questions like these posed immense hurdles.

My backward–forward gaze was not based on an imaginary purity about the validity of early views but rather on a quest for origins and anticipations. Later – in the forward pose – I would become immersed in the works of Dr Thomas Trotter (1760–1832), the intuitive bachelor physician of the Fleet who established 'nervous temperaments' as the cornerstone of his social pathology of medicine in the early nineteenth century.[46] He was struck by modern nerves as the clue to the stresses of urbanization, especially restlessness and its opposites, the indoor sedentary life: nerves' first-generation heir since Dr. George Cheyne and company immortalized them circa 1740 for the gathering empire.

Modern novelist David Malouf's fictional Ovid would have found much in common with Trotter. His Ovid, reviewing his life while in wretched exile, concludes that he has 'been brought up to believe in my own nerves, in restlessness, variety, change...'[47] The strains of motion were opposite to the 'calm life' for Trotter; evident in the sedentary scholar brought up entirely on books, and always living in a soft state of pampering the self through intellectual cuddles. Both versions – radical motion or frozen immobility – produced nervous waste. Trotter also romanticized the literal anatomic nerves he thought the ancients possessed, hence his interest to me then.[48] He viewed them as untainted by the luxury, vice – especially drunkenness – and urban sprawl over-taking England at the end of the eighteenth century. Such purity, Trotter thought, prevented the 'hereditary debility' prevalent among the Georgians.[49] This idealized primitivism informed his whole view of the ancient 'Greeks and Germans' (as he calls the Frankish peoples of middle Europe), especially their two greatest writers, Homer and Tacitus (his view). It was an attitude to ancient nervous physiology drenched in Romanticizing tendencies, yet it demonstrates his reverence for the ancients. The Greeks and Romans did not view themselves in this way, nor did such primitivism invade their own anatomical theories. My dilemma was not Romanticization, diverting me to a mythical ancient world, but insufficient understanding of ancient science itself.[50]

3   Early modern nerves

Few blocks of Western history are more problematic than the transition from paganism to Christianity, especially in their Roman and Christian incarnations. Historians have written libraries on the subject, touching on every aspect; not least in the second century AD when man could sentimentally be said – as Flaubert noted in the 1860s – to exist between different gods: the demise of the Roman panoply and the rise of the Christ-child. Yet in the domain of nerves all was as silent as the jesuitical midnight in Bethlehem. Neither monolithic advances nor declines occurred: man's nervous system just remained dormant. While Lucretius and the later Stoics spewed their atomistic cosmological theories, the nervous system declined and atrophied. Soranus, already mentioned, had generated theories of apoplexy and epilepsy, paralysis and madness, containing genuine insight, and he grasped the nervous foundations of mental theories of illness. But others in decaying Rome and the Byzantine and Islamic lands generated nothing genuinely new. It now appears odd, from the removal of almost twenty centuries, that Stoics – for example – should not have pegged their philosophies of abstinence and balance at least in part on nervous states; as strange to our neuroanatomic mindsets that primitive Christianity did not expend more thought on the nervous resurrection of Christ.[51]

This is no place to survey in detail the history of nerves from the Stoics and Gnostics through Augustine and Aquinas, let alone extend it beyond them to the Reformation. The salient point from the perspective of the synoptic view is that nerves are absent from the early history of passions.[52] If belief and faith swayed philosophical minds prior to Leonardo, it also compelled them to view their bodies as anything *but* neuroanatomical organisms. The gyrating saint or levitating hermit knew what the fate of his blood would be in the afterlife, but it seems never occurred to him that his earthly states of possession were 'nervous.'[53] The confessional Augustine who was racked by guilt over the sins of the flesh never appeals to his fallen 'nervous body' (even the offending nervous genitals) as a possible culprit: even in the state of lust his torment is curiously mental. The sixteenth-century anatomists – Leonardo, da Carpa, Eustachius, Vesalius, Fallopio and company – must do their work before mankind will be able to attribute its earthly plight to the neuroanatomical realities of its corporal body. Only in the seventeenth-century world of Bacon and Descartes will the *corpora fabrica* again begin to make a case for causal relations between anatomy and human behavior, and thereafter more or less in a steady, if unpredictable, line to the transformative period 1650–1700 so crucial for the below essays. The impact of nerves arose relatively late in Western thought.

Even in the crucial domain of the passions, the nerves are absent. And it is only much later – during the mid-eighteenth century – that they will enter the discussion.[54] One reason was the new morality equating nerves and communal sensitivity: the main ethic of the new sensibility was that it led to aesthetic refinement and civic responsibility among all educated persons and was therefore something to be cultivated. Yet the more it manifested itself among the elite in their behavior the more the lower orders aped it, prompting observers to inquire whether the degree of sensibility in any individual instance was genuine or feigned. The aesthetic route rose upwards – almost *gradus ad parnassum* – through

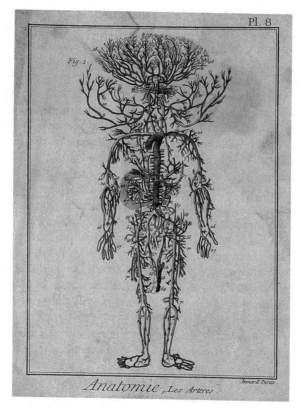

Pl. 8

Fig. 1

Anatomie, Les Arteres.

Benard Direx.

Plate 1 Nervous system after Raymond Vieussens (1641?–1715), no date. Reproduced by kind permission of the Wellcome Library, London.

stages of appreciation to the realms of sublimity. Indeed the youthful Edmund Burke wrote a famous essay about the psychological origins of sublimity that launched his philosophical career: *An Essay on the Origin of our Ideas of the Sublime and Beautiful* (1757). Here he drew on centuries of 'passion theory' to demonstrate what relation fear (in some senses the most interesting of all the passions) bears to the sense of sublimity. Fear may have been the 'most interesting of the passions' for Burke's contemporaries,[55] but even in Burke the nerves aroused in the requisite awe and terror prerequisite for 'sublime passions' are muted though quietly present. The progression is clear: medicine, built on the anatomy learned in the schools, presented nerves first; thereafter fiction imported their medical theories, as novelist Samuel Richardson did from physician George Cheyne;[56] only after early fiction had vivified nervous protagonists, such as Laurence Sterne's inimitable 'Tristram Shandy' and Scottish Henry Mackenzie's tearful 'Harley,' the crying 'Man of Feeling,' did they migrate further afield to infuse philosophy – as in Burke's philosophy of sublimity – and the civic thought of Adam Smith and Adam Ferguson. These fictive male characters (Rousseau's St Preux and Goethe's Werther are others) brim over with 'animal spirits' and are staunchly 'nervous.' They shed tears because their nervous systems are exquisitely calibrated to their innate physiologies,[57] their nervous frames anything but delicate and soft in the way female figures of the period are.[58] They could not have been hewn of such fiber if they had been members of the lower working classes: in this sense it made perfect sense to refer to the nerves of the poor, just as it did – later on – to notice how nerves fed directly into the aims of empire and colonialism. Yet in one category – early modern theologies of decadence and luxury – the nerves appeared well in advance of the eighteenth century as an outgrowth of religious thought.

Early theories of the pursuit of pleasure, as found in the credos of Epicureanism and hedonism long before the European Renaissance, largely condoned luxury.[59] One argument arose from dietary concerns: excess nourishment might make the body soft, but it also increased its ruddiness and well-being, possibly enhanced its longevity. This was the opposite of our contemporary position on obesity. When asked specifically where this vitality existed, replies often focused on the 'vital' or other 'animal' spirits. Hence from the start – Roman and Medieval times – spirits assumed an importance in excess of any visible proof of their salubriousness. Furthermore, the credos of abstinence from the neo-Pythagorean and early Christians to diverse later Calvinists and Puritans made little impression on this attitude. When the proponents of luxury resurfaced in the Renaissance, they drew even more forcefully on the benefits of rich diet and drink to bodily health.

Early modern luxury was placed on a physiological footing despite its religious, even theological, anchors. The anatomical 'facts' of the *corpora fabrica* counted less than the implication for 'spirits,' already seen to be more invasive than the early anatomists dreamed.[60] My own view of this enduring religio-biological heritage had been formed, first by the long biochemical history of vitalism (especially of Georg Ernst Stahl's variety where 'sensitive souls' were construed as more progressive than Aristotleian psyches), and then (working chronologically backwards) by modern neuroscience's homage to Newtonian proofs of cause and effect as offered by mechanists and materialists. But what about the lingering 'spirits'? Were they enchanted humoral 'spirits' of the type Jean Fernel (1497–1558) assumed in his early system of medical pathology and disease?[61] Some type of metaphysical *pneuma* or *geist*? What had been the profile, I wondered, of mystical 'spirits' in the world between Paracelsus and William Blake, van Helmont and Swedenborg; not merely what we would call the 'scientific world' (an anachronistic notion) but in their composite cultural mindsets?[62]

4   'Fire in the soul': the animal spirits

The 'animal spirits' and their complicit nerves constituted a large chunk of the early history of science. Their sheer longevity is astounding. By 1970 I had also realized what a large swath of linguistic history had been captured in the trajectory of the tantalizing phrase *animal spirits*: 'spiritus animalis,' 'spiritus vitalus,' 'spiritus vitae,' and many other forms of this psychic *pneuma* from the Middle Ages to the Enlightenment – a hulk so vast not one but several important volumes could be amassed about it.[63]

Nor is *vitalism* in the life sciences a dead concept today: as recently as 1986, for example, the trope was used to organize a British Academy Lecture on microeconomics.[64] Throughout this long record a delicate balance has always existed between natural and supernatural forces: reason versus magic, observation versus faith, intuition versus superstition. Each component of the phrase – the *animal* part, the *spiritual* part – had permeated thought by 1500. No matter what its precise anatomo-physiology the 'animal spirits' continued to imply a fundamental facet of humanity: a 'nature of Nature' well beyond the physical realms of motion and sensation. The biological 'spirits' were a 'Life Force' – an inexplicable fluid, a 'fire in the soul'[65] taken on faith – long before they were tagged to this Romantic label in the nineteenth century.[66] Eventually they became something more than a linguistic commonplace transformed into a veritable habit of mind and means of perceiving the cosmic world. Fundamentally Ovidian in their metamorphic status, they became the basis for self and identity.[67]

Embedded in the *animalistic* component were all things material, bodily, and susceptible to the physical laws of motion. The implication for living animals was considerable.[68] The *spiritual* part signified the holy residue for the soul – hence its 'fire': everything known and taken for granted but nevertheless unseen, unverified, unknown.[69] Animal spirits pertained to both flesh *and* soul and it is no accident that for five centuries, from approximately 1400 forward, the 'animal spirits' commanded the front rank in all theories explaining life. In the seventeenth century, as blood in human life assumed ever-greater significance, they became the field on which the nervous revolution in anatomy and physiology was fought. By then blood and these unseen 'vital spirits' were practically synonymous. It was only a matter of time until microscopes would reveal their secrets in coded protein forms. None of this development could have occurred outside the sweep of a long lineage, extending over many centuries, which conjoined the superlative mystery of human life: not its physical existence in eternity, but the interplay of mind and body, soul and flesh, in a dualistic riddle incapable of resolution.[70]

The linguistic dimension of 'spirits' proved immensely problematic. I had been trained at Princeton in the American 'New Criticism,' that method of close reading in which no nuance or shade of meaning is overlooked. Yet it was one thing to apply their theoretical models to a short poem or play and quite another to trace a trope as metamorphic as this one – animal spirits – over many centuries and across national and linguistic boundaries. The possibilities, nevertheless, for transculturation encouraged me to push ahead. William Empson's *Seven Types of Ambiguity*, published in 1930 and reprinted almost every year in America after the Second World War, still cast a long shadow in the 1960s. As I worked through prolific primary treatises on the animal spirits published before 1750 I sought to interpret into their tropes the energies of the *opposing* camp: biological into theological, theological into biological.[71] Others before me had made similar attempts without reaching grand hypotheses, yet I would not be deterred.[72]

Intuitively I was searching, without knowing it until much later, for a Western chronicle of mind–body relations.[73] By tracing tropes, decade in and out – the 1650s, 1660s, 1670s, etc. – I imagined myself capturing the growth of both domains, as well as their influences and reciprocities. Still, no synthesis arose: the animal spirits explained everything and nothing. I gained some clarity (even partial closure) on particular thinkers and texts, even decades and generations, but the vitalistic component – the place where the borders between the theological and biological realms meet – remained murky. Vitalism as a protean, living and breathing idea has itself not had a social or cultural history.[74] But its appearances funneled into this concrete trope – *animal spirits* – did not lend itself (the trope) to much greater harnessing. What formed were developing metaphors, few more enticing than 'fire in the soul,' a type of primal Promethean ether waiting to be versified by poets.[75]

I considered the possibility that animal spirits were too difficult to study in this way. The questions were impossible: what were the 'spirits' before they became attached to the *Holy* Spirit? What was the relation of pagan gods to these anatomical spirits? Why was early modern science unable to shed these miraculous properties of flesh and blood in the aftermath of the Renaissance anatomists? What to do with the metaphoric dimension? – a Tower of Babel was decaying there, layers of encrypted dead dialects.[76] There were also byways in magic and astrology: what effect did these have during the Reformation and religious wars? Complexity gripped me: after all, the 'spirits' had been developing since the Middle Ages.[77] Their early appearances in the eleventh century, as we have seen, were shrouded in dark causes hovering around animal heat and motion; their referents – agency, cause and effect, spiritual versus physical force, mechanisms of sensation – remaining as nebulous: principles incapable of naming or defining. Animal spirits were taken on faith no less than other components of the Trinity, the holiest of which was the 'spirit' of love itself: the *Holy* Spirit. Each was embedded in the other, if not an extension of itself then a mirror reflection. This shared space suggested that both types of spirits – animal and holy – resisted clarification and definition; were as indefinable as soul itself.[78]

The long-term visualization of animal spirits was as tantalizing as its discursive profiles and no less daunting than microcosmic cell images prior to 1800 or artists' renditions of inner anatomical transformation before the detection of MRI scanning.[79] The animal spirits had carried three cognitive meanings from the start: (1) sources of sensation; (2) seats of temperament, i.e., especially courage and masculinity; (3) sources of human inclinations, i.e., especially vivacity and gayety of disposition. Set the chronological dials to circa 1700 and all three are firmly in place, and this is before the body will be moralized in the way it was in the long eighteenth century. The physics of the first accreted to itself many hypotheses, each toppled by successive generations. Courage and masculinity (the second) were touted as the spirits' primary attribute in epochs when the two genders – soon to be three – came under cultural strain.[80] Vivacity, the innate trait of a person's most basic temperament (here the opposite of gloom and doom), ultimately proved as indefinite as the other two, assuming that animal spirits were inherited and that persons naturally vivacious had been born so. This last inherent trait proved so elusive that it continued to be metaphorized, as in John Locke, who conceptualized the spirits in his *Essay Concerning Human Understanding* (1690) as 'set a going in motion,' and later in Sterne who opens his opus magnum, *Tristram Shandy* (1760), with a description of how they go 'cluttering like hey-go-mad' as if they were so many trains on tracks. Nothing, it seemed, could be done to alter these markings, the proofs of heredity and pedigree. Even if so, how could you *visualize* animal spirits?

For long stretches I grew impervious to the links between spirits and nerves, forgetting how the brain depended on both, and how it mediated them – so all-engrossing was the *linguistic* dimension of spirits. When I emerged from this quasi-amnesia other types of 'fire in the soul' leapt out at me – medical and moral – not least the way that the nerves were appropriated by the time of William Harvey.[81] One dominant metaphor was a system in equilibrium. Throughout the sixteenth and early seventeenth centuries, ancient views of the nervous system were interpreted along lines of stasis and harmony. Vesalius, Bacon, and even Descartes in his *Treatise on Man* (written between 1629 and 1633 and published in 1664), theorized spatially about a place for animal spirits to roam throughout the body. Descartes reduced the three 'spirits' to one and sustained the belief that nerves were hollow tubes: spaces waiting to be filled up. Others posited valves and other hydraulic parts.[82] Still others viewed the 'spirits' as producing commotion within the body's 'juicy canals.'[83] From 'commotion' it was only a step to murmurs of disruption and cries of anarchy, metaphors invading these discourses by the time Charles I was executed and Cromwell named Protector.

The model of the 'nervous body' as analogous to the political state was inevitable. By the time Hobbes and Milton had died, 'the political Nerves and Arteries' were commonplaces. Notwithstanding the dense metaphoric jungle of seventeenth-century prose, the analogy would still have been prolifically drawn. Such analogies intensified during the English Revolution; afterwards, they were often drawn as encomiums of Thomas Willis: the brain reformer and subject of chapter 5 of this volume. Two generations after Willis' death in 1675, Thomas Morgan, a minor English physician, was still celebrating the master by resorting to the *two* kinds of bodies, nervous and political:

Among ourselves, Dr Willis must be allow'd as a perfect Master...He understood perfectly well the Explosions and Suffocations of the animal Spirits [sic] and the chymical Effects, Changes, and Transmutations produced in the Body...and though many have endeavour'd to imitate, yet none could ever equal him in this way...This Cant Philosophy [of animal spirits] has furnish'd abundance of Gentlemen in the profession of Physick, with Matter enough for Ostentation, and an inexhaustible Fund of Absurdity and Nonsense, which is much admired by those who cannot judge of it. But, perhaps, it might be very well if the [English] Parliament would take this Matter into Consideration, so far as to oblige all Physicians to talk English to their Patients, and not to amuse them with technical Words and Terms of Art [drawn from nervous bodies]...[84]

Here the political disruptions appear as 'eruptions and suffocations,' as they had in Willis' densely metaphorical prose. Throughout the epoch we glimpse this 'nervous body politic' best through minor secondary writings rather than in canonical prose from Milton to Johnson and Hazlitt. Hollow vessels were sitting ducks for 'streams' run amok by 'animal' and other 'pernicious' fluids. The heated fluids roamed the body, armies creating havoc wherever they went; once invaded, the nervous tubes erupted into disarray, warfare and states of siege. Helkiah Crooke's *Microcosmographia* (1618), subtitled 'a description of the body of man' and not to be confused with Robert Hooke's later *Micrographia* (1665), had also analogized the modern state to a 'nervous body politic.' Here was seventeenth-century anatomy in full political blossom, demonstrating what difference it made to the world of Hobbes and Spinoza to have nervous bodies pictured as passive hollow tubes rather than material fibers. The 'succus nervosus,' or 'nervous juice,' was another type of fluid, they thought, running in these bodily rivers. It could be pure or corrupted, but whether it was a fluid or a 'fire' was anyone's guess.[85]

5 'Spirits' and mystical nervous fluids

The discourses of the nervous body in the world of Descartes and Willis are so linguistically charged, following in the aftermath of the flowering of English in Elizabethan and Caroline glory, that there is no upper limit to their close analysis. Once you start 'reading' them as linguistic codes, as literary critics do the drama and metaphysical poetry of the era, the metaphors fly high in all directions. Discourses of the spirits and fibers were especially vulnerable as their tiny, invisible, atomic structures were unseen. All was taken on faith. Here is a typical passage in Descartes that could be replicated in dozens of smaller fry:

The Nerves are nothing else but productions of the marrowy and slimy substance of the Brain, through which the Animal spirits do rather beam than are transported. And this substance is indeed more fit for irradiation then a conspicuous or open cavity, which would have made our motions and sensations more sudden, commotive, violent and disturbed, whereas now the members receiving a gentle and successive illumination are better commanded by our will and moderated by our reason.[86]

'Marrowy, slimy, cavernous, violent, disturbed, gentle, illuminated': the nerves are everything and nothing. They especially give rise to metaphor and lexical neologism. Entire books can be written about their linguistic encodings, rhetorical and metaphoric. After Willis' landmark treatises on the brain the writings of Bernard Mandeville – remembered for his brilliant satire on the paradigm that 'private vices are public benefits' (*Fable of the Bees*, 1714–23) yet trained as a physician who wrote treatises on psychosomatic conditions wherein the nerves play a principal part[87] – and a generation later those of George Cheyne, already mentioned, were especially prone to these stylistic flights of fancy. The close proximity then between doctors and authors – many doctors, like Sir Richard Blackmore, known for his epic poems *Prince Arthur* (1695) and *Creation* (1725), were themselves authors – further mediated an easy intertraffic of anatomical nerves in both technical treatises and widely-read poems.[88] In all these genres, the nerves found themselves at home in diverse discourses. Even the Shaftesburyian *sensus communis*, a doctrine ultimately about the public good achieved through language and Nature, was born in the notion that the civic community, no less than each person, shares a 'sensual body' – a constellation of senses – which presides over its actions and policies.

Still, the anatomical debates of the seventeenth and eighteenth centuries were substantive *apart* from their labyrinthine linguistic mazes. One heated topic as the century wore on was quantity: the precise number of such units or 'spirits' in the bloodstream. As late as 1705 Thomas Fuller – a physician intrigued by exercise and the riddles of nervous anatomy – was claiming that the healthiest blood streams were those profiting 'from the Augmentation of the *Quantity* of the animal Spirits infused into it.'[89] As in later Malthusian principles based on population, numbers counted: the greater the quantity, the healthier the person; to live long you need maximum amounts in your blood, a view anticipating the red-blood cell hypothesis long before cells were discovered.[90] Fuller harbored no doubts. Yet for every Fuller who adumbrated this view, dozens of non-publishing doctors silently adhered to them and prescribed therapies aimed at elevating their count.

Exercise and motion – even more than diet – were thought to raise the count. Here again the representative Fuller pronounced unambiguously, claiming that 'no Paradox [in all medicine] is greater than the more a Man stirs himself, the more animal Spirits are made in the Brain.'[91] The syllogism was seemingly straightforward: hard motion stirs the brain to produce ever more spirits and spirits are the key to health, even longevity. The message endorsed exercise: the rustic's hard labor over the sedentary clerk's calm or priest's inertia. Extend the paradigm and you soon idealize a Christian society, especially in Europe, celebrating noble peasants and denigrating sedentary armchair types (this view accounts, as much as any other, for the decline of the scholar's image after c. 1700, who will be seen as unkempt, silly and obese).[92] Here was the new primitivism anatomically legitimated: the sweaty farmer and laborer extolled, while the scholars and students are displayed as weak, degenerate creatures even if no theory of degeneration is yet in place.[93] A century later, when Swiss anthropological physician S. D. Tissot decided to make himself a world-class expert on the ailments on the sedentary classes in the 1760s, the physiological theory he needed had long ago been generated.[94]

Paradoxically, the physical space available for animal spirits shrank as the anatomy of the nervous system enlarged. By 1653, William Harvey might still cite medieval authorities: 'According to Avicenna, the nerves are like plantings of the brain, and provide ready intelligence for the organs of sensation, like the fingers of the hand; wherefore the brain neither sees or hears, yet knows all things.'[95] Mutedly the brain lords over these millions of slaves who occupy abundant space in the bodily mansion. The nerves are hollow tubes, yet they take up room. Harvey took great pride in his epoch's extensive anatomical studies that identified the optic, auditory, and olfactory nerves. Yet disagreement and disparagement flourished: despite Descartes' insistence on the hollow nerve, the Scottish medical student John Moir, for example, records in his 1620 lecture notes that 'nerves have no perceptible cavity internally, as the veins and arteries have.'[96] Both Harvey and Descartes represent transitional moments during which the animate powers of the nervous system were in doubt, but no fully satisfactory alternative was available to explain how unseen 'messages' were transmitted through the body.

The animal spirits' organic niche also continued to be contested. William Harvey wrote in his *Lectures on the Whole of Anatomy*: 'The nerves only carry down, they do not act, move, nor are they sentient by a faculty, but are organs.'[97] Descartes offered an important revision of this position by responding to 'different kinds of particle motion (we should term these stimuli)' which stimulated the nerve endings, producing a series of successive motions that connected nervous fibers, a new and more physically precise description of the structure of nerves in relation to the brain. Descartes concluded: 'The essence of motor control is, then, the direction of animal spirits into the proper interfilamentous [sic] channels for transmission to the proper nerve.'[98] Philosopher Malebranche, the cautious Cartesian dualist who was more influential than any other in the milieu of Bayle and Locke, also speculated about the nerves and applied Willis' physiology to theories of learning and education.

But Malebranche had no solution about the organic riddles: a further reason for scientists and philosophers, of wide backgrounds, to join in the debate about the sources of sensation and higher associated ideas. The difference between neuroanatomy in their world and ours is that it had yet not broken off – fragmented – from general discourse. As Malcolm Flemyng, an obscure Yorkshire physician obsessed with nerves, would comment in the 1740s in the introduction to his treatise 'on the nervous fluid,' it was a 'cloud of arguments' about which everyone had 'clamoured.'[99] Perhaps so, but the skeptics among them, like Malebranche, wanted proof that this indefinable 'nervous fluid' – this subtle 'fire in the soul' – existed; while others, especially idealist philosophers of mind and entrenched churchmen, were satisfied to 'clamour' on. One fact, however, was plain: it was an epoch as dedicated to the mysteries of the body as of the mind, and to attention paid to the advocates of body, such as Willis was – especially to those proponents of the 'nervous body' who were not recalcitrant to explain why their theories must summon up as much attention as the works of the philosophers of mind like Locke.[100] The body's claims on the soul were also increasing over the elusive mind's. It was the Age of Willis as much as the Age of Locke. Ideas formed in the head before they developed in the mind and then – later on in the process – into higher association as the abstract notions of things. Ideas were impossible without the nerves.

6 Gendered bodies and 'nerve doctors'

By 1681 Willis' translators had coined his English word 'neurologie' to describe the study of neuroanatomy and it would not be very long before the Italian physicians Galvani and Volta were themselves 'clamoring' that the 'fire' was actually electricity, although they hotly disagreed on the source of the current. Stephen Hales, a clergyman who was also a biologist, had already speculated in *Vegetable Staticks* (1727) that it might be some type of current. Yet whether ethereal fire or electric current, what difference did it make to the 'operations' of the brain?

This was one question satirist Jonathan Swift asked – facetiously – in his *Tale of a Tub* (1704) and *Mechanical Operations of the Spirit* (1710). The ironical Dean of St Patrick's in Dublin answered it by ridiculing *all* theories as preposterous. His claim, if one can ever posit overarching moralities in Swift's satires, is the hubris of thinking more than one knows. To demonstrate the folly he concocts the crazy hypothesis, never actually advanced by anyone, that

> ... it is the Opinion of Choice *Virtuosi*, that the Brain is only a Crowd of little Animals, but with Teeth and Claws extremely sharp, and therefore, cling together in the Contexture we behold, like the Picture of *Hobbes' Leviathan*... That all invention is formed by the Morsure [sic] of two or more of these Animals, upon certain capillary Nerves, which proceed from thence, whereof three Branches spread into the Tongue, and two into the right Hand. They hold also, that these Animals are of a Constitution extremely cold;... Farther, that nothing less than a violent Heat, can disentangle these Creatures from their hamated [hooked] Station of Life, or give them Vigor and Humor, to imprint the Marks of their little Teeth. That if the Morsure be Hexagonal, it produces Poetry; the Circular gives Eloquence; if the Bite hath been Conical, the Person, whose Nerve is so affected, shall be disposed to write upon the Politicks; and so of the rest.[101]

So much for Swift's derisory view of 'scientific' animal spirits operating in tandem with the 'mother's imagination' to mark the fetus. Swift's 'spirits' were too proximate to other, holier, types in another incarnation – in the winds of inspiration in *A Tale of a Tub* (1704) – to sit comfortably with him. Even he displays the ambivalence apparent among other divines delivering sermons on the 'Spirit' (Warburton's 'Operations of the Holy Spirit'), and it is too simple to think that all Swift summons is Augustinian modesty before the Deity's creation of the human body. His pen's lash strikes out at those persuaded they have landed on the *definitive* truth of the mysterious nervous underpinnings of emotional life.

The tribe continued long after Swift's death, beyond Newton and Hales, Galvani and Volta, down to Faraday and Helmholtz and beyond to the present day. Nor will it end. Each time a new paradigm evolved (gravitation, electricity, cell theory, proteins, DNA, etc.), proponents of the old approach to 'animal spirits' were debunked, no less ferociously than the Newtonian mechanists who were flaunting Paracelsus. The history of neuroanatomy has indeed been, to some extent, the record of search for this fire, fluid or ether, which mediates between mind and body in the eternal search for consciousness, emotion, and memory: up to the present moment when Professor Antonio Damasio corrects Descartes (*Descartes' Error*, 1994), and – in 2004 – when Damasio's critics, most recently Ian Hacking, upstage him.[102] But the quest for spirits, construed in this monolithic key, could disguise the social and cultural dimensions.

One stark proof of the confluence is found in Mesmerism, the vitalist doctrine holding that electrical shock cures illness and all else. After electricity had been discovered in the mid- eighteenth century it was not long until the nerves were swiftly inducted into its magic. Robert Darnton has demonstrated how Mesmer – charlatan, doctor, raconteur, showman from the provinces from whose name the word derives – summoned electricity to transport entire salons out of themselves.[103] Yet Mesmer pulled off his theatrical act by possessing intimate knowledge of nervous physiology. His followers, especially in Paris, applied neuro-mesmerism to neurohypnosis – the putting to sleep of the nerves through calming them – to assure that devotees were not merely lifted out of themselves, but were also made to forget and remember events deemed impossible in the eras of such 'nerve doctors' as Cheyne, David Hartley, Daniel Smith and David Kinneir.

By the peak of Mesmerism in revolutionary France the 'nerve doctor' had come into his own (only a few were female) already in possession of professional lineage. The engraving of a 'Doctor Wood' is subtitled 'Nerve Doctor,' a surprising appellation for such an early colleague of Thomas Willis who was too savvy a physician to label himself in this way.[104] Later 'nerve doctors' in the long eighteenth century included Nathaniel Highmore, Nicholas Robinson, George Cheyne, David Kinneir and dozens of others: all acknowledged their specialty, most wrote treatises on nerves, fibers and spirits (see the Bibliography). After mid-century the 'nerve doctors' formed a specialty among practitioners. In Britain they appeared in every shape: ex cathedra (Thomas Morgan), scholarly (David Hartley), Newtonian (Nicholas Robinson), fashionable (Cheyne, Hill, Adair), provincial (Flemyng), clerical (in 1768 Daniel Smith was still aiming to prove that nerves could pierce through to the soul),[105] quackish and downright illegitimate (the tribes of mountebanks and empirics peddling their pills and potions on street corners everywhere). France, Germany and other countries were imitating Britain in the rise of this new medical breed. As life became more stressful through nationalism and industrialization, modern nerves found themselves increasingly frayed. Only the 'nerve doctors' claimed to have a ready answer.

They continued to flourish throughout the nineteenth century, after much transformation, and retained their name and identity as nerve specialists even when challenged by psychologists and psychiatrists after the Great War of 1914. A delicious chromolithograph postcard survives in London dated circa 1918 showing a wife sending her forlorn husband away on holiday in order to pursue an affair with a 'nerve specialist,' called as such on the card.[106] Here the wonderfully named

'Dr. Frank Truth Nerve Specialist' has prescribed a medical order to the husband shown with his suitcase packed, cane in hand, bowler hand affixed, depressively on his way to the train station. 'I consider,' nerve specialist Dr Truth has written on note paper to the husband, 'you require an *Entire Change* in your surroundings. In fact you must get *right away* from the Conditions you have been *living under*.' One could think the emphasis here to be adulterous love rather than modern nerves; still, the cartoon loses all effect without public recognition of the medical specialty.

Such 'nerve specialists' had long ago been idealized, despite Swift's excoriations, if not in Thomas Willis' Restoration then certainly by the early eighteenth century when the South Sea Bubble burst (1720–21) and produced widespread depressives and suicides of every variety. Dr John Midriff (an appropriate name for a 'nerve doctor') was an illustrious specimen. He concentrated his London practice on (what we would call) 'depressed patients,' treated their low spirits and intuitively recognized that their woes sprung from their fiscal misfortunes in the Bubble, a pyramid scheme in which almost everyone who invested lost money.[107] Midriff documented his 'nervous' cases – one by one – in reading as fascinating as Freud's 'Dora'.[108] He avoided the mousetrap of thinking their problem either organic or hereditary, nor was he naïve enough to think their 'nervous crisis' without effect on their health. Instead he saw darkly into the situational crisis they faced, as his patients divulged how they had been reduced – overnight – from riches to rags. Within months, great houses and country estates were deserted, only to be occupied by those who earlier would not have dreamed of occupying them. Midriff's patients were undone not by the sudden deprivation of sex, love or fame, but by the loss of wealth: a situation that would replay itself many times after 1720–21 – in 1848–49, the 1870s and 1929. It sounds remarkably modern, even in our post-Prozac era of antidepressants.

The proliferation of 'nerve doctors' continued in Dr Cheyne's era as city dwellers craved them. These residents, not yet all cosmopolites to be sure, began for the first time to travel in great numbers through the countryside and seaside in an attempt to restore health to their (newly nervous) bodies and, equally importantly, harmony to their souls. Urban dwellers, of whom there were now – in the 1740s and 1750s – more numbers than at any previous time in history, found themselves sorely out of joint and spirit. Some, like the Countess of Huntingdon, Cheyne's perpetual patient, concentrated on the spirit rather more than on their joints, as the good doctor's correspondence with her demonstrates.[109] A generation later – when Dr Adair was paid large sums to treat the most fashionable depressives of his day at Bath and Edinburgh – nervous patients had grown accustomed to 'telling all' to their doctor. By the time 'Dr Truth' (above) swept our impeccably dressed lady-in-red off her feet two centuries later, the depressive patient has been diagnosed as splenetic, vaporish, hyppish, nervous, hysterical, neurasthenic, neuralgic, neurotic and in any number of equally colorful labels.[110] The labels indicated the range and nuance the 'specialist' could then expect to interpret. It is historically wrong to think that the nineteenth century was the first era to promote 'nerve doctors' (by whatever name) who commanded high fees and wooed elegant ladies. Mesmer and other 'animal magnetists' appeared at the peak of a long wave rather than the beginning, when nerves had already socially matured.

Hitherto our context for these 'nerve doctors' has been too narrow by virtue of generational tunnel confinement. We have looked at eras in isolation, but in actuality the Cheynes and Mesmers held greater spheres of influence over their patients than doctors in our time. The reasons are hydra-headed, but owe something to the status of the physician. Recently doctors have come under such suspicion that their hypotheses are also viewed skeptically. Contemporary physicians must be Nobel Prize winners to be noticed or must generate such striking theories that they capture attention by virtue of originality.[111] This was not so in the era from Willis to Boerhaave, Cheyne to Adair. For instance, Dr Peter Shaw (1694–1763) was representative of the type, even if he was not specifically a self-styled 'nerve doctor.'[112] He wrote about nerves and propeled himself into the public sector as a cultural spokesman; so blazingly that despite his hectic schedule he was enticed to comment on human affairs in journalistic papers called *The Reflector*.[113] He was not an original thinker, but some saw him committing hubris – as did Alexander Pope (Swift had died in 1745) – and lashed out. Shaw also translated and edited the Latin works of Francis Bacon for the first time – a magisterial undertaking in the history of Western scholarship – as well as editing Boyle's scientific writings and popularizing Boerhaave's medical theories. The point is that the satirists thought Shaw was elevating nerves above their due rank. Had his contemporary Cheyne not usurped the realm of nerves for himself, Shaw may well have discovered a way to write his own *English Malady*.

Yet the 'nerve doctors' disagreed and quarreled among themselves. By 1732 Dr Thomas Dover was publicly proclaiming to the masses that they 'had nerves'[114] and should take cognizance of them. Dover was familiar with the anatomical plates of William Cowper – the distinguished London physician whose folio works had illustrated bodies and brains – but took no notice of Cowper's claim that the animal spirits were too fraught a topic to debate. Cowper had written that animal spirits were not only controversial, but also difficult, impossible to adjudicate: 'there is too much Argument on the matter.'[115] His caveat was pronounced even without recourse to the areas of human perception, psychology and aesthetics. Cowper never fused his anatomical theories with Newtonian mechanics. If he had, he might have recognized how much more controversial spirits and nerves were, especially for the body–mind barrier.[116]

A sense of what nerves signified to educated, well-read persons is gleaned in the millions of words left behind by Horace Walpole, son of the Prime Minister and *memoirist extraordinaire* of his epoch. Nerves, especially 'ruined' and 'shattered,' frequently appear on his pages, much more than spirits: the 'spirit' has waned, secular or holy, the 'nerve' replacing it as the agent of all body states. Walpole gendered nerves according to the sex of their owners – male, female and the alternatives – and was curious about nervous transformation that occurred in private and public settings (such as the 'tea table,' as in the famous scene in the first canto of Alexander Pope's *Rape of the Lock* where this still new, potent drink altered the body's states).[117] Walpole also noticed that travel plays havoc with 'nerves abroad' and observed their altered states while on the Grand Tour, especially in seductive Italy. Most of all they were sensitive barometers of emotional states and consequently permitted the passions to be more nuanced than in previous eras. Walpole would not have commented so frequently if nerves had not seemed so noteworthy – a novelty still to be captured and recorded for posterity.[118]

This is not the place to comment on Walpole's sexuality, except in its 'nervous' context, and to wonder to what degree his bachelordom had sensitized him to the gender component: muscular, male nerves versus soft, flaccid female. Although irritable, even bitter, he was intrigued that nerves were one of the new clues to masculinity, especially as he thought of Lord Hervey: Pope's 'Amphibious Thing' christened by Lady Mary Wortley Montagu a member of a 'third sex,'[119] with its suggestion of the possession of a new species of nerves. It is thus wrongheaded to think 'gendered nerves' a restricted medical category – even in the eighteenth century – reserved for the doctors: the Boerhaaves, Cheynes and Dovers. Already in *secular* correspondence of the 1740–1750s, such as Walpole's, sexual conduct is being assessed along the lines of nervous constitution. It continued down through the century: when Harriet Wilson, who blackmailed King George III and other public figures later in the eighteenth century, exclaimed that 'dates made ladies nervous,' she assumed that the public would be able to grasp her point about gendered nerves.[120] Nerves loosely understood were the body zone most attuned to emotional calibration. Soon – by the mid-nineteenth century – they would be the key to understanding the whole personality. Male brains had already been gendered as being different from women's, although there was no anatomic or biological proof. Cases like Wilson's demonstrated that anatomy had filtered into culture and were influencing its developing notions of gender.[121]

By the 1770s Jean-Louis Jauffret (1770–1840), a French doctor who had written about the plague of 1720 and advocated exercise, explained nerves for children and adolescents in books eventually translated into English.[122] He illustrated these popularizations that contained simple notes in an era when children's books were mushrooming. Lucy Aikin, the English bluestocking writer, thought so well of his anatomies that she translated Jauffret as one of her earliest undertakings in 1810–12. Jauffret's 'wonders of the body for youth' take the form of a dialogue between 'Mr Vermont' and his children. Opening with an exposition of William Paley's 'clockwork body' derived 'by design' from his *Natural Theology* of 1802,[123] the children learn that nerves underlie every part of the skin and fire up the body's five senses. The nerves are recounted as if 'miracles.' Richard eventually surmises – logically after the demonstrations – 'it appears that it is the nerves which perform the most important part in the action of our organs.'[124] His 'anatomy lesson' is primarily about skin, a body part that will continue to recur in relation to the mystery of nerves.

Others who wrote similarly, demonstrating how eager their readerships were for information about the miraculously nervous body. Child readers were especially susceptible to this wondrous dimension because so many more actions could be explained by recourse to nerves than had been possible without them; not merely the differences between boys and girls, but also matters as diverse as weather and joy and evil. The miracle of nerves became something of a minor *topoi* by the 1790s, drawn on variously by children's writers and visionaries like Blake, who celebrated states of childhood innocence. A contemporary of Jauffret's – Hugh Smith, a physician himself who was the son of a family of doctors – also expounded on nerves for popular audiences.[125] Smith's immediate agenda was the discreditation of 'foreign teas' as pernicious to the nervous system. He may have been set up by the British tea-lobby (see their advertisements at the back of his book). Yet Smith's wide-ranging view of the philosophical basis of human life as 'nervous' suggests that he believed what he espoused. His argument amounts to the view that the caffeine in foreign tea overexcites the nervous system and throws it into disarray. James Parkinson, the pioneering geologist and practicing physician, endorsed Smith's 'philosophical books' as sound,[126] but not without commenting on the contemporary debate about the 'nerves' inside rocks, a controversy then still unresolved among Neptunists and Vulcanists.[127] Through these minor figures, rather than the giants of the era, we gain further insight into the diffusion of nervous sensibility.

7 Passions, emotions, affections: writing out the nervous sentiments

Pressing questions remained. How did nervous constitution translate into passion and emotion? The answer would color one's view of the dim past: Eden to the Fall and the rise of civilization. As the century progressed, the riddles of mind also increased.[128] Those exploring its operations in the decades when the third Earl of Shaftesbury wrote early in the century were far fewer – and less troubled – than those at mid-century who had been bitten by the bug of the new neurophilosophy.[129] Even Hume established an identity of selfhood based not on one's coherent thoughts, organized logically and sequentially, but, rather, on a physiology of bodily passions, even if it was a counter-physiology ultimately uninformed by the nerves.[130] After the Haller–Whytt debates of the 1750s over 'nervous irritability' and 'nervous sensibility' it seemed that physiology would invade yet another domain, the realm of *affect*. Their learned dialogues advanced theories of the nervous system by asking how signals sent to the muscle-fibres stimulate them into sensation and movement. The riddle was less the exact anatomical pathways than the contractions and expansions in the muscles themselves. Haller preferred to conceptualize these contractions as 'capacities' rather than definitive motions and called the phenomenon 'irritability.' Whytt, on the other hand, an early pillar of the world-famous Edinburgh School of Medicine, focused on voluntary and involuntary movements to generate a theory of the 'reflex action' that contradicted Haller's theory of irritability. The neuroanatomy of these positions is immensely complex, but both physiologists were nevertheless searching for explanations of *feeling* rather than delimiting themselves to movement. Feeling – especially feeling in relation to *sensation* – was the genuine source of their explorations, as it would be in the next generation for the Viennese neurophysiologist George Prochaska and the Scottish surgeon Charles Bell in his landmark work, *The Anatomy of the Brain, Explained in a Series of Engravings* (1802).

Laurence Sterne's *Sentimental Journey* (1768) made this influence – what I have vividly called invasion – apparent in the changing moods of his inimitable traveler who laughs and weeps by turns, and suffuses himself in emotion at every bend of the road. Whether stalking impecunious street urchins or flirtatious pretty lasses – 'Maria' – Yorick responds from the heart where all his animal spirits seem to have charged in one fell swoop. So far has Yorick's 'nervous sensibility' carried him in a decade when the sensibility craze was ascending towards its peak. Yet Sterne submerges whatever intellectual origins his fictive strain may have had: unlike the much more diverse *Tristram Shandy*, this is primarily a journey through the kingdom of *affect* despite its chapter divisions by place ('Calais,' 'Montreuil,' 'Amiens'). By turns Yorick's nerves are strained, shattered and stretched, the basis for his huge mood swings. Still, his maker – Sterne – withholds any didactic commentary about them. But this does not prevent him from composing a 'sentimental journey' through the land of feeling and replacing the more adventuresome type Fielding had cultivated in his 'new species of writing.' The reader's dilemma is that it is never clear what Sterne's moral is: whether this is a novel of sensibility – i.e., one that explores the kingdom of feeling without a priori condemning it – or a satire on sensibility aimed at exposing its excesses in an attempt to curb or even obliterate them.[131]

Sterne's didactic silence notwithstanding, precisely how did neurophysiology account for character, feeling, and the sentiments among people? The epoch began to refer to itself as an age of sensibility; but did it mean sensibility of the anatomico-physiological variety? Slavery and racism were then burning topics;[132] did black slaves have different nervous systems from white Europeans? Why were some people extremely sensitive? – was this also the product of certain nervous states? We have already broached matters of gender and sex; the sexual domain was beginning to ask pinpointed questions about (what we would call) sexual desire. The male penis has approximately 6,000 nerve tips ending in this organ, the female clitoris 8,000; given – as Bienville and other doctors noticed by mid-century – that sexual intercourse occurs in this anatomical environment was it then principally a 'nervous' act? The classical poets had thought so, calling the penis the 'nerve' (as in John Dryden's 'limber Nerve').[133] Were nerves also implicated in blushing, crying and the tears produced? What about war and commerce: gendered activities then thought to be tied to masculinity and bodily hardness?[134] Were women excluded from business and commerce as the result of weak nerves? Did gendered bodies have different kinds of 'spirits' and 'brains' as well as nerves?

The list was extensive. And lunacy, insanity, madness? The first treatises theorizing its genesis explicitly resorted to nervous explanation.[135] Even before the French Revolution sent chills through Europe it was clear that madness was the product of flawed nerves that might never be put right.[136] But nerves also had their hale and comic side. The treatment of lunatics included ridiculous therapies: filling the bladder with bubbling liquids, implanting the nerves with gas, straitjacketing the body so securely that it became distended with the suffusing gas. This link to political upheaval was not missed as commentators in nostalgic poses glanced back at the 'revolution' in nerves they had witnessed. Scottish John Armstrong, poet-physician, had written his most imaginative poetry as a young man.[137] But when collecting his medical essays (1773) near the end of his long life he referred to the revolution in nerves he had witnessed.[138] He had written poems, *OEconomy of Love* and *Art of Preserving Health* (which more than one Victorian critic considered to be the most masterful blank verse of its kind in the whole century), which should be set side-to-side with Akenside's *Pleasures of Imagination*. As a young buck in Edinburgh and later a doctor to the British troops in the West Indies, the nerves signified little to him despite a quasi-philosophical interest in animal spirits;[139] by his last decade in the 1770s, the crest of nerves was reached and he could see the toll it had taken on medical theory. Armstrong baffled his contemporaries: bachelor, miser, misogynist, would-be poet, quondam physician, scourge of his medical profession. Yet on the 'arrival of nerves' he was lynx-eyed, nowhere more so than in love, as his poetical essay of that name (*OEconomy of Love*) suggests: love sans nerves was unthinkable to an Enlightenment man like Armstrong. He intuited this when he began to write his poem in the 1730s. Four decades were necessary to articulate it.

A magisterial presence in Edinburgh was William Cullen, whose view of neurophysiological life stemmed from 'nervous energy'. All through these decades the mind–body debates continued, often reinvigorated by considerations of the emotions, now (in the 1780s) gradually replacing the passions in a paradigm shift. But the course of nerves in the nineteenth century would also have been different if Cullen had not developed his systems of taxonomy giving nerves a prominent role, for which he was dubbed 'Cullen the neurologist.' This was not of Willis' variety: the world had changed since the 1660s. Besides, nerves had come of age and infiltrated Scottish society by the time Armstrong died (1779) and Cullen professed in the 1770s and 1780s (he died in 1790).[140] An influential doctor and charismatic professorial lecturer in Edinburgh, he was well placed in the hub of Enlightenment culture to shape the course of nervous knowledge.[141] Working under the shadow of Haller and his theories of sensibility, Cullen generated a system of classification based on 'nervous sensibility' and added a branch of pathology called 'neurotic,' named for those with disorders of the nervous system rather than purely for the mental disturbance it has denoted since Freud's psychoanalysis.

Similar to the earlier Newtonians who applied laws of mechanics to geographically dispersed realms – aesthetics, music, painting, the laws of government and society – Cullen's disciples extended his nervous taxonomies. They claimed that much behavior was 'nervous': civic, economic, social, urban.[142] Some of the applications were old, but what has been less clear is how Cullen's far-flung students gradually formed the backbone of a generation in the revolutionary world of the 1790s. They came to Edinburgh from far afield – from Russia to the Iberian peninsula; after hearing the Haller–Cullen line they were prepared to rethink mankind as if converts to a new sect. One of these was John Brown, a Scot who created his own Brunonian theory based on the balance of external stimuli, which had a long life after 1800 and which Romantic philosophers and poets considered so vital that it must be construed as an intrinsic part of its sensibility. Some were young and impressionable, well-aware of the international distinction of their medical professors in Edinburgh (James Gregory and Joseph Black were others), and they listened with awe to the new 'lines of physic and physiology' predicated on 'nervous man's body.' As the century altered to 1800 Cullen's effect seeped into the tissue of Western culture, if not penetrated its marrow. 'Nervous' was now applied to collective groups of urban dwellers living under stress, as well as to workers radically adjusting to the new sense of accelerated time.[143]

The growth of 'nervous civilization' had not been shaped by a string of prominent figures, but their presences demonstrate in a nutshell how nervous theory was altering to keep abreast of ever-more agitated citizens. Hartley, Haller, Whytt, Cullen and company were all theorists who fundamentally changed some essential aspect of nervous knowledge. Another was the already mentioned Thomas Trotter, although his contributions were more perceptive than authoritative. He had been a student of Cullen and practiced in Newcastle. He was in the lecture theatre when Cullen expounded the strange instance of a 'nervous lady'.[144] Cullen taught using actual case histories, thereby imbuing his students with a profound sense of the implicated nervous system and brain. In one of his lectures he presented a lady whose whole life was uprooted as the result of shattered nerves. Trotter never forgot the case, and he went on to amplify Cullen's theories with particular application to alcohol and other substance abuse. He was struck by the degree to which 'nervous patients' are 'uncommonly fond of drugs,' a strain as true in 2005 as it must have seemed to him two centuries earlier.[145] His major leap was the intuitive sense that society became more 'nervous' as the marked stresses of modern life – at home, in the workplace, within the community – increased.

Every decade following Cullen witnessed waves of heightened interest in nerves. By now the category had long since been constituted and brought into the public domain, but it would be impossible to say in which sectors it most thrived: medical, artistic, the practical arts of living, economics. Furthermore, Haller and Cullen were theoretical giants who made huge strides in nervous knowledge. Yet we also derive a sense of the dissemination of nervous knowledge from small fry. For example, George Rowe (1792–1861) was a successful army physician, widely traveled and *au fait* with tropical climates, who practiced for most of his life in the London area. He was not academically distinguished and contributed nothing new to the theory of medicine. But his treatise on nerves, which he conceived at the end of the 1830s as he was being elected to the Royal College of Physicians, filled a huge 'practical' gap and went through two dozen editions in about twenty years.[146] Its contents contain no new research but emphasize the Haller and Whytt debates. The link to hypochondria and hysteria, claiming that all nervous cases are fundamentally 'hypochondriacal,' includes practical directions for 'calming the nerves' with liquor and 'loosening the nerves' with exercise and pure air, and he writes as if these categories could be understood by common folk.[147] Who can have constituted his readership? Even if the book became a textbook for medical students, the number of reissues raises questions about the extensiveness of nervous inquiry.

And hysteria circa 1790 or 1800? It had not yet been construed as the 'neuromimetic disease' it would soon become. Historian Mark Micale has written:

> As most commonly understood in nineteenth- and twentieth-century medicine, hysteria is a 'neuromimetic' affliction. It is the masquerading malady that has no essence but rather emerges, chameleon-like, by aping the symptomatological forms of other organic diseases, most often neurological diseases.[148]

This 'neuromimetic' synopsis is accurate and it continued to bewilder doctors, and even inquisitive patients, trying to understand hysteria. The Victorian Harley Street 'nerve doctor' Sir James Paget had read a passage in Coleridge's *Table Talk* where hysteria is called 'mimosa' for its ability 'to counterfeit so many diseases, even death.'[149] This led Paget

OH! THERE'S SOMETHING IN MY EYE!

I wish you'd take it out,
There's always something the matter with me!

Plate 2 Aquatint drawn by Ego. Engraved by F. C. Hunt. No date. The man is surrounded by objects denoting his neurosis, including a table lader with 'nervous cordial'.

to improve on 'mimosa' and coin 'neuromimesis' about which he lectured to medical students. Coleridge captured Paget's attention; the Coleridge who – of all the canonical figures of English literature – has had the greatest affinities with medical theory. Even so, hysteria's recent past in the world of Cullen, long before Paget, amounted to a type of intensified genderization. Debates raged not merely about its neuroanatomical origin (how could men without wombs contract it unless their nerves were as frayed as women's?) but also suitable cures. These formed the background of the new 'moral therapy' that would be applied in the era of Pinel and Esquirol, when the Paris doctors gained their reputation as the most advanced in the world. Long ago Sydenham had noted that hysteria could be chameleon-like, 'neuromimetic' in slyly aping other diseases, but it had not yet completely broken off from 'low spirits'.

Brain theory had also come a long way since Willis and the 'nerve doctors' had made their strong case for the connection of anatomy with human cognition and perception. Poetry and early fiction absorbed their teachings and continued to.[150] The nerves were seen as increasingly important not merely to the 'general inclinations' – as one philosopher had called human disposition – but to smaller realms, the senses, especially smell and touch. Novelists 'Smelfungus' Smollett, so-called for his extraordinary olfactory sense, Trollope (in *The Three Clerks*) and Elizabeth Gaskell (in *Wives and Daughters*), were especially keen on the nervous route to offensive smell. Smell and touch were also gendered, along the lines of brain genderization. Offensive smells and touches even led, allegedly, to illness as they do the prickly clerk in *Wives and Daughters*.

Social class, especially in the British Isles, absorbed the new neurology even more quickly than imaginative literature. I have described in an entire essay (chapter 5 below) the 'conversion to nerves,' according to Dr Dover, in the 1730s, and others whose way was paved by Dr Cheyne.[151] What continues to be minimized is the toll such conversion (if it was that) took on rank and privilege. This psychological process was not merely cosmopolitan: occurring in the vast metropolises that London and Paris had become by 1800. Decades of social exclusion and rules for fashionability had consequences further afield even than the countryside among the rustics. Travelers and explorers newly read in nervous anatomy began to apply the theories to foreign natives, as did the wealthy Humboldt brothers, Alexander and Wilhelm. Climate theory added to these accounts of national stereotypes, as did topography and vegetation. For centuries, explorers had been exposed to races other than white, but before approximately the mid-eighteenth century their observations rarely commented on racial difference in the light of nervous constitution. Nicolas le Cat, a French physician who applied the new theories to race, never traveled widely but shrewdly speculated about the *probable nerves* of blacks in the heart of the jungle.[152] Scientific traveler Robert Dallas had voyaged extensively in the Caribbean Maroon Islands.[153] His theory of race centered on climate in relation to anatomic organization, and he generated a fantasy of how 'Marooners' might thrive in northern countries: 'It may be well imagined, that at first the pinching of the frost will not be agreeable to [nervous] fibres accustomed to the full flow of blood produced by the rarefaction of the torrid zone; but time, the nurse of habit, corrects this acuteness of nervous sensation, and accommodates corporeal sensibility to the influence of climate.'

By the 1820s a preliminary social anthropology of nerves developed: of wide interest to the Germans and based outside the recognized disciplinary specialties of anatomy and medicine, geography and history. It formed in the 1780s and 1790s, when civic life itself was being radically rethought in the run-up to, and then in the aftermath of, political revolution and the great scientific debates over mechanism and vitalism. For example, in Germany the young Samuel Taylor Coleridge heard these theories, especially Brunonian medicine, in Blumenbach's lectures and was galvanized for the way in which it uprooted the old Lockean associationist philosophy and Hartleyan vibrationism.[154] As cell theory took hold and the new protoplasmic biology based on microscopy rather than morality gained sway, the body's nervous organization asserted itself. Booksellers cashed in on the thirst for new knowledge and began to produce compendiums like Christian Friedrich Ludwig's *Scriptores neurologici minores selecti* (1791–95); anthologies gathering snippets about spirits and nerves, as in this German collection which Coleridge may have seen while studying in Göttingen.

Nervous theory and practice also awakened the curious. By the 1760s plans had been drawn for a machine to measure the strength of the nerves with the aid of electricity. The dream instrument would ascertain tonic health or its opposite among patients, especially women.[155] In the optimistic decades before the French Revolution, when instrument makers thrived and ordinary folk believed measuring machines would soon be invented, a 'neurometer' (as it was called) was thought to be imminent. Two generations earlier – in the 1720s – Stephen Hales had devised a way of measuring blood pressure (although the 'mercury manometer' was not invented for another century). It would be another generation before Galvani and Volta performed their experiments on 'animal electricity,' but electricity provided the necessary impetus for the 'neurometer.' By 1818 Robert Southey, the writer-traveler, claimed in his letters that he had become intrigued by the possibility of measuring the tension of the nerves in this way. A neurometer instrument could be predictive as well as diagnostic: augur one's manliness and vivacity, become a guide to personality rather than aid to failing nervous constitutions. One imagines how promising it must have seemed in eras when military victory alone ensured national might. Commanders may have been less selective in peacetime; during war, when the outcome of battles counted for all, the last thing they wanted to recruit were sickly males of nervous constitutions such as Eugene of Savoy, who was declared by Louis XIV to be completely unfit for service.[156]

8   The 'nervous' style

By 1740 the nerves had been widely versified across Europe, especially in England, which assumed the lead: if not yet having found their single muse, as Newton had in James Thomson, Pope and the Newtonian poets,[157] nerves were nevertheless widely disseminated through a host of lesser writers. Comb the vocabulary of English poetry, decade by decade, from the 1680s forward and you discover how this accumulation intensifies. From the allusions in Dryden and Swift, to those in Pope (especially in *An Essay on Man* and *The Dunciad*) and the poetasters of the 1740s,[158] the poetry of nerves amplifies itself down through the century. Its appearance further intensifies in Smart and the early visionaries (Collins, Cowper et al.) as faith itself turns 'nervous' under the spellbound trance of the beholder; continuing through the lyric poets of mid-century down to Lunar Society Erasmus Darwin in his various 'Temples of Nature,' where 'Next the long nerves unite their silver train,/And young Sensation permeates the brain.'[159]

More must be said of *this* polymathic Darwin who versified nerves so acutely. First it must be noted, however, that in the long eighteenth century the writers themselves claimed to be composing in a mode they called *nervous* – it is not a modern imposition. They *themselves* used it and there is sufficient evidence to warrant that they had a sense of its vitality even when ignorant of their Willisian heritage.[160] But it is chronologically faulty, and historically false, to think that the new anatomy of the *eighteenth* century gave rise to a *nervous style* in literature. It enhanced its existence but did not invent it. The contexts are more varied, the arrows more complex than those of science influencing literature.

Glancing backward at authors read by Collins, Cowper, Erasmus Darwin and their contemporaries, a case for nervous style can be made among the ancients who harbored a shrewd sense of 'manly' and 'virile' modes. Its effects appeared in prose and poetry, although 'nervous' did not denote the style. Cicero comments in his famous essay on old age (*De Senectute*) on the main function of orators and teachers as instruction of the young; it should be in 'nervous and manly eloquence' in the English words of translator William Melmoth. The translation is noteworthy because it occurs in his English translation of 1773 precisely at the moment when nervous style peaked in popularity in Britain.[161] Later in the treatise Cicero configures old age as a time when the body begins to decay; Melmoth renders the Ciceronian idea as 'old age has not totally relaxed my nerves and subdued my native vigour.'[162] Here the nerves function as a metonymy for the body of man, vigor the chief sign of his maximum efficacy. Yet long before Melmoth translated Cicero the word 'nervous' was commonly being used to describe styles of expression of a particular type. Joseph Mede, the Cambridge polymath active in the early seventeenth century and author of *Clavis apocalyptica,* referred to 'a nervous, close and well-composed Discourse,' when awarding high marks to didactic treatises of which he approved.[163]

The opposite style – weak and flaccid – was the diffuse, usually as a term of opprobrium. It too was in place by the time Willis published his landmark treatises on the brain during the English Restoration. Both styles matured by 1700, the energetic 'nervous' overtaking its flaccid 'Anti-Christ of Wit' by far. Nervous writers garnered praise for the briskness of their arguments and exposition. 'Nerve' was also conjoined to innovation: the triumphal sense that what may have been pathological in a 'nervous person' was brave and laudatory in writing. Such composition was, more than anything, 'brisk' and 'robust,'the two words most often used to describe it, but it was also 'muscular' and 'tight.' As such it required all sorts of compression and 'musculature,' which energized it further by dint of pithiness. If the Enlightenment, as we have seen, constructed subjective 'nervous selves,' it also encouraged the 'nervous style' that was a hallmark feature of such persons. For truly masculine selves did not write in a soft, flaccid, effeminate style. Thomas Hale, a naval writer in Pepys' Restoration London, practically bristled with delight when reading the work of another male 'projector' because 'the Author hath in so nervous a Manner given [his] Directions.'[164] These could not have been *female* 'directions.'

A generation later, in the high Restoration of Dryden and Rochester (poles apart), English poetry was defining itself not merely by style, but also by the nervous bodies of its writers. It was unclear whether nervous styles had prompted nervous selves or vice-versa – they probably developed in tandem. The route was indirect and no literary history has yet explicated it.[165] Some of the context of the new 'nervous' group of poets was its quantum of 'nervous rapture,' which ensured the requisite degree of imagination or 'fire' for inspired poetry.[166] They were not merely suffering from mania, and thus 'nervous.' One type of fire – the old Willisian 'fire in the soul' – migrated to replace the poet's selfhood. A new 'nervous self' formed whose persona the poet aimed to dramatize, if not consistently celebrate. By the high tide of Restoration literature plays were being described as 'nervous,' and – in the next generation – English poetry. Added to this new dimension was a developing Newtonian aesthetics of perception that relied heavily on the brain for its operations. It posited close links between the condition – or state – of the brain and the language the mind generated. Hence 'nervous selves' were unimaginable apart from their physical bodies; and the literature they (these selves) imagined informed by a new constellation of brain–mind–language, all this against a backdrop – later at mid-century – when nerves and nervous selves more than anything were being gendered along lines of weak and strong, female and male.[167]

The nervous style was also gendered, the assumption being that men would write taut muscular literature and women in the flaccid or weak style reflecting their selves and larger cultural identities. Lack of vigor or energy was construed as 'nerveless,' hence soft and feminine. Applied to the aesthetic realm, and especially directed to style, it implied a gendered ethic which gathered – more or less continually – down through the eighteenth century. 'Nervous style' was a widely applied label, equally broad in recognition. When Bishop Warburton wrote in his *Doctrine of Grace* (1763) that under the pen of certain authors 'Western Eloquence' could 'appear nerveless and effeminate,' he assumed his readers knew what he meant by this shorthand conjoining gender and style.[168] No less than a broad range of Victorians, a century later, knew what Thomas Arnold meant by 'muscular Christianity,' which also built upon this nervous heritage. Between the two, in the Regency, a standard critical aperçu about Byron's style was that it was 'nerveless': a pointless kind of blank verse, as if (shockingly) to suggest that the soft female sex could not follow an argument through to its logical conclusion (i.e., pointless).[169]

Comb critical commentary between 1700 and 1820 and dozens of similar appraisals are found. Nor did the habit wane after 1820. If anything, it intensified all through Victorian sensibility, culminating in the view that 'nerveless' art – whether literature, painting, or music (as we shall see) – was somehow defective and 'pointless' merely by virtue of gender difference. Any positive valence attached to the 'nerveless style' remained an undercurrent, not to alter until the twentieth century. When asked specifically how the anatomy and physiology of the genders differed, the doctors were mostly at a loss. What they had been taught was what they knew; it amounted to knowledge that the two nervous systems differed.

Erasmus Darwin – already mentioned – was one such medical doctor. He was far more enlightened than most of his contemporaries on these nervous matters. Trained in medicine he also wrote a great deal of poetry, some of it versifying the conflicting theories of nerves. His last poem, *The Temple of Nature*, was written just before his death in 1802 – and published posthumously in 1803. It followed on the heels of Wordsworth's and Coleridge's *Lyrical Ballads* (1798) and incorporated the neuroanatomical views he had recently (1798) put forward in his 'Mechanism of the Human Body.'[170] Here Darwin reasoned hierarchically upwards from the plants (which were especially 'nervous') and smallest creatures to 'nervous human beings' capable of sustaining poetic selves, like his own – even to ideas formed through series of 'nervous links' in the body, each link consisting of successive trains of motions conducted through the extremity of the nerve tip.[171] Poetically Darwin also aggrandized this nervous self by claiming that the sensation nerves permitted had been the great *evolutionary* step: if not an outright gift then a giant stride in human progress:

> Next the long nerves unite their silver train,
> And young Sensation permeates the brain.[172]

This was a point his grandson, the famous evolutionist, would tackle with greater scientific sophistication than his ancestor. Yet his grandfather had posited that down through the ages, the animal kingdom had evolved by the development of sensation; each sensation leading to ever-greater capacity of movement and awareness of the surrounding environment. New sensations evoked new passions and degrees of self-reflection; these – in turn – gave rise to diasporas of emotion that defined human beings as the creatures of sensibility they were becoming. Much of the opening canto of his *Temple of Nature* deals with this nervous physiology of emotion wherein associations and thoughts combine to produce the full range of passions:

> Through each new sense the keen emotions dart,
> Flush the young cheek, and swell the throbbing heart.
> From pain and pleasure quick Volitions rise,
> Lift the strong arm, or point the inquiring eyes.[173]

Darwin combines evolution and emotion in a manner that the Romantics will elevate to a full-blown aesthetic, and the late-nineteenth century Romantic Evolutionists (Samuel Butler, Bernard Shaw, Henri

Bergson, Julian Huxley et al.) still further. Even in the miraculous human hand, with its indispensable opposable thumb that makes musical performance of the type discussed in the opening of this essay, nervous physiology holds the key:

> Nerved with fine touch above the bestial throngs;
> The hand, first gift of Heaven! to man belongs;
> Untipt with claws the circling fingers close,
> With rival points the bending thumbs oppose,
> Trace the nice lines of Form with sense refined,
> And clear ideas charm the thinking mind.[174]

Such human neuroanatomy inevitably took, Darwin suggested, a toll on literary style. Samuel Johnson and other literary critics had noticed the point much earlier and they found no contradiction between theory and practice in the application of its tenets: taut expression, everywhere muscular, without any trace of excess or rhetorical flab.[175] But after circa 1820 aesthetic pronouncements about the nervous style diminished (not the application of the style itself) and were replaced with a new concern for anatomical nerves themselves as the basis of modern life. This was not so much a continuation of evolutionary position (Darwin to Lamarck) in the aftermath of the French Revolution, as the sense that the strains and stresses of modern life were taking a *new* toll on nervous anatomy.[176] Yet it was not necessary to wait for the novels of the 1790s, or the poeticization of these views in works such as Thomas Peacock's in *Nightmare Abby* (1818) and Jane Austen's extended treatment of Marianne Dashwood's nervous sensibility in *Sense and Sensibility* (1811) – they were present in the eighteenth century, perhaps nowhere more vigorously than in the fictions of Tobias Smollett (1721–71).

Like Darwin, Smollett had trained as a physician in Scotland, but he eventually joined the London tribes of professional hacks and journalists, constructing his fictions around the idea of these nervous stresses and strains, and one of his six novels on the notion of modern life as a shattering of the mind–body relation. In *The Adventures of Lancelot Greaves* (1762), the first serialized novel in the English language built around the idea of a 'Romance about English Quixotism,' Smollett imagines a renaissance of quixotism, and suggests that all rebirths, such as the Spanish Quixotism, imply nervous consequences. Hence heroine Aurelia Darnel's nerves are totally shattered. Her father's disapproval of her marriage to Greaves forms the engine of the plot of endless separations from her lover. One of these incarcerations is in a private madhouse owned by 'Dr Kawdle.' Sweet 'Kawdle' turns out to be thoroughly benevolent – as in so many of Smollett's fictions where being and seeming are at odds – and all ends well for the lovers, Darnel and Greaves; but not before Aurelia's nerves are restored, amounting to an arduous process requiring the author (Smollett) to quote lengthy passages from a contemporary treatise on nervous derangement as the basis for insanity.[177] This thick content of 'nervous derangement as mental affliction' would be less noteworthy in Smollett's fiction if his prose had been less 'muscular and nervous' than it is; so energized that it was often said to be the result of his own excessively wrecked nervous system which had left him irritable to the degree that others bristled. Readers of his fictions from Walter Scott and Hazlitt noticed this massive 'energy' (their term) in his style. It was as if vitality poured from the joints of his nervous passages.[178] A theory of energy and the nervous style had developed in tandem, often prompting savvy early readers to inquire about the sources of his terrific degree of energy.

Even so, the nervous style could not have developed without a vocabulary of newly minted words. The developing glossary was crucial to the rise of a style deemed to be muscular, masculine and beneficial to the developing state and gathering empire.[179] Style and state thus entered a new phase; and the old 'body politic' as 'nervous' – as we saw – was transformed into something exceeding mere metaphor. The opposite applied to effeminate, soft and flaccid styles, implying a crude gender-based homology of 'as is the style, so is the state.' Another introduction would be needed to list and comment on the rise of these qualitative antitheses and their coinages even for the period 1660–1850. A single example from the sexual domain makes the point by tapping into many of these tensions at once. The Renaissance 'pathic' was a passive catamite in the service of a sadomasochistic master who abused him anally. The word arose, in part, from its Greek root in *pathos* or suffering: the self-evident pain inflicted from behind. By approximately 1600 'pathic' had migrated from highbrow books about the Ancient world into popular discourse, especially commonplace on the stage. However, within a generation or two a new variety of 'pathic' was added, the '*neuropathic*': a creature not merely passive and clearly derived from the old passive sodomite, but now *compulsively* passive at large (physically and emotionally), hence now the possessor of a 'neuropathic constitution.'[180] The diagnosis took a large toll for several decades, especially in the 1860s forward, until its aura was demystified. This was satisfied by the new neuropathic, born in the eighteenth century. Dozens of similar words also exist.[181]

Style and neuroanatomy would seem to have little in common. Nothing could be further from the truth. After the 1860s, the rise of sexology brought them closer still, and they continued into the twentieth century as writers began to calibrate their own creative states to their momentary nervous disposition. This was not so much a fleeting emotional flutter – a state occurring only once – but the writer's sense of herself as a nervous creature on a large scale of perceived nervous others. Woolf, Beckett, Genet, Yourcenar and dozens of others have commented on themselves in this sense.

9   Nerves and music: the aesthetic debates, 1740–1890

The role of the nerves in music has loomed as large as the nervous style, especially since the Renaissance forwards and in relation to questions about music and the emotions. For example, is there a distinctive aesthetic emotion? How, neuroanatomically speaking, does music evoke emotions? What are the pathways? More closely related to the brain and its cognition, does music evoke emotions or does it represent or portray emotions? And are there some emotions that cannot be evoked or represented by music – such as shame or envy? Why does some music – the purest language of melancholy known to mankind – make us nostalgic? Why do people willingly listen to music that makes them sad – is there a nervous dimension to this tendency or is it primarily cultural and habitual?

The range of questions is large, few of them without any apparent nervous profile. What cognitive function makes us hate some songs? How is it that some drugs can transform the emotional experience of music? Are the moods we experience from music related to what the classical world called the 'sentiments' and our more modern, the 'personality'? Closer to the sensory side of this experience, can we experience musical emotions in the absence of sounds? We saw that nervous style was firmly gendered by the mid-eighteenth century: do women and men also experience music differently? According to their nervous anatomy and physiology? Or is the body irrelevant to these perceptions?

The intimate connection between the body and music is ancient. It was already old when the Ancient Greeks correlated the musical modes (Dorian, Phrygian, Lydian) to emotional moods (see Brocklesby (1749), Gouk (2000), Nutton (2004–)). More recently, the eighteenth-century Sephardic Jew Isaac Disraeli commented on the degree to which his own intellectual curiosity was piqued when pondering precisely how – through what anatomic processes – music healed the body. He thought he had hit on something new, but later concluded that he had entered the debate rather late. His trilogy of books entitled *The Curiosities of Literature* had propelled him into the limelight in the 1790s, offering him the amplitude to range over the terrain of numerous pressing puzzles ('curiosities') in history. But he was not scientifically or medically trained. The little he knew he discovered through reading. Once apprized of the literature about healing he changed his tune: 'I since have found that it is no new discovery.'[182]

His main source seems to have been an article in the May 1806 issue of the *Philosophical Magazine*. It could as well have been any number of other sources, for by the 1790s the idea of musical healing was pervasive. Disraeli also found a theory delineated in the early music-historian Charles Burney's essay titled 'On the Medicinal Powers Attributed to Music by the Ancients,' which Burney had included in his pioneering *History of Music* (1776–89). Here Disraeli discovered, for the first time, the source of the mystery: 'Music relieves pain by ... occasioning certain vibrations of the nerves ...'[183] This explanation stunned him: he paused on nerves and pondered them. Once enmeshed he found himself wondering about the network of nerves, spirits and fibers in relation to the larger nervous system and governing brain. He was mystified, as any well-read person would have been then – even someone medically trained, which he was not. He turned to philosophers and physician-authors for guidance, locating an informative source in 'Mr Burette and many modern physicians and philosophers [who] have believed that music has the power of affecting the mind, and the whole nervous system ...'[184]

Pierre-Jean Burette (1665–1747) may now be an obscure figure, but he had become something of a legend by the late eighteenth century for his views on ancient Greek music. He was a medical professor in Lyon, a well-read classicist and member of the recently formed French Academy for Belles Lettres. He also frequently published articles in the *Journal des sçavans*, of which he was an editor. His vast medico-musical library was catalogued at his death in 1747 and created a bibliophilic sensation.[185] Fluent in most aspects of ancient Greek life, he focused on ancient athletes, gymnastics, musical modes (Doric, Lydian, Phrygian, etc.) and the Renaissance revivals of musical composition executed according to Greek acoustical principles. Burette also translated into French Plutarch's *De musica*, in which the ancient writer expatiated on his theories of Greek music.[186] Burette's focus is on individual musical instruments[187] in an attempt to explain their specific healing powers through a discourse of voice: flutes, lutes, pipes, horns, drawing some of his evidence from the

Renaissance lament tradition in which *malades* are assuaged by listening to singers accompanied by just *one* instrument.[188] The figures Apollo and Pan, medical healers and musical instrumentalists, haunted him. He even wondered what Apollo knew about the nervous system.

Had Disraeli reckoned backwards in time – proceeding from the 1780s to Burette in the 1740s and then to generations prior to the 1740s – he would have found what steep controversy had surrounded this debate about the healing properties of music. John Usher and William Brocklesby, the influential Scottish physician, had touched on it in the 1740s and 1750s, and a generation before then it was debated in Dr Richard Browne's *Medicina Musica*.[189] Browne wrote during the heyday of Newtonian mechanics in the 1720s, when many explanations for neurophysiological transformation were being ascribed to the mathematical forces of heat and motion on bodily matter. These were the 'iatromechanical operations of the human anatomy.' Browne and other die-hard Newtonians stuck to their mechanical credo: the belief that 'matter in motion' could explain the body's altered states.[190] Before the 1720s – still working chronologically backwards – music's healing properties had been debated in almost every generation: especially by doctors and philosophers who exercised themselves over the precise mechanisms of arousal. Many were persuaded of music's stranglehold over the anatomical nerves, but the brain's role was more elusive. Patients like John Usher, who had healed themselves, legitimated the abstract theories by confirming music's restorative properties.[191]

The main point, however, of these examples of theories of musical healing from Browne to Disraeli is their chronological sweep: from the Greeks to the late nineteenth century and beyond. In our world the mind–body divide in musical healing has asserted itself in new and far more complex ways.[192] But if Disraeli could have lived beyond 1848, he would have witnessed the new crest of confidence rising in the late nineteenth century. Neurophysiologists then revived the old Greek beliefs in surprisingly new ways: arguing, for instance, that particular classical composers (Bach, Mozart, Beethoven, etc.) deployed sound in specific acoustical ways to affect the nervous system *directly*. That is, particular harmonies and rhythms have predictable effects on the synapses, their stimulation capable of creating specific moods.[193] Mozart was a favorite, and as the centenary of his death approached in 1891 the neuroanatomical approach was increasingly invoked as validation that he had been the greatest composer who had ever written.[194] Musical genius was thus calibrated, at least in part, to its neuroanatomical potency. The medical implication for healing was equally impressive by 1900:

the notion that one could elicit *specific emotions* – appeal to specific nervous mechanisms that would affect the brain to produce particular moods, ease, joy, abandon – through the music of these composers only. Cure would eventually arise through the process. The implication for the production of (our current) mood swing was considerable in this model in which music virtually competed with pharmacology.[195] Central to these musical debates in the pre-1900 world had been the role of the brain: how it participated in these processes and what part, if any, it played in breaking through the mind–body barrier. No matter how the musical question was framed – and historically it was usually not configured in these terms – the argument revolved around the nerves and nervous system. There was another factor in these discussions: the nerves as minuscule wonders. For centuries, knowledge about them had depended on the microscope. Before the seventeenth century nerves are often noted in anatomical treatises as the most remarkable parts of the body structure. But the naked eye wanted proof: hoping to see them and witness some proof of their pathways. The simple microscope provided scant evidence, but it was better than nothing and confirmed the view of early music theorists like Browne (above) of the major role the nerves played. But nerves were also configured in the popular imagination before 1700 as technical anatomical parts – too difficult for the ordinary person to comprehend. Thus it is no surprise that they hardly figure in the 'little wonders of the world' tradition, as found in Nathaniel Wanley, or in equivalent works in the discourses of curiosity about the exiguous universe.[196] Nor did the Willisian revolution of the brain do much to alter this profile until the 1740s, when poets and scientists alike began to celebrate nervous anatomy. Once that leap was made, it was predictable that nerves would enter the musical debates as principal players, as Kevin Barry and others have demonstrated.[197]

This aesthetic boundary of the 1740s is crucial to the means by which Willisian neuroanatomy was absorbed into the arts. By 1705 (for context, a year after Locke's death and the appearance of Newton's *Opticks*) the Boyle Lectures in London – an interdisciplinary series aimed to bring religion and science close together – began to celebrate nerves, spirits and fibers in published physico-theologies.[198] The prolific Bishop William Derham's deistic lectures of 1711–12, delivered to packed audiences, were particularly explicit on the mysteries of the nervous fluid.[199] After a generation of such descanting on the marvels of the body's interior netways the idea took hold: something new in nervous anatomy, it seemed, had recently been discovered. But it had not. Nevertheless, the public's perception was enthusiastic, credible of almost anything pertaining to the newly *nervous* construction of man, which the poets would celebrate in the 1740s. Poet Alexander Pope had already anticipated the mood in his *Imitations of Horace* (1733–37) when constructing a myth about himself as having been most his most creative, and original when 'enraptured': the word through whose nominative forms he described his bodily state in the time of high imaginative power. 'You grow *correct*,' he lamented in maturity of his lack of nervousness, 'that once with Rapture writ.'[200] The didactic poets of the 1740s followed the lead, celebrating at great length the aesthetic consequences of man's anatomy: composed of more than blood and organs, the human fabric and its component parts also had to be taken account of by poets and artists.

In the 1740s poets began to versify these wonders of the nerves, especially in the long poems already mentioned by Akenside, Armstrong and Flemyng, and in prose treatises such as David Hartley's *Observations on Man*.[201] All these thinkers refer to musical aesthetics as a correlative of the body's expressiveness in the other arts. The Boyle Lectures, however, had not ceased in the 1740s, a decade of sweeping aesthetic and scientific demarcations: they continued to be delivered, disseminating further knowledge about nervous anatomy among other scientific subjects. The model of a meticulous clocklike perfection required anatomical nerves filled with the body's most vital fluid. It alone enabled man to perform perfectly, most of all in the expressive arts. In that epoch of desired progress, there was also room for nervous advancement: if the nerves were not wholly adequate now, they could be educated to perform better. In this sense, they were more plastic than all other parts of the body: organs, solids and fluids. Plasticity was in their nature, a position that almost dovetails with our presentist views of 'the emotional brain' (LeDoux and company) learning to be shaped by its experience rather than being hard-wired from birth with no possibility for alteration.

Nowhere was nervous anatomy more colorful than in the role nerves were said to play in the automata of the mid-eighteenth century. Jacques de Vaucanson's diverse automata – defecating ducks, flying flautists, three-horn pipers and many others – were ample proof of the sunny progress many expected.[202] Salient now were not merely the engineer's ingenious inventions, but also ingenious attempts to create artificial life in the century culminating in Frankenstein (whose 'nerves' merit a chapter in themselves). The internal nervous system, newly translated into motion and movement, became a pillar-mainstay of this invention. And if the automata's 'nervous life' was a topic inevitably pursued by the instrument-makers – the Vaucansons and their brethren charged with demonstrating *nervous* activity among their automata – these projectors were also seen as doing for 'artificial bodies' what God had accomplished for the universe and Prometheus for fire.[203] As for the nerves in life – natural and artificial – so it seemed it would be for them in death and *beyond*; and questions about the survival or demise of the nerves *after* death were becoming inevitable.[204] However, medical research produced from the corpse-world in the late eighteenth century did not catch up with the dancing automata. That arose, as we have seen, in the next century – the nineteenth – when musical debates vis-à-vis nerves became unnervingly sophisticated after the 1850s.

The listener featured prominently in these debates spanning a century (1750–1850). The audience had always been a main player in aesthetic commentaries about music's mysterious language, specifically the listener's perceiving body as distinct from the mind (again the mind–body divide). The quintessential question revolved around the ability of the genius-composer to stir the nervous system. Stephen Jay Gould, the late writer on natural history who wrote so perceptively about the popular under-standing of science, has commented that 'A complete neurological analysis of the listener ... will not explain ... the ravishing beauty and emotional power that I experience in Handel's three great Old Testament oratorios of tragic figures felled by their own incubi of madness, bad judgment, or rash vows (*Saul, Samson,* and *Jephtha*). And a complete neurological analysis of the composer (unfortunately impossible) will not explain why Handel was a genius ...'[205] This approach seems to argue contra the neuroanatomical explanation and to delimit its value. Gould suggests that nothing can explain why Thomas Arne and Salieri cannot rival Handel and Mozart. Yet he comes perilously close to Bishop Berkeley's idealist position about the sound of the tree in the forest if no one is present to hear it. Would Handel's music retain its power if a listener with a defective brain and nervous apparatus heard it? We can imagine Gould moving toward the Romantic view of genius as finally inexplicable, yet if the gap between genius composers (Handel) and the next rank (Arne) is so vast – at least according to Gould – is music then a special case, or is this also true in the case of Shakespeare and lesser literary fry?[206]

The disabled listener may constitute a special case proceeding by exception, but it makes the point about the auditor's body nevertheless. His neuroanatomy cannot explain much, but was fundamental to the aesthetic experience. Part of the reason is that music was said to communicate directly with the soul through a so-called 'language of the soul.' But it soon became apparent that a deaf or mute listener, no matter how 'soul-full,' would experience no overwhelming Wordsworthian joy when listening to a Beethoven symphony. Besides, it may be that music speaks to the body in 'languages' (if this is the right designation) as yet unknown: through 'awareness neurons' of the kind proposed by the late Francis Crick.[207] Either way, the aesthetic debates about the listener's body moved forward after the 1860s. The Romantic revolution in music, the inexplicable genius of certain composers to the detriment of others, and creeping Darwinism after 1859 (i.e., the notion that ideas and institutions as well as organisms were evolving) all took a toll. The strides of neurophysiology in the 1880s, especially in Germany and conjoined to the developing discipline of psychology, were also influential, as they reified the old question of the emotions during the musical experience, bringing the listener's body to the foreground again.[208]

The 'Wagner phenomenon' in this era demonstrated what was at stake in these debates. Controversial German composer Richard Wagner symbolized diverse forces to different groups: cultural icon, patriotic nationalist, deviant usurper, decadent artist, political anarchist and revolutionary, the ultimate Arch-Satan. But among music critics and singers – those living voices who performed his operatic roles – he was also the subject of controversy for the way in which his music was said to derange, and even corrupt, the listener. The routes to corruption were various and included Wagner's harmonies and melodies, as well as the sheer volume of his music and – not least – the stories he dramatized in specific mythical landscapes and settings. Many critics then agreed that his music represented something novel capable of causing 'nervous deafness,' 'diseases of the ear and mind' (toward which end the nerves lying between the brain and ear were analyzed and measured), thereby propeling listeners into the madness that deranged them. This was an approach in counterpoint to Gould's delimitation of the neurological explanation.[209]

The basic idea was that sounds affect the body to create emotion and mood. Beneficial sounds evoke healthy feelings, strange sounds degenerate passions. Music could thus heal as well as corrupt and derange. The anatomical site of these physiological changes was said to lie in the auditory nerves. The sounds would leave the ear and creep down the ganglia of the spine to the rest of the body. In this way pernicious sound could attack the nervous system and derange the listener. It was a view bearing affinities to the old Romantic 'moral therapy' practiced earlier in the century in which bodily health could not be conceived apart from healthy minds: *mens sano in corpora sano*. But which were the good sounds, which the bad? How could the critic – or conversely the medical doctor – tell?

This was the type of question critics such as Eduard Hanslick faced and Hugh Reginald Haweis braved in his *Music and Morals* (1876), in which he built on the post-Schopenhauerian philosophy of his generation and the burgeoning neurophysiology of his time.[210] He harbored no doubt whatever that sound impinges on the nervous body by producing 'moral consequences' (his phrase) in it. For Haweis, the whole nervous system was implicated, especially the brain. The listener was thus the crucial component in music: responding to the composer's sounds and transforming them through nervous responses (auditory, synaptic, cerebral). In an era ripe with hysteria diagnoses and the murmurings of psychoanalysis, this tack was a persuasive antidote to the vague notion that sexual longing deranges the psyche. It was much more persuasive to claim that perverse sounds could derange the ear, brain and the rest of the nervous apparatus, thereby producing 'diseases of the mind.'[211]

Nor was the composer at work in his studio imagining these sounds immune to moral degeneration – indeed he was prone to it. The conjuring of such noise was concrete proof that the composer's mind was *already* degenerate and – if not checked – would derange his listeners. Sound and mental illness were thus brought into closer association than they had been at any time since the Renaissance. The Wagner phenomenon was proof in a nutshell: a composer whose own nervous breakdowns had attested, it was thought, to the impaired state of his psyche. His black sounds, sustained over multiple hours in operas, would steer his listeners into mental illness. Little wonder, then, that it was also claimed he had deployed his operas as vehicles for an attack on the main institutions of Western civilization: social, economic, religious, marital. Hence Max Nordau, the philosopher of degeneration, thought Wagner to be living proof of his theory. And when George Bernard Shaw later interpreted *The Ring* as a veiled attack on capitalism he amassed his evidence based on a tradition of moral criticism of just this nervous physiological variety. Similar charges would be made against Gustav Mahler in subsequent decades, but by then the 'music and morals' ethic was loosening its stranglehold as German scientific positivism gave way after 1900 to Gallican Bergsonian vitalism: the view that all great art moves its viewers by cultivating the unnamed – je *ne sais quoi* – and that it is impossible to pinpoint how sound creates emotion.[212]

These turn-of-the-century thinkers underestimated the complexity of the questions raised about cognition in relation to emotion. Their agendas were often isomorphic: strange sounds producing deranged minds, then sick minds subverting the established order in their symphonies and operas. Nevertheless, they overlooked fundamental relations between the sympathetic nervous system and cognition. In an age when Nietzschean morals were being celebrated, they generated their linear homologies of sound and sickness (pandemic 'shattered nerves') or visual image and derangement (equivalent theories that 'sick' pictures and paintings also corrupted the mind and propelled it into nervous illness). In these pursuits they were not so different from nineteenth-century dream theorists intent on identifying what role the nerves played in dreams. Pre-Freudian dream interpretation often relied on explanations based on changes in the sympathetic nervous system function rather than the prior physiological view locating the dream in the midriff.[213] The old Rabelais-to-Coleridge premise that indigestion caused nightmares yielded to a newer brain-centered view. But it had not yet arrived at the position, for example, of George Mandler. This recent philosopher-scientist has demonstrated how emotions act, in effect, as signals to consciousness that alert an individual to re-evaluate or assess the meaning and significance of current events. Changes of activity in the sympathetic nervous system function as 'computer interrupts' that commandeer cognitive processes. Today we carry the baggage of all these views. We continue to vex ourselves about the relation of sound to cognition and emotion, always invoking the brain. But the degree to which we implicate the sympathetic nervous system varies widely, and many of us agree with Stephen Jay Gould about the limits of the neurophysiological explanation. Something else is necessary (a species of Crick's 'awareness neurons'?) to explain why Handel and Mozart move us as they do. These are not empty arguments about words when we reflect on the unique power of music.

10   Nerves and 'modern life'

From the moment when Georgian 'society doctors' such as Cheyne and Adair described the nerves as the barometers of modern life, the nervous system became the battlefield on which civilization and its discontents would be played out. Their treatises were mainly directed to 'refined readers,' especially female, and the bodies they described primarily 'fashionable bodies,' but the implication for communal nerves could have been applied to all classes. Increasingly, the working classes – even farmers and rustics – began to ape the upper classes; it was only a matter of time before nerves – especially damaged and shattered nerves – would become mankind's common lot.

Cheyne and Adair had written intuitively; without the backup of a social anthropology of mankind, as it were, which would have permitted them to compare savage and civilized nerves, as well as different national varieties. Had they had possessed even the intellectual tools available to Lord Kames in his studies of comparative man, they would have framed their positions differently.[214] But they searched in the dark and still produced remarkable results. The next two generations – those of Trotter of 'modern nerves' fame and the Scottish empiricists (Cullen and company) – configured the nerves as the anatomical guides to modern life. They did so *ne plus ultra*: first as the product of late comparative Enlightenment thought of the type on which Kames drew, which measured civilizations and their progress according to their cultural products and material goods;[215] secondly, while building on a new post-Enlightenment anthropology that compared races and nations, and ranked them according to virtually every category of human endeavor from creativity to productivity;[216] and finally in the light of the doctrines of progress which permitted thinkers to historicize the nerves in ways they had never been before.[217] This was a tall order. It could therefore be said that the nervous explanation of human behavior peaked in the early nineteenth century. It certainly crested when scholars then charted the progress of mankind from the dawn of time to the present day; from the wildest savage in the forest or jungle – so the argument went – to the most civilized European in the most cosmopolitan capitals. In practice all of this amounted to a new view of history based on physiological advancement.

There had been numerous versions of what I am crudely calling 'nervous social anthropology' before the development of 'proper' anthropology. Between the late eighteenth century and Darwin's evolutionary writings, there were many: not merely Kames and the Scottish philosophers, but others as well.[218] Nowhere is it clearer what was at stake than the full-length (perhaps too complete!) treatment Robert Verity gave to nerves and civilization in the early nineteenth century. Like Cullen, his teacher together with Dr Monro, he was a Scot trained to be a physician in Edinburgh at the peak of 'Enlightenment' in the 1780s.[219] There he first heard about the 'physiological view of mankind.' But he soon headed for France and Germany, as had Coleridge, Wordsworth and other idealistic Romantic thinkers. In Göttingen he sat at the feet of Blumenbach and heard his lectures on the races of comparative man. And he read widely, and in different languages, while there and – later – in Paris: Sir William Temple's reflections on the Dutch, Vico's comparative study of peoples, Kames on developing societies, and his Göttingen teacher Blumenbach for his anthropological views of comparative man. He also perused B. G. Niebuhr, Herder and Humboldt for their travels and the French naturalists from Cuvier and Cousin to Broussais, Guizot and Michelet.[220] Kames was the major influence: beginning with the long Kamesian epigraph of his book to his conclusion about the 'nervous way of life' among modern peoples, Verity looked to Kames to buttress a physiological conception of human history.

Verity's argument amounts to this: savage man possessed *small* brains (what we would call brains with few neurons) and underdeveloped nervous systems. As populations developed into tribes and races, and nutrition improved over the centuries, both brains and nervous systems became more highly developed. Even before the flowering of ancient Greeks, barbarians and civilized races could be distinguished by their nervous apparatus. Nutrition was the great differentiator: the more protein nourishment a civilization had, the more *nervous*, and hence more prosperous, it became. Abundant food increased body size and strength, which in turn strengthened the mind and its intellect, leading to ever more secure systems of commerce and wealth. Verity's sequence errs, but nevertheless is thoroughly a product of late Enlightenment thought: based on logic and reason, and predicated on the belief that nutrition and capital were essential for progress and social amelioration. As Verity repeatedly states, it was Kames who first stumbled upon the first principle of the system he (Verity) would generate: the recognition that civilization proceeds upwards according to 'the extent and magnitude of corresponding changes taking place contemporaneously in the interior of the nervous structures.'[221]

It is nevertheless an odd view, even if Verity has absorbed the writings of the French evolutionists. While the purpose of his exploration is historical – 'to recognize the existence of the predominating nervous organization of modern individuals compared with those of former times' – Verity's grand hypothesis aims to explain how the future will be determined by the nervous system. Verity develops this point principally in two chapters (33 and 36), which leads him to the conclusion that all great men and women – the Jane Austens, Brontës and Wordsworths of his time – 'have developed nervous systems.'[222] By 'developed' he means *more* nerves, *more* neural pathways, *more* exquisitely developed fibers and muscles capable of ever-finer gradations and distinctions than the average person, and a larger brain size.[223]

Verity never uses the word 'plasticity' (in any of its grammatical forms) but his analysis clearly implies it. Modern life, he claims, echoing Thomas Trotter before him, presents nervous man with all sorts of new stresses. The more highly developed his brain (especially through protein nutrition), the more he will be able to cope with increasing these public and private strains. It is a sort of Mandevillian myth of the beehive – the idea that private vices are definite public goods – applied to the evolving anatomical self. Furthermore, the more successfully mankind copes, the more his brain gradually habituates itself to these stressful conditions. But it is always *civilization* that alters the nervous system, not the reverse. The hypothesis Verity constructs is seemingly airtight – or so it appeared to him – placing high value on the flexibility (our plasticity) of the synaptic nervous system to adjust to its cultural milieu. And if not precisely 'plasticity' in our current neuroscientific sense, then he suggests a nervous system sufficiently malleable to the extent that the brain learns from experience.

This was Verity in 1837, revised and expanded in his much later revision of 1870, when he also brought out his second book on 'the double brain.' But even he sensed that not all persons would cope well. There were bound to be those whose evolving nerves and brains spelled doom for them: ostracism and marginalization from the masses. If the 'nervous type' were evolving, only the fittest would adapt to civilization's stresses. The rest would not cope, and would fall by the wayside and die. Verity was still alive in 1859 when Darwin explained to the world how species transform themselves for survival, some unable to adapt, and was still spinning his theories about the brain in 1870.[224] Among the slowly evolving 'types' surveyed by both thinkers were the sedentary species – scholars, students, clerics, literati, all those who sat for long periods without the regular activity of their muscles – and they were especially susceptible to nervousness. What remained to be shown was that their *nerves*, rather than other parts of their anatomies, constituted the vulnerable sector. 'Shattered nerves' came to be seen, in part, as 'literary nerves': a temperament hewn out of naturally inherited melancholia coupled to a forever failing nervous system.

Only the educated of high Victorian England can have aspired to these sedentary professions. Thus 'literary nerves' was de facto a class-bound condition, as Adair and his cohorts had recognized almost a century earlier.[225] But how – we wonder – did Verity's theory validate itself in routine life of the time? Who are the suppliant examples of his nervous coping mechanism and, conversely, of the nervous decay he sought to define before a theory of degeneration was in place? If we could understand a test case, we could fathom how Verity – an early pioneer of this theory despite its sundry failings – persuaded himself of its legitimacy.

11 'Literary nerves' and evolutionary biology

A case in point is found in the life of the idiosyncratic John Addington Symonds (1840–93).[226] He is a particularly useful biographical case for Verity's 'nervous civilization' hypothesis because he was a lifelong patient, as well as a remarkably able classicist and scholar. He was also an overt homosexual man who agonized and reflected abundantly on the origins of his sexual nature. And he was a lucid commentator on the condition of his nerves, and the plasticity of his brain, at a time when the connection between the two was not yet entirely explicit. Some of this awareness arose in his childhood. The son of a prominent doctor in posh Clifton Village, outside Bristol, nothing medical was alien or repugnant to him: indeed, he was morbidly fascinated by his 'nervous history' to the point that he wrote it out in his letters and memoirs.[227] He was a 'patient' in many senses and he conceptualized his deviant sexuality – which he construed as a branch of 'nervous medicine' itself – before there were ready-made labels or words to convey it. Biographically, he had been the lynchpin of a series of sexual scandals at Harrow and Magdalen College Oxford that irrevocably damaged his health, or so he thought they had, and propelled him to abandon his Oxford fellowship, alter his profession and permanently leave England.[228]

The scandals at Oxford were 'epistolary': another Oxford classicist had written to the Fellows of Symonds' college (Magdalen) that their classical scholar was in love with boys, which prompted the College to hold lengthy investigations in the autumn of 1862. The Fellows formed an internal court, held proceedings chaired by the President, requested sworn statements and legal depositions, and – in the end – found that while Symonds' erotic desire for choirboys was reprehensible it could not, in itself, be the basis for termination of his fellowship. But the proceedings were too much for the fragile Symonds, who survived to the end of the term and then resigned his fellowship. Nervously ill with a consumption that had returned when his lungs caved in under the pressure of his inquisition, and hoping to recover under sunny Italian skies, he left the country. But he paused in Davos, then a hamlet high up in the Swiss Alps, where his sister was already recuperating, took to the place, and spent most of the rest of his life there.

Symonds' upbringing indoctrinated him with an understanding of the new 'nervous' medicine few others of his generation enjoyed. While still a pre-pubescent boy his doctor-father had been intrigued by abstract questions about the nature of sleep, memory, dreams and the role of the imagination during sleeping states. Dr Symonds had pondered what part the nervous system played.[229] His son John heard these ideas and commented on them later in life in his massive correspondence and in carefully composed memoirs that annotate what I am calling 'literary nerves' and evolutionary biology.[230] One passage, chapter 14 of Symonds' memoirs, entitled 'Intellectual and Literary Evolution,' is especially valuable in this regard.[231] Here he explains how, at an early age, his nervous system and brain colluded to stage the demise of his health and set him up to be persecuted in the Oxford scandals. From then on, he retreated into a private world in remote geographies where he could spend his life reflecting and writing. The course from nervous breakdown to imaginative writing was smooth, the one segueing into the other almost preternaturally. What was his rationale for this extraordinary conclusion in so young a man?[232]

'Irritable nerves and a morbid condition of the reproductive organs,' he writes, were 'due to the particular erethism[233] of my sexual instincts, and the absurd habit of ante nuptial continence, rendered me physically a very poor creature.' This self-diagnosis is not merely the work of a physician's son, but also reveals someone who is widely read in the medical-scientific literature of his time. Symonds turned himself into an authority on 'nervous erethism' on the grounds that it might explain his sexual preference for other males. Even before he went up to Oxford, he combed libraries for books capable of explaining his constant sexual excitability: his 'particular erethism,' as he calls it. The idea of excessive sexual excitability – erethism – had developed in the aftermath of Haller's theories of sensibility and excitability, already mentioned above.[234] By approximately 1800 medical doctors were exporting Hallerian notions of 'excessive sensibility' to the new 'moral therapies' aimed to restrain sexual excess.

Jean Amédée Dupau is an example. Trained at the famous medical school in Montpellier by medical professors vigilant to the mind–body barrier and theories of sexual excitability, Dupau specialized in erethism.[235] By the 1830s the French doctors, especially those concerned with sexual aberration, invoked erethism as a cornerstone of their explanations: the patient is predisposed – anatomically and psychologically through his nervous state – to it through 'moral pollution.'[236] The nerves are in a perpetual state of inflammation owing to sought-after sexual stimuli. The more the patient restrains his appetites, as Symonds did, the more voracious the erethism, or excitability, becomes. Inflammation escalates so quickly that the mere thought of, or fantasy about, anything sexual arouses the patient and causes emission – so delicate and exquisitely sensitive are his nerves, especially in the genitals. The process depends entirely upon an overly aroused nervous state. Thus the genesis in Dupau's generation of the word 'nervine' – the state of the nerves in extreme sexual excitability – coined to assist 'nerve doctors' like himself to describe erethism. Remove the nerves and 'erethism' ceases to exist. It was one of several erotic moods predicated exclusively on nerves that – glancing backward from our vantage point – seem to have been conceptually generated to account for perpetual arousal.[237]

Symonds also attributed his nervous sexual sensibility to an exquisitely delicate dermis. The result was involuntary sexual stimulation, even when not consciously aroused. An overly sensitive dermis and a keen imagination worked in conjunction to excite him into states of sexual arousal. He considered these 'morbid' specifically for the vivid way they crept up his spine and ganglia to alter his brain: '...in some obscure way my brain became functionally disordered. They called the affection hyperaesthia, and gave it all kinds of names.'[238] But Symonds considered it derivative from his nervous apparatus. This neuroanatomical constellation drives his self-diagnoses and leads him to conclude 'morbidity' or – more accurately – '*homo*morbidity.'[239] He reasons that the 'morbidity' extends to the plasticity of his brain responding to, and becoming altered by, erethism.[240] Moreover, the neural damage in his altered brain also causes a state beyond mere erethism: to 'hyperaesthia,' or excessive sensibility in the brain itself. This was his idiosyncratic explanation, based on the best medicine of his time, of how he had been brought down at twenty, driven by nervous pathology but aggravated by pulmonary tubercular consumption.[241]

Elsewhere I have argued for the need to refine this term from mere 'morbidity' to '*homo*morbidity': not merely for Symonds, but for homosexuals in history who tell themselves stories about the way in which their strained nervous systems have altered their bodies and brains.[242] This linguistic refinement notwithstanding, mind and body functioned, Symonds thought, to predispose him to an alternative sexuality: 'I lived in fermentation.'[243] Larger than life and more nervous, he thought he burned at a higher temperature in all activities: intellectual, domestic and sexual. His lifelong chronic tuberculosis was proof of this nervous fatalism. Both served to shield him from challenges he could not have faced and duties he could not have met. Yet working against the grain of this nervous hyperactivity was a repressive Victorian moral order, heightened in fastidious Oxford where the mere notion – let alone the reality – of sexual longing for other boys was grounds for expulsion. All these streams fed into the massive river that would be neurasthenia by the 1890s in the milieu of Ellis and Krafft-Ebing. The developing homosexual was inherently neurasthenic, which further served as proof of his degeneration: sexually uncontrollable, debased, 'erethistic' and thoroughly 'nervous.' No wonder Symonds became so 'morbid' about the way the Magdalen College authorities had turned on him.

This disentanglement of literary and medical nerves may seem technical and involuted to us, but we must not forget that 'erethism' of Symonds' variety was the archetype in Foucault's mind when expounding on the repressive discourses of Victorian sexuality in the *History of Sexuality*.[244] The 'repressive hypothesis' was originally a Nietzschean concept necessary to demonstrate the degree to which the Victorians had engendered an inimical discourse. Foucault amplified the Nietzschean notion by emphasizing the *discursive* part: i.e., sexual arousal occurred primarily by discursive stimuli. Victorians reading about Arcadian shepherds in love, or the loves of Corydon, were as likely to be aroused by them as by the glimpse of boy choristers in their college chapels. Hence the readerly path to sexual freedom – Foucault pronounced in a now much-quoted passage – could be found in direct *resistance* to 'discursive erethism':

> Since the eighteenth century sex has not ceased to provoke a kind of generalized discursive erethism. And these [Victorian] discourses on sex did not multiply apart from or against power, but in the very space and as the means of its exercise...From the singular imperialism that compels everyone to transform their sexuality into a perpetual discourse, to the manifold mechanisms that, in the areas of economy, pedagogy, medicine, and justice, incite, extract, distribute, and institutionalize the sexual discourse, an immense verbosity is what our civilization has required and organized.[245]

Symonds is living proof of Foucault's 'discursive erethism' and would have sympathized with Foucault's view, I think, if he could have known about it, for the way it had ruined him and his health. He spent the rest of his life writing out – 'discursively' – his nervous health: creative states, moods, breakdowns, loves, the lot. He was not alone: one thinks of Carlyle steeped in his own morbidities, and – at the other side of the nineteenth century – Fréderic Amiel nervously obsessed in Switzerland, the bed-ridden Proust in Paris (although not entirely from illness), not to mention dozens of Italians and Spaniards who have also been strung out on their nerves. Writers as diverse as Woolf, Beckett and Genet have each incorporated nerves into the fabric of their artistic visions: Woolf explicitly in her sense of the heritage of sentiment and sensibility when writing about Laurence Sterne's cognitive states of mind, as well as the condition of her own nervous body;[246] Beckett for the nervous laughter that captures the basic horror of his black humor; Genet in the way flesh – sinewy, nervous, nerveless – becomes an *arrière pensée* of his dramatic textures. Genet empties himself into a sensible world and, in turn, incorporates this world back into himself, maintaining his own perpetual state of erethism – nervous sexual excitability – even if the condition is less edgily 'nervous' than Beckett's. He folds and unfolds his flesh like a game of cadavre exquis. The grotesque results are unexpected and monstrous, as if in a game: a juxtaposition of body parts organized around intervals separating each fleshy segment. The versions of 'literary nerves' presented so far were not the only types. Nerves also played havoc at the limits of discourse: playful and alternately bitter, each encouraging their own discursive practices – as Foucault had intimated in his analysis of Victorian erethism – against moral interdiction. Others in the decades leading up to Symonds' death in 1893 – one thinks of Russian composer Tchaikovsky who also died in that year of mysterious causes, a possible suicide wrecked by his own 'shattered nerves' after a failed homosexual infatuation with a young corporal[247] – followed suit. Many of these artists, composers and writers sought out the Philosopher's Stone, so to speak, to explain the mysterious workings of their own nerves and brain. But by the nineteenth-century *fin-de-siècle* neuroanatomy had moved on. Even classicists like Symonds resorted to neurons and dendrites, the interconnections of various nervous systems and the ganglia, to denude the mysteries of man. When Ramon y Cajal (1852–1934), the pioneering histologist of his era who won the Nobel Prize in 1906 for his neuroscientific research on the brain, mounted his own campaign in the 1890s for the role of regeneration of the brain which implicated synapses, he intensified a nervous view of life that would crest in the twentieth century's synaptic revolution.[248]

This transformation could not have occurred without the aid of language – language in the service of science. We have already commented at large on the rise of a 'nervous style.' It was not the only linguistic domain to codify the gains of the new neuroanatomy. Late nineteenth-century theory continued to generate concepts from nervous coinage as well as neologisms derived from the nerves (momentarily we shall have examples). An argument can even be mounted that the nineteenth-century hysteria diagnosis thrived on it. One case was that of the influential surgeon Sir James Paget, knighted for his services to the teaching of surgery.[249] He never claimed to be an authority on nervous diseases and was first and foremost a general surgeon (rising to principal surgeon to Queen Victoria and King Edward VII) and a prominent professor of medicine in London. But he wrote abundant articles intended for the medical community where he was scrupulous in the use of medical terminology. He considered hysteria to be the most mysterious ailment of his era and, in common with many other physicians, was intrigued by its murky disguises. To help him explain to students what occurs in the hysteric's body he generated the term 'neuromimesis.'[250] He intended this as a synonym for hysteria itself while searching for a word dramatically emphasizing the nervous dimension of the malady's protean transformations – one word capturing hysteria's imitative components (its ability to 'mimic' other diseases) without excessive concern for mimesis' different profile in art and aesthetics.[251] Paget was resolute to summon attention to the fact that all forms of hysteria arose when dire changes in the patient's nervous system occurred. Paget supported his claims by glancing back at earlier nineteenth-century views that the nervous system itself is inherently disposed to mimic local disease, leading to symptoms that are 'disguises' of other conditions. His vivid neologism – neuromimesis – allowed him to system-build:

> Neuromimesis cannot be found in all persons alike, or in any person at all times. It may be regarded as a localized manifestation of a certain constitution; localized, that is, in the same meaning as we have when we speak of the local manifestation of gout or syphilis, or of any other morbid constitution which we regard as something general or diffused...[252]

His lectures continue for dozens of pages divagating about 'nervous mimicry': the 'essential hysteria' itself – and cautioning his medical students that 'there is no greater fallacy than to suppose that nervous mimicry, or hysteria, or any of the allied forms of disease, can be referred to any malady or any other part than the nervous system.'[253] His *cri de coeur* was genuine (what he keeps calling 'real'): hysteria is a disease that attacks in any part of the body. However, the localized condition, apart from hysteria, must not be minimized. Hysteria, the disease par excellence of 'neuromimetics' (the dynamic process of nervous transformation), is *real* precisely because it is constitutionally capable of generating genuine injuries to every organ in the body and diseases of every kind known to men and women through the nervous system.

Doctors drawn to 'rare and new diseases,' as Paget was, understandably gravitated toward these nervous glossaries and odd neologisms.[254] Hysteria in particular lent itself to be crowned at the summit of the neuromimetic conditions for the way in which it could insidiously disguise itself (almost like a virus). The ultimate trickster, it masqueraded as other diseases to fool patient and doctor alike. In Paget's day, two centuries after Thomas Sydenham and his colleagues had made the first neuromimetic diagnoses in the late seventeenth century, the contexts had altered. Darwinism had taken its toll a generation earlier on nerves as nervous reorganization of the body had permitted weaker species in evolution to emerge. Then, in the 1870s, war and widespread economic upheaval left European society (as well as American, in the pathetic aftermath of the Civil War as Americans tried to 'reconstruct' themselves) inert. It was as if their nerve had given out. Waves of migration in the 1880s and the displacement it caused created further alienation of mind and body. By the 1890s tides of scientific positivism sweeping over Europe while Western societies rapidly industrialized themselves and the pace of life, especially in cities, accelerated into the quick-paced rat race it became in the twentieth century. As mobility increased, and people found themselves hundreds of miles further apart at night than they had been in the morning, the sense of time – literal and psychological – changed.

These shifts took a huge toll on nerves in daily life and produced a counter-response in the realms of imagination. No wonder that 'new and rare diseases,' in Dr Paget's words, were cropping up in life and in literature. Some appear in the obscure Francesco Canaveri's *Neuronomia* of 1836.[255] This is an odd prose fantasy commingling literature and medicine, the new 'French' hysteria and (even) Italian nationalism. Two generations later and across the English Channel, George McIver seems to have been as obscure in England as Canaveri was in Italy. Another 'one-book man,' he is the more able satirist, if also less preoccupied with 'new nervous conditions.' McIver transformed his anxiety about the unrelenting acceleration of time into a Swiftian satirical utopia called *Neuroomia: a New Continent* (1894).[256] His 'new continent,' whose name translates literally as 'nervous place,' takes the reader to an imaginary country called 'Atazalan' for its Aztec associations. The narrative proceeds by recounting the reflections of a 'Captain Periwinkle' who – like Swift's Lemuel Gulliver named for his chief attribute – visits Neuroomia and Atazalan only to be struck by their difference from his native England. Here the average lifespan spans between one hundred and two hundred years. The absence of 'labor strikes' and 'inflammatory speeches' leads to perfect health. Disease is absent, hospitals unnecessary in this 'country of perfect health,' and – remarkably for the 1890s as Europe's cities swelled in filth and grime – the health of urban dwellers is as perfect as the ruddiest cowhands in the country.[257] The 'new threatening diseases' of McIver's era – cholera, hysteria, neurosis, sexual 'perversions' such as Symonds' – are unknown because in mythical Atazalan the nerves exist in equipoise, sans excitement or irritation. In both city and country the anatomical nerves oscillate harmoniously as time marches slowly and the pace of life is peaceful. McIver's aptly named continent – 'neuro-omia' or the 'nervous place' – is an imaginary country, a new Atlantis or Houhynhymn land, lorded over by relaxed nerves. How timely McIver's utopian fantasy seems in 2004.

McIver's parodic utopia coexists with other 'fantastic literature' – however different it may be – inspired by fanatic religion and driven by sectarian evangelicalism. For example, a discourse about the body of Jesus had developed among enthusiasts in the nineteenth century making much of the savior's nerves. It is not altogether clear why such an arcane topic should have interested them but it did and was part of a larger (not always scientific) discourse about the bases for the resurrection of the body of Christ. Henry G. Waters, an American Quaker calling himself 'Salvarona,' a pseudonym meaning savior and not to be confused with Savonarola, the sixteenth-century Italian religious reformer, produced a whole treatise about Jesus' nervous system.[258] His intentions are as spurious as his various names on the title-page, not merely 'Salvarona' but also 'Jesus Christ.' It is the fact that he tapped into a sub-discourse about the nervous body of Christ that arrests.

Salvarona's voice in this strange little work is difficult to locate, but he is serious and less than ironic. The five chapters have titles such as 'An Analysis of the Nerves of Jesus' and 'The Nature of the Nervous Forces of Jesus' and Waters' point, perhaps predictably in light of the nineteenth-century heritage of nervous sensibility, is that Jesus' nerves were more exquisitely sensitive than anyone else's in history – this is specifically what distinguished him. There can have been few readers of this work other than believers, nor is its message credible. But it demonstrates how far earlier doctrines of nervous sensibility had progressed. If their validity was waning among the educated classes, or at least altering to a different doctrine about neuroanatomy in the generation after Ramon y Cajal's influential one at the turn of the century, the fringe literature absorbing its ideas continues to be written.[259]

The language of nerves was also imported into ordinary critical parlance after the turn of the century and frequently appears after 1900 as a descriptive convenience to convey the domain of the excessive. Coinages such as nerve-center, nerve-like, nervine, now taken for granted, begin to appear with such regularity by the Great War that it is unnecessary to document any particular instances. When, for example, French novelist-commentator Marguerite Yourcenar (1903–87), author of the best-selling *Memoirs of Hadrian* (1951), wrote in her juvenescent *Diagnostic of Europe* (1929) that it was certain 'our successors will pay a price for our nervous expenditures,' she invoked the word without any of its anatomical connotation.[260] Their expenditures are 'nervous' merely because those granting them, the miscreants of the Crash of 1929, are profligate. It is a psychological attribute, as well as an indictment, of a generation, not any quality inherent in the expenditures themselves.

'Nervous words,' as another commentator had noticed shortly afterwards in 1930 on the anniversary of the Crash as its effects were hitting hard, are 'through and through psychological.'[261] They had slowly been moving that way for a century. Even Yourcenar's context (above) remains that of a decadent 'Western civilization' teetering on the edge: having peaked, fragmented and now in postwar gloom newly chaotic, tumbling toward precipitous decline, broke and penniless. Ultimately Yourcenar's coinage – 'nervous expenditures' – is not a fiscal trope but a cultural one; personal shorthand for imminent doom – as so many of the medical doctors' 'nervous bodies' had been before psychological collapse. 'Nervous words' are performative: they enact what none others would or could, denoting spheres of delicate uncertainty and reckless excess. If the 'expenditures' of the generation of 1929 had been confident rather than ruinous they could have been justified. But they were not, anymore than the nerves of healthy (male) bodies had been taut and tonic in the generations of Georgian Doctors Adair and Cullen.

By 1900 nervous words also begin to appear with some frequency in drama criticism. Phrases such as 'nervous warmth' and 'nervous tension' are routinely invoked to capture a *je ne sais quoi*: the actor's emotional attachment to the audience combined with a constantly moving edge. 'Warmth' and 'tension' are commonplace qualities in actors often remarked upon by the critics. The inclusion of 'nervous' attempts to heighten the effect of these qualities while conceding that the perform-ance radiated energy defying precise description. But 'nervous words' are not confined to economic diagnoses or dramatic performances. They have been diffused so widely throughout the language, in most Western languages, and imported into so many domains beyond the literary and the medical that it would be tedious to compile a glossary of types – let alone usages and nuance. Deconstructionists as well as postcolonial novelists also now select from this broad reservoir of 'nervous vocabulary' to test boundaries and explore limits that would otherwise be more difficult to capture discursively. When Tsitsi Dangarembga, a contem-porary Zimbabwean novelist, calls her novel *Nervous Conditions* she draws on this sector of contemporary vocabulary to enable her to interrogate the borders of current 'body discourse.'[262] Her plot focuses on the alienation of Shona women from restrictive traditional practices. Barred by custom from disagreeing, verbally or actively, with their families, one woman bravely rebels by feigning paralysis (African version of the old European hysteria?) while another starves herself into a Western-style anorectic body. In both cases, the women rebel explicitly by acts of bodily denial, hence their dire ensuing 'nervous conditions.' Dangarembga's novel illustrates how the acquisition of education and the adoption of Western practices (hysteria, anorexia) can have painful, if nervous, consequences for modern African women. Anti-colonialist critics have condemned the colonial education that alienates African intellectuals, such as Dangarembga, from their roots; but this stance bases itself on the posture that males alone may rebel against colonialism, and reduces women who do so to such 'nervous conditions' as madness and hysteria. Both of the Shona women in Dangarembga's *Nervous Conditions* rely on solid colonial educations to escape from their subordinate states and discover their identities, but not before their bodily states – their 'nervous conditions' – have been altered. Meghan Vaughan is right to claim that 'body novels' such as *Nervous Conditions* can only be understood as cultural anatomies of colonial power dissected through the grid of 'African illness.'[263]

Alternatively, there is Michael Taussig, a hard-to-nail-down theoretical anthropologist who titles his recent study of the place of the tactile eye in both magic and modernity *The Nervous System*.[264] Taussig was trained in medicine and has held major professorships of anthropology. Based on anthropological fieldwork in Australia and Colombia, his fascinating essays draw upon on the workings of the human nervous system to illustrate concepts of culture, especially its fetishistic and shamanic elements, as well as the role of mimesis and the magic of the state. Here the central theme is conveyed by the title's double meaning: on the one hand portraying the human nervous system as a controlling – even deterministic – force; on the other, as a system that is not at all systematic, but intensely nervous, and therefore on the brink of collapse. Tacitly he cajoles us to ask whether 'all systems are to some degree nervous.' Taussig has always been intrigued by the senses, especially the visual sense, and the way in which the five senses combine to produce *nervous* terror. In his gaze the political state too has long tentacles, intruding into the anatomical ones so menacingly that paranoia is inevitably produced.

I have tried to follow suit in these 'nervous acts.' Mine amounts to a lifelong Apollonian attempt to test the limits of the borders between literature and medicine – the arts and the sciences – through the provision of a textual body for nervous discourse in history. But the approach is also meant to be 'realist' in every sense, not least because the 'nervous body' is not merely a fiction any more than the 'brain' is an imaginary construct. All over the world millions of people suffer from their combined effects every day – nervous diseases, neurological disorders – despite their inability to be as articulate about the drastic fallout as Zimbabwean novelist Tsitsi Dangarembga or Australian theorist Michael Taussig.

Exploration, like so many of these prior 'nervous acts' from the Greeks forward, necessarily dwells on the broad neuroanatomical lineage construed contextually and discursively.

12 Coda: discursivity and the pharmacological future

We have traveled a very long way in Western civilization from Greek Galen and English Evelyn to the advent of this century. Two centuries after Evelyn's death in 1706, by the time high Victorian culture came to a final close and the fruits of empire had long set in, the 'nerves' had finally arrived; having been anatomized, literalized, medicalized, patholo-gized, and now – more recently, as we have seen – colonized and absorbed into the discourses of empire. At no time was the economic dimension of 'nerves' overlooked or soft-pedalled. Entire books can be written anticipating our contemporary billion-dollar industries predicated on pills and potions – books that chart the drugs-for-profit trade in 'calming the nerves.'

Too easily perhaps we imagine that the arrival of the large pharma-ceuticals, the Glaxos and Pfizers, were necessary historical conditions for the growth and profitability of drugs, but these million-dollar businesses long antedated the twentieth century even if their profits were then smaller. They flourished at large in the mid-eighteenth-century world in which now thoroughly obscure self-made entrepreneurs such as Sir John Hill, to name only the most illustrious of this tribe, quickly pounced on 'new nerves' and made himself rich through his 'nerve panaceas.'[265] They intensified in the Regency and Victorian eras. Someone, for instance, signing himself 'I. Lewis' who continues to defy biographical identification was as unabashed about the efficacy of his nervous panacea as any author of the eighteenth century.[266] Lewis wrote a full-length treatise in the 1870s serving up the whole, he thought, of 'modern nervousness' construed 'medically, philosophically, religiously.' But his goal was lucre and profit and he exploited Edward Jenner's Victorian legacy as the founder of vaccination by advertising that his (Lewis') potion was as efficacious for the nerves as the great beneficial Jenner's 'Neuropathic Remedy.' 'This muscular elixir,' Lewis claimed, is every bit a 'great discovery' as Jenner's vaccination for the smallpox:

> Dr Jenner's Neuropathic Remedy is sanctioned by experience, and originated in the labors of a scientific and philosophic mind, whose basis was nature. Those who are depressed, hopeless and despairing, who look on life as a burthen, and, whilst surer its load, have not yet

been tempted to lay it down... whose nights are as wretched as their days, and who unknowingly are suffering from neuropathic maladies, affecting the nerves or muscles, or both, will, by taking this invaluable medicine... feel once more those ecstatic emotions, which the diseased brain in vain hopes to find in intoxicating beverages, narcotics, or sensual gratifications.[267]

Substitute new names and nostrums, render them recognizable in 2004, and you grasp the range of continuities in this nervous heritage. Lewis' words ring true almost a century later for all those in search of a quick fix: depression, hopelessness and despair; life captured as burdensome, stressful days, sleepless nights; distraught bodies afflicted by poor diet, lack of exercise and obesity. This possession of 'the diseased brain,' in Lewis' confidently bloated prose, which often makes life appear impossible to live. In bleaker and more sculpted dying words, they could be Beckett's – so nervous is their content, as was his. So fraught that only our latest pill or potion, Prozac or Ritalin, or whatever will replace these, which in turn replaced Hill's 'wild valerian for nerves,'[268] sustains us in the hope that we will ever again 'feel once more those ecstatic emotions' which we have long forgotten. Or which we thought – in Lewis' construction – could be restored by 'intoxicating beverages, narcotics, or sensual gratifications.' The chemical recipe of Jenner's 'nervous panacea' may never be found, nor the profits it made for those, like the obscure Lewis, who later touted it as *'Jenner's universal nervous remedy'* and then superadded their own brand. Jenner's was merely one of the most successful of many such nostrum makers. There have been hundreds of others since then.

In my earlier essays, as well as in this re-evaluation, I have been suggesting that the lineage of the nerves – the basis for my 'nervous acts' – has had a discursive, literary, rhetorical, metaphorical, epistemological, ontological and even theological profile as consuming as the more apparent pharmacological heritage. All my efforts have been directed in that direction and it would sustain me to think, more diachronically in the long eighteenth century, that it may not be possible any longer to view the rise of the European Sensibility movement apart from its scientific and medical moorings. Indeed my plea for retrieval has been almost entirely discursive and pitted against the realist pharmacological claim that proceeds by pretending that puffing like Lewis', no matter how disguised, is not performative and rhetorical. Yet the die is cast: as contemporary life grows ever more stressful in late capitalism, and as personal depression of many protean shapes disguises its earlier versions, the pharmaceutical claims on nervosity, which hold out great promise for all of us, will gradually replace all these.

The transcendental nerves – metaphoric, symbolic, subjective, trans-historical, neo-Kantian in their essential idealism – have another profile in recent history. Their trajectory, qualitatively not dissimilar to the one implied in the disjunction between the anatomical penis and the subjectively signifying phallus of contemporary theory, has been different. The transcendental nerves were born in the late neuro-anatomic milieu of the last century and often spoke, as it were, in different tongues from the physical nerves. My research over four decades was grounded in the historical, primarily non-transcendental nerves, despite inviting these literal and physical nerves to speak for themselves. When theory took off in the 1970s I naturally saw the symbolic conjunctions of the two types and often made links between them, as the below essays aim to do. Indeed it was impossible not to recognize, and be struck by, the velocity with which phrases such as 'shattered nerves' and free-floating concepts such as 'the nervous body' had become unmoored from their anatomical anchors in such a short time. One could think it was a disjunction, again, between 'two cultures' speaking different languages containing different vocabularies, as we saw in the opening of this essay. But the disjunction was less that than the collision of ideologies in conflict, or – at the least – the struggle between the epistemological profiles of different academic disciplines: the materialist history of science and history of medicine dwelling on bodily organs versus the more transcendentally signifying – and immaterial – literary and psychoanalytic theory. 'Essentialist nerves' exclusively in the service of theory must, of course, be configured ahistorically and transcendentally – are necessarily loaded ideologically in the way Kaja Silverman and others have reconstructed the 'phallus of malice', as in Silverman's *Male Subjectivity at the Margins* (1992). Yet even Silverman knows that the symbolic immaterial phallus had a life prior to its existence in the blissful transcendental state of neo-Kantian abstract transhistory.

So too the nerves, despite no '*nervocentric*' ideology having developed in any parallel linguistic universe to the '*phallocentric*' one. I am not the first to concede the abstract capability of transcendental states of 'shattered nerves,' especially in late capitalism when theory has become such a necessary and intrinsic part of our critical lives. But here, in these equally 'nervous acts,' I have not sought to retrieve that more transcendental, if also far more recent, tradition. That labor awaits someone else, or at least awaits me in another lifetime.

Notes

1. Evelyn (1850), I: 54.
2. This was symptomatic of the conventions of periodization in literary history.
3. Francis Harris (2003), for example, is indicative of the new Evelyn.
4. I have found no other use of the phrase 'originated neurology' anywhere in Evelyn's generation.
5. For knowledge in that decade see D. O'Malley (1996), Robert G. Frank (1990).
6. For Thomas Willis, one of the main figures in this collection, see chapters 2 and 5; also H. Isler (1968); Conry (1978); Robert G. Frank (1990); J. T. Hughes (1991); A. N. Williams and R. Sunderland (2001) and the items under Thomas Willis in the Bibliography.
7. I mean Sensibility with an upper-case S and shall have more to say about this word in relation to the rise of neuroanatomy.
8. See the works in the Bibliography under Gerald Edelman.
9. See J. Z. Young's works listed in the Bibliography.
10. Edelman (1987), p. 3.
11. See C. P. Snow (1959) and F. R. Leavis' replies, collected in Leavis (1972).
12. The Ancient–Moderns debates provided further impetus after 1700; more recently our disparate fields have been fused in the work of historically-oriented theorists: Canguilhem, Foucault, Habermas, Feyerabend, Kuhn, later on Roy Porter and many others.
13. See Stanley Fish (1992) and the reply of G. S. Rousseau (1999).
14. For example, I remember endless conversations with my contemporaries Arthur Donovan and Arthur Quinn, both of whom have become distinguished historians of science, about these matters; they were then self-professed 'Kuhnians': students of Thomas Kuhn and definitely in the know.
15. Rousseau (1965).
16. See the poeticization of the thumb's neuroanatomy in Erasmus Darwin below, p. 44.
17. Charles Rosen in *Piano Notes* (2003) has discussed these and other related matters; there were no such books in the 1960s.
18. We worked then under the sway of books about the 'edge of objectivity' such as G. C. Gillespie (1960).
19. Important books such as Corsi (1989, 1991) and Wedlin (1989) did not exist then; one had to rely on one's own naively compiled histories.
20. The 'Romantic instauration' was a phrase that still resounds thirty years after I was using it in the late 1960s to denote the coming of 'imagination.' Later, I had read and reviewed Charles Webster's magnificent book of that title (1975) and began to think the 'instauration' of anatomical and physiological supremacy parallel to the Baconian instauration of learning.
21. M. H. Abrams had not yet published his magisterial study of *Supernatural Naturalism* (1971), which interrogated some of these areas along interdisciplinary lines.
22. There were not then the illuminating writings of Stephen Jay Gould, Oliver Sacks, Richard Selzer and Richard Dawkins to shed light on these issues.

23. By 1970, I had read Foucault but had not yet digested him on the origins of discourses and genealogies of thought.
24. See Nutton (1979), Nutton (1981), Nutton (1999), Nutton (2004), Cunningham (1997).
25. Thomas Laqueur's book *Making Sex*, studying the differentiation of two sexes out of one in the Renaissance, was an exception but would not be published until 1990. See also Hillman and Mazzio (1997).
26. See Singer (1957).
27. For the history of the brain see O'Malley (1952, 1996), L. R. Lind (1975), C. Singer (1957), Conry (1978), K. B. Roberts and J. D. W. Tomlinson (1992).
28. Thomas Brown seems to have appreciated this fundamental fact as early as 1827; see T. Brown (1827).
29. Oppenheimer (1971); Wedlin (1989).
30. Galen (1968); Nutton (1979, 1981, 1999).
31. Galen (1968).
32. Nutton (1979, 1981, 1999).
33. For the European reception of Galen's theory see Edwin Clarke (1968).
34. During the 1960s and 1970s I combed the brilliant works of E. R. Dodds for their late Roman applications of Galenic anatomy, but they offered little help.
35. Circa 1970 the explanations of Vivian Nutton had not yet been published; had his books been available my bewilderment would have been less than it was.
36. See Bekker (1695) for an example of how commingled these concepts had become by 1700.
37. F. Rahman (1952) on Avicenna's *Canon of Medicine* and other works of medical psychology.
38. See Cusanus (1650) from whose work the passages in this paragraph are taken.
39. For the background see Roberts and Tomlinson (1992).
40. See William Harvey (1961).
41. See Dewhurst (1972).
42. See Benedetti (1528).
43. See O'Malley (1982); Vesalius (1973).
44. See C. F. Ludwig, *Scriptores neurologici minores* (1791–95) which collected and anthologized many of these works; Romantic philosophers and thinkers, such as Blumenbach, Lichtenberg and Samuel Taylor Coleridge, commonly learned what they knew about the nerves in this collection.
45. See Cotugno (1775), especially the plates facing p. 102.
46. See Trotter (1804–5, 1807, 1988); Porter (1992).
47. See David Malouf (1978), p. 39.
48. See Trotter (1807), especially the chapters on the Ancient Greeks.
49. Trotter (1807), pp. xv–xvi.
50. The reader now sees why the interdisciplinary sub-theme of this essay is not fanciful.
51. Later sectarians did; as late as 1907 they still were expounding Christ's nervous system; see A. Waters (1907) and p. 65 below.
52. See Fisher (2002), Dixon (2003), Rousseau (2004).
53. See Camporesi (1988, 1994).
54. The point is confirmed in Thomas Dixon's study (2003).

55. Tobias Smollett, whose phrase this is, thought it was; see his *Adventures of Ferdinand Count Fathom* (1753) composed in the same decade as Burke's treatise. As further proof see John Baillie (1747).

56. See chapters 2, 5, and 7 for extended discussion of this influence.

57. For the physiology of tears in crying and sensibility see R. O. Allen (1975) and A. Vincent-Buffault (1991).

58. Their makers were young when they imagined them: Burke and Mackenzie in their twenties, as were most of the others.

59. For its history from Edenic time forward see Sekora (1977).

60. P. Medawar (1988) noted this as well, as had Mikhail Bakhtin long before in the 1920s; for which see G. S. Rousseau (1992).

61. See Fernel (1554).

62. J. D. Spillane's (1981) approach on pp. 53–109 represented the narrow approach I sought to avoid.

63. This has never been undertaken but see G. S. Rousseau (2005), 'The decay of scientific theories'.

64. R. C. O. Matthews (1986).

65. Thomas Willis, the physician, uses a version of this phrase in several of his treatises on the brain.

66. For the long history of the Life Force in relation to nerves see G. S. Rousseau (1992).

67. The below essays were an attempt to document that record about which no full-length book still has been written.

68. As Keith Thomas has shown for the natural world (1983).

69. As late as 1768 physico-theologians aimed to demonstrate how the nervous system invaded the soul; see Daniel Smith (1768).

70. Recently the amalgam has been historically demonstrated; see Roy Porter (2003).

71. Typical of the works I analyzed linguistically were David Kinneir (1737); then I would take my analyses and bring them to bear on texts by great writers such as Jonathan Swift and Alexander Pope.

72. See, for example, Caroline Thompson (1935), who wrote under Empson's spell five years after his book appeared; also W. V. Reynolds (1933, 1935).

73. I continued to, with Roy Porter, in the 1980s; see our collaborative introduction in G. S. Rousseau (1990).

74. For Mikhail's Bakhtin's neo-Kantian contributions see the Appendix in G. S. Rousseau (1992).

75. Romantic poet Shelley, building on neo-Platonic attractions, did just that.

76. For the survival of this nervous rhetoric and metaphor in classical Johnsonian prose see W. K. Wimsatt (1948).

77. For the blend of religion and science then see R. D. Stock (1978).

78. For further discussion of the complex connections see chapter 2 below and Roy Porter (2003).

79. By the time I had read B. Stafford's *Body Criticism* (1991) circa 1993 it was too late to be useful for the essays in Part II.

80. When witty Lady Mary Wortley Montagu quipped in the 1730s about Lord Hervey that there were 'three sexes,' she drew on a hermaphroditical type that would soon be subjected to the analysis of nerves.

81. See Robert G. Frank (1990).

82. Dutch microscopist Jan Swammerdam (1637–80) denied the valves and claimed that his dissections showed that animal spirits do not increase in volume.
83. Galileo's friend Giovanni Borelli (1608–69) applied sophisticated mathematical principles to the animal spirits, claiming that they flowed in the nerves by producing 'commotion' according to geometrical laws.
84. See Thomas Morgan (1735), p. xv; the treatise provides a mini-history of animal spirits. For Willis' place also see J. Slotkin (2004), p. 94 and chapter v, 'Eighteenth-Century Social Anthropology: Man's Nature,' pp. 244–356; for the animal spirits and the literary satirists see Rousseau (2005).
85. For a modern neuroscientist's view of this long record see Ian Glynn (1999); also W. Clower (1998).
86. Descartes (1985), p. 101; F. C. Copleston (1963); A. R. Damasio (1994).
87. See Bernard Mandeville (1730).
88. Blackmore was crucial for Pope's aesthetic development; see Blackmore (1725, 1725, 1726, 1797); for the tradition of the physician-author in the early modern world see Randolph (1941).
89. Thomas Fuller (1740, 2nd edn), p. 21 who capitalizes all his 'Spirits' but never the animal portion.
90. Village doctors in Victorian England still touted these views, extolling numbers for ruddiness; see Langbourne (1842).
91. Ibid., p. 21; for literary repercussions of theories of exercise and motion see Carol Flynn (1990).
92. In this paradigm shift from romantic Luthers to deluded Isaac Casaubons, Swiss Dr Tissot was a principal theorist; see Tissot (1768, 1784). Developing prose fiction also played a part in dissemination, as in the much-read novels of Trollope and Balzac.
93. That will rise after 1850. For the role of economic luxury in these matters see J. Sekora (1978).
94. S. Tissot (1768, 1784).
95. For Harvey's 'nervous anatomy' see Robert G. Frank (1990).
96. See John Moir (1975).
97. See William Harvey (1961); for the contexts see F. Duchesneau (1996).
98. Descartes (1985), pp. 99–102.
99. M. Flemyng (1751), pp. 1–2.
100. The prefaces to Thomas Willis' various treatises on the brain make this point perfectly clear.
101. A. C. Guthkelch et al. (1958), p. 277.
102. As he has in June 2004 in the *New York Review of Books* 37: June 2004.
103. See Robert Darnton (1968).
104. A generation later a nerve quack called 'Dr Wood' had set up by 1740 in Covent Garden; see the caricature of him in the British Library, Prints and Drawings catalogue number 2475. There were many other similar cases.
105. See Daniel Smith (1768).
106. See Wellcome Library Catalogue no. 544301I; the postcard was made by Birn Bros.
107. See J. Midriff (1721).
108. Cheyne curiously never refers to Midriff in his post-1721 works, of which there are many; can this be the case of one nerve specialist not wishing to recognize the competition?

109. See Mullett (1940).
110. A full glossary of the terms is found in Midriff's treatise (1721).
111. Contemporary examples today derive mostly from the realm of nutrition, as in Dr Atkins' diet.
112. The only recent study is by Jan Golinski (1992).
113. Peter Shaw (1750).
114. See below pp. 262–3 and Thomas Dover (1733), Dewhurst (1957).
115. Thomas Fuller (1705), pp. 21–2 reconsidered these debates a decade later; see also William Cowper (1695), I. ii–iii.
116. See J. Bellers (1714), p. 23; Bellers and Cowper debated some of these topics.
117. See Hugh Smith (1780) who provided the scientific explanation during Walpole's adulthood.
118. The only reliable means of locating his passages is by trudging through the fifty volumes of his correspondence.
119. As I show in chapter 6, the Enlightenment commentators on sexology also noted it, as had the still anonymous author of Satan's Harvest Home (1749).
120. Frances Wilson (2003), p. 1.
121. T. Laqueur (1990) downplays this aspect.
122. Jauffret (1810).
123. Jauffret (1810), p. 18; Natural Theology (1802) had been translated into French during this decade 1802–10.
124. Jauffret (1810), p. 97.
125. H. Smith (1780; rep. 1795) who must not be confused with his contemporary Daniel Smith.
126. James Parkinson (1780).
127. The Neptunists believed that water was contained at the center of the earth, the Vulcanists fire; each group had considered that fissures deep inside the earth were analogous to nerves carrying the 'vital spirit,' a type of ethereal fire that expanded when heated and caused eruptions and earthquakes; for these debates see Roy Porter (1978).
128. See Thomas Browne (1827); M. D. Wilson (1980); M. Gazzaniga (1998); R. Carter (1999); A. Richardson (2001).
129. For its Enlightenment phase see P. S. Churchland (1986).
130. The phrase 'counter-physiology' to describe Hume's position is Stafford's (1990), p. 192.
131. A survey of the critical literature since 1950 shows it is divided on the matter, an even number of critics coming down on either side.
132. See chapter 4 below.
133. John Dryden, The Satires of Juvenal (1693–97), x: line 262.
134. See Roy Porter (1992).
135. See William Battie (1962), published in 1757.
136. See William Perfect (1787); W. Bynum (1985); Roy Porter (1987).
137. Adam Budd in Toronto and Edinburgh is undertaking the most important work on poet-physician Armstrong.
138. John Armstrong (1773); see also John Moore (1786).
139. See Armstrong's 'Alcalescent Disposition of Animal Fluids,' Sloane MSS, no. 4433.
140. See C. Lawrence (1979, 1985).

141. For Cullen see Thomson (1832), though old still the most reliable study; Cullen (1780, 1785).
142. For the social implications see C. Lawrence (1979, 1985).
143. Jon Mee (1992). Cullen's revolution also enabled readers of works such as Bienville's to understand how his nymphomania had been configured; see below chap. 6.
144. See Trotter (1807), p. 230 ff.; Trotter (1804–05).
145. See Trotter (1807), p. 104. The most perceptive summary of his work is Porter (1992).
146. See George Rowe (1840).
147. Rowe (1840), pp. 82–4.
148. Micale (2004), p. 90. Micale attributes the phrase 'neuromimetic' to Sir James Paget (1902), chap. 7: 'Nervous Mimicry.'
149. Paget found the passage in the 1852 edition of Samuel Taylor Coleridge's *Table Talk* on p. 81.
150. See Ian Jack (1966); J. Nadelhaft (1968); H. Small (1996); G. Sill (1997).
151. For Dover's dramatic account see pp. 263–4 below.
152. For Le Cat see chapter 4 below; for an anthropology of nerves H. P. Wasserman (1974).
153. Dallas (1803), I: 200–01, where the cited passages appear.
154. See G. S. Rousseau (2003), pp. 246, 264. An anthropologist at heart, Blumenbach was professor of medicine.
155. George Rowe, already cited above, mentioned it in his 1840 treatise on women's infirmities.
156. The King later regretted his decision not to recruit when he proved himself a valiant soldier.
157. M. H. Nicolson (1949).
158. G. S. Rousseau, 'In rapture writ' (2005).
159. Erasmus Darwin, *Temple of Nature* (1973), 24, I: 269–70.
160. For background see W. K. Wimsatt (1948); W. Kenney (1961); P. Parker (1996); C. Hawes (1996).
161. Cicero (1953), p. 232.
162. Ibid., p. 233.
163. Mede (1637), p. 847.
164. Hale (1691), p. xlii.
165. Nicholas Sweeney (2002) has recently made a sustained case that draws substantially on the history of science and medicine and adopts a similar approach to the one taken here.
166. See G. S. Rousseau, 'In rapture writ' (2005).
167. See the further examples of nervous style below in chapter 8.
168. Warburton (1763), I. ix.
169. Roy Porter (2001) takes a different approach: by 'nervous style' he denotes economic preferences, i.e. luxury or labor.
170. See Erasmus Darwin (1798); J. Slotkin (2000), pp. 296–301.
171. For Darwin's passages on this intricate process see J. Slotkin (2000), p. 301.
172. Erasmus Darwin, *The Temple of Nature* (1973), Canto I, lines 269–70.
173. Ibid., lines 271–4.
174. Ibid., Canto III, lines 121–6.
175. For another avenue to nervous style see Patricia Parker (1996).

176. A century later 'Dr Phyllosan' explained how drugs were necessary to take the edge off these stresses; see Phyllosan (1937).
177. See William Battie (1962); for the larger connections see Clement Hawes (1996).
178. For contemporary comments on Smollett's 'nervous style' see the criticism of him in Goldsmith, Charles Churchill, William Kenrick and Samuel Johnson among others.
179. Roy Porter has commented on these historical links in economic terms in Porter (1991).
180. For the contexts of this development see Vernon Rosario II, *Science and Homosexualities* (1997).
181. Examples include neuromaniac, neurohysteric and neurohypochondriac, demonstrating the crucial significance of labeling and naming. See also the neologisms discussed below on p. 253.
182. Disraeli (1881), I: 269 in the main Victorian edition of his works.
183. Ibid., 269.
184. Ibid., 270.
185. Edited by Gustave Martin, it ran to four volumes; the British Library copy annotates titles with the prices each one fetched.
186. Burette, *Dialogue sur la musique* (Paris, 1849); the work was reissued a century later.
187. See Burette (1746) for the expressiveness of individual instruments.
188. Did the listeners of heart-rending lamentation sermons in the mid-eighteenth century expect their nerves to vibrate as strings as David Hartley (1749) had suggested they would? David Lockman's 1736 sermon on the lament of David for Saul suggests so; see also Burette (1748).
189. Brocklesby (1749), Browne (1729); see the discussion of Browne's extraordinary treatise in Rousseau (2000).
190. See Gibbons and Heller (1985).
191. See Brocklesby (1749), Woof (1994).
192. See Penelope Gouk (2000).
193. For a broad perspective on the topic see Malcolm Budd (1992).
194. Eduard Hanslick (1891) was no scientist, but he anticipated these restorative properties of music.
195. German neurophysiologists of the 1890s were particularly influential in this realm.
196. Nathaniel Wanley, *Small Wonders* (1678).
197. See Kevin Barry (1987); Lucy Hartley (2001).
198. For the Boyle Lectures see Nicolson Jacob.
199. See William Derham (1711–12), which work continued to be reissued for decades together with his others.
200. See G. S. Rousseau, 'In rapture writ' (2005).
201. Richard Blackmore (1797) had anticipated them in his poems; see also pp. 27, 35, 50, 89 for Hartley.
202. Jacques de Vaucanson (1742); see R. G. Mazzolini (1991).
203. See Gordon W. O'Brien (1966) for the instrument-makers and the cult of genius at the end of the eighteenth century.
204. Daniel Smith had addressed these matters en passant when exploring the interface between the soul and nervous system; see Smith (1768).
205. Stephen Jay Gould, *The Hedgehog and the Fox* (2003).

206. Tuberculosis, especially common among composers and other artists, was the superlative sign of this genius; see Lewis Moorman (1940).
207. See Francis Crick (1994) and the epilogue at the end of this book.
208. Some of the aesthetic debates are discussed in this light in George Mandler (1984).
209. Gulliver Ralston (Oxford University), to whom I am in debt here, is developing these ideas in a book about Wagner; see Ralston (2004).
210. See Eduard Hanslick (1891) and H. R. Haweis (1876), which was reissued in many editions by 1900. Earlier aesthetic philosophers, even John Ruskin, had speculated about these matters, although with less focus than Haweis.
211. Haweis uses this phrase and other similar ones in an attempt to show how degeneration occurs.
212. For these complex scientific-aesthetic developments see Frederick Burwick et al. (1992).
213. For a view of 1851 see John Addington Symonds (1851); for a more recent approach see George Mandler (2002).
214. See Henry Home, Lord Kames (1781), Home (1788).
215. See J. S. Slotkin (2004), p. 357ff.
216. See Emmanuel C. Eze (1997) for a synthesis of racism in the Enlightenment.
217. For ideas of progress and science see D. Spadafora (1990).
218. For example, the writings of Daniel Smith (1768, 1778). See also Christopher Lawrence (1979).
219. See Verity (1837).
220. Verity refers to all these writers in his treatise of 1870: he may not have read them, but he felt sufficiently confident about their views to discuss them in detail.
221. Verity (1837), p. 10.
222. Ibid., p. 130.
223. For discussion of the history of brain size see D. C. O'Malley (1996).
224. See Verity (1870).
225. Adair (1786); Roy Porter (1992).
226. Elsewhere I present Symonds' Oxford scandal in the context of blackmail; see G. S. Rousseau, 'Sodomitical Scandals and Choristers' Blackmail: Sex in the Oxford Colleges' (forthcoming).
227. This section draws on Symonds' own writings, especially his journal, as well as Grosskurth (1962), Jenkins (1980).
228. The story has often been told without attention to his pronounced self-conscious 'nervous states,' which constitute my contribution.
229. Symonds (1851).
230. See Grosskurth (1984), still the most reliable edition for Symonds' memoirs.
231. Ibid., pp. 230–1, the source of the passages in this paragraph.
232. Other writers in the twentieth century would follow suit, connecting nervous states to the need for writerly retirement; Virginia Woolf and Tennessee Williams were merely two among many.
233. This technical word etymologically means *extreme* sexual excitability, often aroused during nocturnal dreams and causing frequent discharge, but also during waking hours and producing fantasies of consummation.
234. See above, p. 32 and below pp. 96–7, 158–9 and 164–73.
235. See Dupau (1819).

236. Dupau's phrase functions as the cornerstone of his theory of nervous disease.
237. See Rosario (1997, 1997), as well as his review of our *Hysteria Beyond Freud* (1993) in Rosario (1995).
238. Grosskurth (1984), p. 230.
239. The word was coined by G. S. Rousseau; see Rousseau (2002), pp. 1–50.
240. If Symonds could return from the dead and read the books by J. Ledoux (2000, 2002), he might be in agreement with their basic tenets about the plasticity of the brain *vis-à-vis* sexual arousal rather than competing models of the brain as a permanently hard-wired organ.
241. By 1860 tuberculars had been advised for a century that they possessed excessively 'sensible nerves': see Moorman (1940); J. A. Meyers (1960); J. H. Williams (1973); Dalrymple (1999).
242. See Rousseau (2002), pp. 1–50.
243. Grosskurth (1984), p. 230.
244. Foucault (1978), pp. 32–3.
245. Ibid., pp. 32–3.
246. See Virginia Woolf (1986), pp. 78–80; for the contexts see Thomas Caramagno (1991).
247. He may have died of cholera. The late nineteenth-century homosexual Russian composer Scriabin was also acutely 'nervous' and needs to have his extraordinary career and early death deconstructed from this vantage of 'creative nerves.'
248. The best introduction to his work is Ramon y Cajal (1982). See also Ramon y Cajal (1928). There has been a massive effort to translate his works in the 1990s. Comment about the history of nerves abound in his vast *Autobiography*, originally published in 1937.
249. There is no proper biography but see his son's detailed sketch in J. Paget (1902).
250. See Paget (1902), p. 73ff. He had been influenced by British rather than Continental theories of hysteria, especially those of Doctors Tuke, Gull, Anstie, and Reynolds.
251. For the aesthetic profile, as distinct from the medical, see Martin Jay (1998), chap. 11.
252. Paget (1902), p. 75.
253. Ibid., p. 95.
254. See Paget's chapter 18 (1902) entitled 'Rare and New Diseases.'
255. See Canaveri (1836).
256. See George McIver (1894), copies of which are in the British Library and the Bodleian in Oxford.
257. Ibid., pp. 48–50.
258. Henry G. Waters (1907). One wonders whether he also puns on the Renaissance Italian.
259. Literature similar to McIver and Waters abounds after the 1860s, but there is no space to present it here.
260. See Yourcenar (1929), p. 751.
261. *The London Times*, 29 October 1930, p. 16.
262. See Tsitsi Dangarembga (1988); analysis along the lines of the nervous body is found in Bahri (1994) and Vizzard (1994).
263. See M. Vaughan (1991) who makes a case for the African nervous body.

264. See Michael Taussig (1992); for discussion of his *Nervous System* see Martin Jay (1998), p. 206.
265. See John Hill (1758); Hill's prose assumes that his contemporaries are 'newly nervous' (his phrase) without expanding at large on the subject.
266. See I. Lewis (1864), now a scarce item.
267. Ibid., pp. 6–7.
268. See John Hill, *The Virtues of Wild Valerian in Nervous Disorders* (1758).

Part II
Essays, 1969–1993

2
Science and the Discovery of the Imagination (1969)

The aftermath of my book This Long Disease, My Life: Alexander Pope and the Sciences, *written in 1964–67 with Marjorie Hope Nicolson and published in 1968 by the Princeton University Press, was the desperate recognition that we had accumulated 'facts' in the void. The book's reception bore out the suspicion. The reviewers were 'Pope critics': specialists, experts, those with vested interests who wanted to know what this book could do to fill out the poet's life when there was still no reliable biography, how it extended the annotation of individual passages of his poetry, and what concrete case it made for the influence of the sciences on Pope's abstract poetic imagination. Distinguished, even great, historians who cared about the contexts of Enlightenment thought were, of course, then at work. Peter Gay immediately comes to mind for the way in which he was then (1964–68) studying and presenting the diversity of Enlightenment culture. But the disciplines then were much further apart than they are now, often working in isolation of each other. We forget the immense progress of interdisciplinarity in just one generation (1970–2000). In any case, the corporeal 'imagination' itself was far less abstract than it has become in almost 300 years since Pope's time.*

At this time I had just taken my first 'real' job at UCLA. Los Angeles' great libraries – Clark, Huntington, the collection at UCLA itself – were remarkable. Harvard, where I had been briefly before then, also had vast libraries. The stimulation for this essay was not to be found in books but in the availability of UCLA's Brain Research Institute where great scientists such as those mentioned in my opening notes were willing to talk to someone from the humanities about neural pathways in history. In those days, long before internets and electronic chat boards, you had to contact people individually; ring them up, visit them, engage them intellectually and establish a working relation: skills that have now gone under in the Age of Electronic Mail, when anyone can send an e-mail to anyone else and even work together without ever meeting the actual person. I explained to them that I was searching for a discourse loosely combining imaginative literature and theoretical science in an attempt to understand what role pre-1800 neural networks played in the human understanding of cognition – all this as a prelude to entering the history of consciousness before the Romantics and Victorians.

Marjorie Hope Nicolson's work had created a revolution in modern thinking about Newtonian aesthetics. But what did the poets in the epoch 1660–1820 themselves think of their imaginations? What did they read in those decades before academic knowledge began to be divided up in the modern way? What were their contemporaries telling them? Our book about Pope had delved deeply into all the sciences except the human ones: into archeology, astronomy, botany, geology, medicine and physics but not psychology, religion, consciousness, dream states, memory – the poet's sense of an awareness of his own mind.

A paradigm about consciousness was forming in this milieu of Newtonian aesthetics and Lockean associational epistemologies and challenging the old position about the presence of innate ideas. It began to coagulate, almost like the blood: cognition, awareness, consciousness, and especially the complex workings of 'the imagination' (with the article, as if it were an organ of the body no different from the others) – all these were mechanistic. The wars of truth about consciousness were fought out in this early modern period on a battlefield dominated by sensation, nervous transmission through the body to the brain, and then higher processing in the cortex – all assuming nervous networks (nerves, spirits, fibers, etc.) rather than among abstract ideas implanted in the mind. In these processes the 'imagination' functioned no differently from the brain or mind or even the soul (seat of the affection and will and possible site, to Pope's contemporaries, of mechanistic operations) – so far had the incursions into mechanistic explanations proceeded.

These bodily processes ultimately depended, or so Pope's contemporaries thought, on the constellation of nerves and animal spirits operating in coordination with the brain. These functions were mirrored in healthy as well as pathological states: what differed in the two states was the physiological condition of the nerves. The nuanced degrees between the two poles were 'exquisite', a term then often used, especially in scientific literature, to designate the response of awe to this varation of nerves: a form of body sublimity perhaps no less finely graded than the degrees of mathematical infinity Newton postulated. The neurologists of Pope's world (Willis and Boerhaave to Cheyne and Hartley) may have had no notion of the modern synapse or nerve cell. Nevertheless, their intuitions drove them to theorize neural networks through a dense linguistic jungle calculated to cope with similar types of plasticity in the nervous system in the 1970s. (I used to refer to this forest of words as 'fire *in the soul', and thought I would write a book by this title, as some of my footnotes note, but never did.) In all this welter of speculation these early neurologists postulated the imagination as the body's most exquisitely turned out organ, the one bridging the body–mind barrier: if not entirely mechanistic and material in its configurations, then still so firmly dependent on this constellation of three systems – animal spirits, fibers, nerves – that it amounted to a type of neural basis of consciousness. Here was the Enlightenment's version of Harvey's 1627 'silent music of the body.' The difference was that no one had any idea in those days of single microscopes how these neural constellations functioned.*

The disciplinary implications of this essay were as problematic in 1969 as its content: namely, that the historical development of the workings of imagination (especially consciousness superimposed on intuition) could not be grasped without taking deep-layer accounts of the history of science and medicine. It seems a cliché to say so now, a generation later, but things have vastly moved on from 1969, and demonstration of some of the changes is one of my aims in gathering this collection. But in the 1960s the humanities were far more insulated than they are now despite far-flung fallout from the 'two cultures' debate of C. P. Snow and F. R. Leavis which rocked the Western academic establishment after 1959 without instituting immediate reforms. To me then, the more baffling state of affairs than this disciplinary one pertained to mind and soul: especially the age-old mind–body divide and where they stood in relation to imagination and the human neural constellation. I was a beginner and could not confront all these fronts in 1966–68. It was enough to bridge the 'two cultures' in working practice.

Science and the Discovery of the Imagination in Enlightened England (1969)

LATE IN the seventeenth century man discovered his imagination[1] and, as Densher in Henry James's novel *The Wings of the Dove* says, 'we shall never again be the same.' When they tell the story of this discovery, historians of science will probably insist that it was part of a long, continuing process of discovering the body. Not until man discovered a considerable portion of his anatomy could he earnestly formulate a scientific model by which to account for the physiology of imagination. The discovery of the imagination does not imply that man suddenly realized he had one; for example, that he had now located the faculty which permitted him to dream at night and daydream,[2] or to envision objects and places not actually present before the retina. What he did discover – with the help of scientists and philosophers – was that the imagination was a real essence, as material in substance as any other part of the body, and that it therefore could be medically described. An important consequence of this development in physiology is that leading thinkers of the next century defied the new 'organ,' endowing it alone with the means of salvaging the soul of man while on earth. Such deification was restricted to philosophers and artists, men like Kant and Coleridge, who felt compelled to reject the modes of explanation of the physiologist, not upon methodological grounds but because it thwarted the expectations it raised. Science and literature were perhaps never closer in their ultimate aims than in the century (1680–1780) that discovered imagination.

I

Such discovery cannot accurately be said to have occurred over a long period of time as in the case of the chemical composition of the blood. It occurred precisely in the second half of the seventeenth century in

Western Europe, particularly in England and France, largely the result of certain medical and physiological experiments and at a time when the scientific spirit, to use the language of the day, was at its zenith – as is evidenced by the establishment in 1660 in London of a Royal Society for the Advancement of Natural Philosophy and the hitherto unprecedented number of experiments undertaken. It was discovered, moreover, at that moment in history when Aristotelian scholasticism could no longer hold water and when the new science, particularly the new corpuscular physiology of Harvey, Boyle, Willis, and Newton, created a revolution in scientific thought.[3] This revolution was not so much a methodological one as it was an awareness that psychology was of paramount importance in the realm of ethics; and that moral conduct was ultimately predicated upon the passions and not the innate ideas of men. The thinker to whom the *literati* in England owed more than to any other single man for their knowledge of psychology is the monumental genius John Locke. And it is reasonable to conjecture that without Locke's theory of associationism (primary sense impressions combining and commingling), the whole course of eighteenth- and nineteenth-century British literature would be different.

The Platonic notion of the imagination as a source of suspicion and distrust (according to the *Timaeus* located in the liver), as a pander to the passions and the appetites, was so soon superseded by the Aristotelian conception in European thought that it can be overlooked as inconsequential in its influence. In Aristotle's ontological world of real and less real essences, the mind and its sub-members (fancy, the imagination, the appetitive faculties, the passions) play a small role in the observer's perception of these essences.[4] Trees, rocks, and drops of water exist in an objective and absolute sense; they ought to appear the same to the baker, bricklayer, or candlestick maker. If they do not, that is the fault of the perceiver (not the perceived), who suffers from a privation of the correct perception of matter. Descartes, Hobbes, Malebranche, and, more significantly, Locke toppled an aged empire of thought when, in their various ways, they introduced for the first time in European thought the possibility of a real imagination: substantive, existential, working physiologically through the mechanical motions of the blood, nerves,[5] and animal spirits.[6] The revolutionary thought of these men is sometimes referred to as the greatest advance ever made in the history of the 'body–mind' problem, but it is more accurately described as the 'discovery of the imagination' since the former was as old as the pre-Socratic philosophers and could not radically redefine its terms until the imagination was physiologically created.

From the literary historian's vantage, the most significant aspect of this discovery was the decline of mimetic art, that aesthetic preference for imitative art that had governed and guided art from ancient times through the late seventeenth century.[7] If the imagination is physically non-existent, then trees must be represented as trees, rocks as rocks, and water as water; but as soon as the imagination is acknowledged as real substance containing matter, it may then transform the perceiver's sense of trees and rocks and permit him to represent these materially tangible objects in artistic shapes that are not immediately recognizable. A belief in the physical existence of the imagination implies a belief in psychology – the science of the psyche – and belief in psychology substantially alters the number of possibilities for imitation. If the imagination contains substance and is material, then, like trees, rocks, or drops of water, *it* may be imitated in art. An observer gazing at Correggio's 'Leda and The Swan' or Poussin's 'Et In Arcadia Ego' can instantaneously recognize true nature in the likeness of the painter's forms, but who has ever seen an 'imitation of the imagination' represented in a form of art that could be called realistic – true to nature?[8] Such non-realistic art (in painting, literature, music) would pose insurmountable difficulties of interpretation. For this reason there was much resistance by philosophers and artists in the eighteenth century to the theory of the physiological existence of the imagination; resistance, however, eventually gave way, and symbolic art (art that imitates, if anything, the life of the imagination) inherited the throne formerly occupied by mimesis.

It was not possible for symbolism to flourish until mimesis died, and the decline of mimesis occurred only when psychology was established as a science. In the late seventeenth and eighteenth centuries psychology was more neurological than it has been at any time until the last few decades (when it appears to have reverted to its original state). Medical theories of physicians like Sydenham, Willis, Charleton, Hooke, and Boyle must be invoked, for without their bold ideas the imagination must have lingered for a longer time in an inchoate state of universal darkness. Although John Locke is famous today as an empirical philosopher and the author of the ethical treatise *An Essay Concerning Human Understanding* (1690), his reputation in his own time was as a leading physician. Scientifically trained at the University of Leyden, which then contained the most advanced medical school in the world, he had studied medicine with Boerhaave, the eminent physician and Professor of Chemistry, and learned physiological theories that later were to enable him to base his ethics and epistemology upon current physiological principles.

The aspect of Locke's *Essay* most important for our purposes is that his deepest questions are ultimately physiological. To be sure, he investigated the science of the mind in order that he might advise men better regarding virtuous conduct. And it is true that he succeeded *most* in his ethics, *least* in his physiology, and not without abundant reason, for he never undertook the kind of scientific investigation and experimentation (as had Boyle and Newton) necessary to explore adequately the neurological aspects of his questions. And yet, his questions were physiological from the start.[9] The imagination, he argued, *must* exist; observation and induction teach that no two men behold and describe a tree similarly; they cannot and indeed are not capable of expressing it alike or suggesting an identical connotation; therefore, while the tree exists, it is not existential in the sense the imagination is; in fact, the tree can exist only in the eye and imaginative faculty of the beholder. So important is the formation of this 'mind,' its impressions and ideas, Locke asserted, that it deserves more attention than the tree. In focusing his attention, Locke's problem was always 'how,' rarely 'why.' He was not a moralist asking *why* men are as they are, nor why they perceive as they do; he knew they are creatures of passion and ruled by irrational desires; he was the physiologist asking 'how' they perceive. Locke was perennially interested in the formation of the imagination as a complicated network of secondary ideas in various associative patterns and in the organic relation of the imagination to the nerves, blood, and animal spirits.

Although it was not possible for Locke to answer the physiological questions he asked, his *Essay* influenced the path of science, not merely epistemology, for more than a century. George Berkeley the Irish philosopher, David Hartley the London physician and associationist philosopher, David Hume and Joseph Priestley, and many others followed in his footsteps, asking the same basic questions. If eighteenth-century medical-philosophical thought was unable to explain the physiology of imagination, that failure should be regarded, first of all, with humility (the most advanced brain theorists today know virtually nothing about the physiology of memory),[10] and secondly, in perspective. No mechanical model could satisfactorily explain the associations of the imagination until the controversies raging over the 'animal spirits' had subsided and ended in less troublesome waters. Precise chemical composition of the substance in the blood giving motion to life eluded definition. It was not until the early nineteenth century, when the reputation of the theory of animal spirits had diminished,[11] that scientists were inclined to believe the problem had always been definitional. And yet, the eighteenth-century physiologists and neurologists conducted their experiments on the nerves in an attempt to understand the secondary associations of the imagination.

The literary effects of the numerous physiological investigations of the century 1680–1780 were considerable, and scientific exploration as significant as this in one segment of the culture would not be expected to remain self-contained. There are more metaphors of the mind and references to mental institutions in the literature of this century than literary critics have noticed – indeed the artist's personal struggle to define himself in relation to the world of sanity or insanity is so pervasive that it may be called a leitmotif. The Restoration and eighteenth century battled out questions relating to the role and place of the imagination in works of art with an ardor that is rare in literary controversies, and the latter part of the century carried on the argument just as ardently. In so doing, thinkers looked to Locke's psychology again and again, and often quoted or commented upon his now celebrated distinction between 'wit' and 'judgment.'[12] In the first half of the eighteenth century, poets like Richard Blackmore in *The Creation* (1712) found a storehouse of opulent poetic images in the physiological terminology of the scientific controversies over the definition of imagination; images taken directly from the language of 'the learned,' who, in Blackmore's words, 'with anatomic art/Dissect the mind, and thinking substance part.' For the second half of the century, Dr. Johnson's chapter 44 of *Rasselas*, 'On the Dangerous Prevalence of Imagination,' with its insistence that 'all power of fancy over reason is a degree of insanity,' is emblematic of the thinking of the day and of the mode in which physiological thought was creeping into non-scientific literature. Writers like Addison, Swift, Pope, James Thomson, and Dr. Johnson were persuaded – however strongly they may have agreed or disagreed with Locke's system – that the role of imagination in literature was perhaps the most vexing aesthetic problem of their time. Without the stirrings of the new physiology these aesthetic questions would never have been asked in the first place.

II

If the 'imagination' exists, then it must be described by anatomists and its functions must be delineated by physicians and other theorists. Obviously this could not be done in the eighteenth century; nor could it be partially completed without a scientific, molecular model that includes neurons and protons. Yet, it was attempted again and again.[13]

The clearest verbal descriptions of the physiological dynamics of the imagination took a mechanical form: the imagination does not work through and by itself although it alone causes the physical motions and chemical processes within the body that result in passions (pleasurable and odious) and secondary, associated ideas. One of the many comprehensive summaries of the physiology of the imagination is found in the *Medicinal Dictionary* (1743–1745) by Dr. Robert James, inventor of the famed 'James's fever powders':

> The whole Bent of the Soul is to court and embrace it [the imagination], earnestly endeavouring to be united to it. She is, as it were, expanded in Pleasure; while the animal Spirits, in a kind of Ovation, being carried within the Brain, are constantly exciting the most pleasing Ideas; and, acting in a lively Manner upon the nervous System, causes the Eyes and Countenance to sparkle, while the Hands and every Member exult for Joy: Besides, the Influence of the Brain affecting the Praecordia, by means of the Nerves, they propel the Blood with more Rapidity, and pour it with Vigour on every Part of the Body... Such are the Effects of the Power of Imagination; Effects which are sometimes almost incredible, and which have been thought sufficient to restore and renovate, to ruin and destroy, the human Structure.[14]

Such definition immediately implies two kinds of imagination: healthy and sick, normal and diseased, and the distinction was increasingly discussed as the eighteenth century progressed. It is an ironic contrast that the supposed 'Age of Reason' should have produced so many cases of insanity among its writers, in England William Cowper, Christopher Smart, and others. That the so-called 'Enlightened Age' should have concentrated so much of its energy on such distinction should not, however, seem strange to us in a post-Freudian age in which the effects of both kinds of imagination are visible daily: the healthy imagination exhibited in creative art and science, and the diseased imagination of the hallucinating schizophrenic.

The distinctive mechanical operations of each type were of concern to eighteenth-century neurologists, although they had more to say about the latter. Disregarding for the moment the distinction, the imagination was explained by various models, most often as an image-producing aspect of the 'mind,' sometimes as the lens of a camera, and sometimes as a multi-motion process of the frontal lobe which throws up images of things or places not present to the mind. Eighteenth-century English poetry, particularly scientific and didactic poetry, possesses a large vocabulary for images describing (analogically) the imagination: cameras, lenses, mirrors, lamps-of-alabaster and lamps-of-reflection, optical instruments then popular, and even the *camera obscura*. A long list of didactic poems explaining the physiology of imagination can be compiled for the period 1750–1820; a late example is L. F. Poulter's *Imagination, a Poem* (1820). There was a tenacious belief that memory was crucial in accounting for the associative modes and chronological order in which these images stream forth. In maintaining this belief, many physiologists were palpably 'modern,' for it is still firmly adhered to by brain theorists.[15] Intensity of imagination was felt to depend upon the size of the vestigia, or tracks, through which the animal spirits flowed, and the animal spirits themselves to depend upon 'the lines or strokes of those images.'[16] The medical writings of Dr. James and his contemporaries explain that displeasing objects are recalled from the memory without being present by a process of contraction and, then, by relaxing of the fibres.

Not until philosopher-physicians like Locke had addressed themselves to questions relating to the physiology of the imagination could the mechanics of ideas (whether innate or acquired) be explained. And yet, almost immediately after Locke formulated questions regarding the association of ideas – a form of 'madnesse' in itself, as he explains in the *Essay* – medical men were quick to distinguish the associative processes under normal and abnormal conditions. The neurological bases of *dementia* had come a long way from Malebranche's indefinite definition, 'L'imagination est la folle du logis'[17] – imagination is the madwoman of the house, less literally the mad creation of the brain. The fact that Enlightenment physiologists centered their attention on the diseased rather than healthy imagination is of tremendous consequence for the development of European poetry; for it was not until culture scientifically defined the very same madness it wished to condemn that poets turned to the writing of 'mad verse' for catharsis and relief. Stated otherwise, the imagination had to be scientifically authenticated before it could be declared ill by physicians, and in turn cured. One cannot cure an unknown disease. In this connection it is interesting to note that the confinement of lunatics to asylums was an institutional creation peculiar to the seventeenth century and unheard of before then; reformation and attempted cure of lunatics occurred only when physicians of the eighteenth century had shown that the disorders of madmen were physiological, not religious. Madness was thereby torn from the imaginary freedom which permitted it to flourish in the Renaissance at that moment historically when scientists could demonstrate that body and mind worked hand in hand in a mechanistic and organic fashion.[18] Such demonstration – by seventeenth- and eighteenth-century physicians like Willis, Sydenham, Locke in England, Boerhaave in Holland, Stahl in Germany, La Mettrie in France – ripped religion out of madness and left it (madness) hovering in an orbit of mechanical cause and effect; as in the case of Newtonian gravity, whose forces everyone could calculate but which no one could explain away, madness was defined as a mechanical disorder of the animal spirits relative to their speed of flow and density, but in almost all cases without the slightest indication of an apparent external cause, i.e., a horrific object or alarming circumstance.[19]

Society, in discovering the 'diseased imagination' within the imagination at large, thereby created the notion of a psychological condition: a melange of similar symptoms constituting a 'condition,' or as it is called today in medical parlance, a neurosis. By so doing, it also rendered the theoretical possibility of man's functioning well in certain areas and poorly in others. For the first time in the history of medicine it was possible for man's body to be sick and his psyche healthy. Never before the rise of psychology in the eighteenth century had this been true; if man was ill, that was because of some radical disorder in the arrangement of the 'humours' which had been caused, in the first place, by a ruptured relationship to God. The sacred causes of illness were once and forever made profane.

Michel Foucault, the French structuralist philosopher and opponent of Jean-Paul Sartre, who has written an award-winning book *Reason and Civilization: Madness in the Age of the Enlightenment*,[20] notes magisterially that hysteria is the true eighteenth-century disease, far more typical of the age than gout, dropsy, or ague, because it alone was explained by the new dualism that replaced Cartesian dualism, the mechanical operation of the *imagination* in relation to the *body*: '[Hysteria was] the most real and the most deceptive of diseases; real because it is based upon a movement of the animal spirits, illusory as well, because it generates symptoms that seem provoked by a disorder inherent in the organs, whereas they are *only the formation* [underlining mine], at the level of these organs, of a central or rather general disorder; it is the derangement of internal mobility that assumes the appearance, on the body's surface, of a local symptom.'[21] The writings of eighteenth-century physicians – works like Mandeville's *Treatise of the Hypochondriack and Hysterick Passions* (1711) – support such a theory of hysteria or hypochondria as a condition, not a disease (a condition embraces many diseases), but not until powers of triggering every kind of physical illness had been delegated to the imagination, could the general malaise of the polite, refined age have been called a disorder of both the imagination and the animal spirits. This malaise was, of course, the well-known 'English Malady,' a neurotic type of melancholic hysteria.

The etiology and taxonomy of the science of 'diseased imagination,' a phrase that by 1720 was common in medical parlance,[22] was unable to rupture the 'holy alliance between science and religion,' although it eventually did. 'It is evident,' Dr. Robert James wrote in the article on 'Mania' in his medical dictionary, 'that the Brain is the Seat...of *all* Disorders of this nature [i.e., madness]. It is there that the Creator has fixed, although in a manner which is inconceivable, the lodging of the soul, the mind, genius, imagination, memory, and all sensations.' And although madness was explained, by James and other physicians, by a mechanical model in which there is an irregular agitation of the spirits, it was now also the obstruction of the body and the imagination, not merely the obstruction of one or the other; an obstruction, moreover, which had grave consequences, as James noted, 'causing stagnation of the humours, immobilization of the fibers in their rigidity, fixation of ideas and a kind of manic concentration on a theme or idea that gradually prevails over all others.' To be sure, some of this explanation was residual Cartesianism, but it was more than that and incorporated the new psychology of Locke and the new physiology of the doctors. Much of eighteenth-century medicine reads like a commentary on Locke's definition, 'Madnesse seemes to be nothing but a disorder in the imagination, and *not* in the discursive faculty.'[23] Science never became so mechanistic as in her attempt to explain 'the obstruction' suggested by Locke and mentioned by James. From the theory of the diseased imagination arose a new conception of the madman. This in turn gave rise to a whole etiology of illnesses created by the 'diseased imagination,' which in turn spurred a series of associationist and sensationist controversies[24] that were to leave indelible scars on literature. The astonishing thing is that Enlightenment medicine dismissed as uninteresting the healthy imagination and concentrated exclusively upon the diseased. It arrogated powers to the diseased imagination in its influence on the body that earlier had been reserved for the Deity himself; imagination, in obstructed and consequently diseased forms in the female or male, could destroy the seed of life, the foetus at any stage of conception or gestation. The last vestiges of Cartesian dualism (mind and matter) were now reduced to a form of tyrannical monism in which all bodily functions – especially those of the fibres and animal spirits –

were enslaved to the will of the imagination. This imagination, still undefined, was recognized as an irrational 'super passion,' as significant for the body as gravity was for the earth, and as the totality of an infinite number of sensations associated and combined in patterns as yet unexplained.

'To the Virtue of the Mother's Fancy [alternate word for the Imagination] have been ascribed the Lineaments of the Embryo or Foetus,' Dr. James wrote in his *Medicinal Dictionary* in the article on 'Imagination,' 'with the Marks imprest upon its Body, both at and after any Time of Conception. . . . Transplantation of Diseases, the Strange Alterations of Bodies by the Virtue, Reliques, and the Invocation of Saints, are all imputed to this Power of the Imagination.' All nascent insanity was believed to result from the mother's diseased imagination, although some theorists thought the father could transmit it during copulation. Medical insistence on the validity of this belief gripped literary and artistic sensibility in a profound way, in a manner that transcended mere influence of science on literature in the form of allusion or demonstrable awareness. Mrs. Pickle's perverse craving for 'pineapples of the finest sort' during Peregrine's gestation in Smollett's *Peregrine Pickle* (1751) raises the medically valid fear that Peregrine may be born deformed and – to carry the idea one step further – there is no shadow of doubt that Tristram Shandy's disordered sense of time is caused by his mother's ill-timed ejaculation, uttered at the very moment of sexual coition with Mr. Shandy, 'Pray, my dear, have you not forgot to wind up the clock?'[25] Eighteenth-century literature is permeated with examples of women whose imaginations are 'diseased,' thereby illustrating more than causal interest of the literary man in medical theories of the day. No science throughout the century was more influential on philosophical thought than medicine, and no science did more to unseat the *literati* from a lingering medievalism based on hierarchy and order, which was now finally obliterated. In the hundred years from 1727 to 1827 there were no Newtons in astronomy, or for that matter in any other science, and while vortices, rainbows, and gravity had enthralled an earlier generation of poets like Pope, John Hughes, and James Thomson, the organic sciences now prevailed and were causing the largest ripples in that ocean of science we have come to call 'the Enlightenment.' As early as 1726, one year before Newton's death, Pope wrote to Swift that the possibility of the 'Rabbit Breeder woman,' Mary Tofts, giving birth to seventeen rabbits had stimulated his imagination more than anything since he indulged in 1713 in 'astronomical dialogues with William Whiston,' whose explications of planets, orbits, and worlds-upon-worlds without end had sent Pope soaring to a fanciful cloud where he remained for some years.[26]

A survey of the vast medical writings of eighteenth-century England and on the Continent, in Russia and the East European countries, shows a preponderance of works concerned with madness and the malfunctioning imagination.[27] The history of the medical concept 'imagination' in this epoch is therefore also the history of madness. It is hardly surprising that the diseased imagination rather than the healthy, should have claimed most theoretical attention: Once derangement was deprived of its former freedom, it was the task of physiologists and neurologists to place still stronger limitations upon it; limitations in the form of incessant redefinition. Madness became a subject of greatest interest to methodologists in science, who made it conform to technological constructs then emerging in the organic and inorganic sciences. Indeed, it is not much of an exaggeration to note that the most creative eighteenth-century medical theory appears to be one long record of reconsiderations of the concept madness: In almost every case an attempt was made to demonstrate the precise dynamics of the new dualism, interaction of the imagination and the nervous system. It is as if theoretical physicians were looking for Newtonian laws of motion of the disordered physiology. Their search was for a calculus, as their spokesman Dr. Thomas Morgan made perfectly clear in his treatise on the *Philosophical Principles of Medicine* (1725), in which imagination (force) is the product of nerves (mass) and animal spirits (acceleration).[28] Why should it be thought extraordinary that recently discovered mathematical laws – the Newtonian calculus – would be applied to psychology at a time when other sciences (post-Newtonian astronomy and physics, French chemistry, Priestleian electromagnetism), as well as Leibniz's 'science of morality,' were turning to them? Albrecht von Haller's experiments in the 1740's and 1750's on animal 'sensibility' and 'irritability'[29] stimulated further laboratory research on the nervous system: Here was a physiologist of the highest caliber who appeared to be on the threshold of discovery of the sensory dynamics of the imagination. His experiments on the sensory perception of animals provided partial answers to Locke's associationist questions, and his separation of all sensations into categories of 'irritability' (impressions which do not reach the cortex) and 'sensibility' (those which do) added new fuel to the fire of theoretical psychologists. Whether mania or melancholy were in question, the cause of mental illness initiated by the 'diseased imagination' is always in the movement of the animal spirits. Haller's explorations into the nervous system confirmed the suspicions of empirical scientists like John Gay and David Hartley.[30] What had been put forth by these men as philosophical speculation was now given the stamp and seal of the medical world. Haller was not only the Professor of Medicine at Göttingen (the most esteemed German university of the day); he was also one of the most reverenced physiologists of the century. Associationism and madness were now wedded, and what had earlier passed for mere hypothesis was now proved scientific fact.

There had been much talk about derangement, of course, in medical circles at the turn of the eighteenth century. In fact, there was so much speculation that Jonathan Swift in 1704 alluded abundantly in *A Tale of a Tub* to one of the current theories, that of 'rising vapors,' and fully expected his readers to grasp his allusions without spelling them out in detail.[31] But at that time, the etiology of madness was an unknown province. In clinical terms, at that time (ca. 1700), the etiology was also the sign and the symptom, and the manifestations of the diseased imagination were thought to be the cause. By 1800 the picture altered considerably. Society at large had heard much about scientific theories of the sick imagination,[32] had observed social reform attempting to improve the lot of the madman, and had even grown somewhat weary by the inability of medicine to cure what it professed to understand. Some of the dissatisfaction was owing to a general suspicion of doctors – 'quacks, empirics, and mountebanks,' to use Jane Austen's invective – but much was more particularized. There was little visible evidence that the new physiology was better than the old: The layman praises theoretical science only when it abets his daily life. We know today that the neurology of the late eighteenth century had advanced to a point far beyond that of the previous century's, but there is little evidence that laymen of the time knew it. To them, discussion by the 'doctors' regarding madness often appeared pedantic and petty, as one long never-ending controversy. There is, thus, an aspect of staleness and a tone of weariness in the creation of Matthew Bramble, Smollett's ailing hero in *Humphry Clinker* (1771), who suffers from 'a natural excess of mental sensibility' and 'whose every attendant disorder of the Body arises from one originally in the Mind.' So, too, is there in Goldsmith's kindly Dr. Primrose in *The Vicar of Wakefield* (1766) whose path of vicissitude proceeds from 'an imagination in discord with its body.'

Greater literary artists than these men absorbed in a curious and sometimes profound manner the findings of contemporary medicine, and began to write in new empirical modes that implied a marriage between 'the diseased imagination' and the normatively accepted 'creative sense' – between sickness and health. The origins of romantic and, later, symbolic poetry were probably as intrinsically involved with this marriage as they were the decay of the so-called 'rules' and 'genres'; or, again, with the decline of mimesis, a decline that had occurred in the first place as a result of psychology and the empirical view of human experience.

Of the various sciences studied in the eighteenth century, geology has been shown to have played a significant role in the development of romantic nature imagery, particularly in poems like Coleridge's 'Kubla Khan' and plays like Shelley's 'Prometheus Unbound.'[33] It has been less clear that the organic sciences generally and medicine particularly played a greater role in the formation of the temperament in art we now call 'romantic.' Without a century of controversy about the 'diseased imagination,' Wordsworth could not have composed his *Prelude* as we know it, the 'Growth of A Poet's Mind.' He was not, to be sure, a scientific poet but if his reaction against physiological law is difficult to pin down, his epic-of-the-imagination – a poem that is now considered to be the greatest achievement of English romantic poetry – is a direct answer to the inadequacies of associationist medicine and sensory physiology. His language was new, but there was nothing new about his subject, the imagination.

III

It is a paradigm that art does what science cannot. Whitehead has written that the great defect of eighteenth-century science was its failure to provide for the deeply felt experiences of man or to tolerate the sense that nature is organic. Enlightenment scientists, try though they did, were unable to explain adequately the interaction of the imagination and the animal spirits.[34] If they had tackled a less important matter, they might have been forgiven, but theirs was the single most important problem of physiology, and physiology was the only science of the eighteenth century in which serious research was conducted. Some aspects of their failure were overlooked because culture, particularly in England, was prepared to accept these failures. The inability of the neurologist, for example, to devise a calculus for this interaction was viewed less disappointedly because society at large had not as yet become quantified in the sense we have known technological quantification in this century.

Had it been apparent to the eighteenth century that the physiology of imagination was tearing God from his seat and creating a new kind of deity – a Godhead of the machine: impersonal, predictable, artificial, thrifty – resistance might have been greater. As it was, the resistance was feeble and inconsequential. There was considerable opposition from all segments of the Church, particularly to La Mettrie's theory of 'man as a machine,'[35] but set in proper perspective and evaluated on balance, this resistance was not influential and gradually gave way. The greatest disappointments of eighteenth-century science were medical, and are most widely viewed in the natural philosophy and romantic poetry of the late eighteenth and early nineteenth centuries; in the poetry of Blake,

> O Divine Spirit, sustain me on Thy wings,
> That I may awake Albion from his long & cold repose;
> For Bacon & Newton, Sheath'd in dismal steel, their terrors hang
> Like iron scourges over Albion: Reasoning like vast Serpents
> Infold around my limbs, bruising my minute articulations.
> I turn my eyes to the Schools & Universities of Europe
> And there behold the Loom of Locke, whose Woof rages dire,
> Wash'd by the Water-wheels of Newton . . .[36]

or that of Coleridge, who was convinced that the materialist tradition of Boyle, Locke, and Hartley had 'untenanted creation of its God'[37] and substituted

> a universe of death
> From that which moves with light and life informed,
> Actual, divine, and true.

For too long it has been thought that the 'revolt' of the Romantic poets was essentially a reaction to a decaying empiricism that could not satisfy the promises it made and the expectations it raised with regard to human progress, a reaction that begged for a new empiricism regarding human experience. Some part of the revolt was certainly due to unfulfilled promises. But much was also the inability of science to formulate laws of organic relationship between the imagination (healthy or diseased) and the animal spirits; that body of physical laws which, if they had been satisfactorily formulated, may have permitted man to be free once again – free in health and free in madness, free as the spirit had been in the Renaissance. The failure of Enlightenment physiology was a genuine, scientific disappointment. Judged by expert standards, it was more than the consequence of an immature science wallowing in a Sargasso Sea and hindered by an absence of genius.[38] It extended beyond the artist's desire that science fulfill in the same way as art; and it is most probably correct to speculate that if science had not taken the turn it did – towards an ever-greater impersonality in the march towards technology – Romantic thinkers would not have rebelled as monolithically and strenuously as they did. Surely the point to be gathered is not *only* that eighteenth-century science actually forced the human spirit into a strait-jacket, but that the Romantics were convinced it had. There can be no doubt, additionally, that eighteenth-century science damaged the artist and impelled him almost against his will, as it were, to indulge his desire for a new variety of empirical experience based on moments of transcendental truth. Claims of impersonality, predictability, artificiality, and thriftiness – claims all made in the name of science and technology – were not enough to satisfy the ilk of a Blake, a Wordsworth, or a Coleridge.

There is, however, still another aspect of the 'revolt.' When science makes promises she must fulfill them however inchoately. In this sense, experimental science as a cultural force is radically different from that of the humanities. Once imagination was created by physiologists, it had to be nurtured, permitted to mature, to evolve into adulthood and ripen into old age. The abortive attempts of eighteenth-century physicians and theoretical physiologists were perfectly visible to all who could see; by late century the record of neurological research read like an antiquarian's journal: Of interest to historians only, these neurological works often repeated, sometimes undigested, the research of earlier physicians, particularly those at mid-century, men like Boerhaave, Whytt, and Hartley.[39] Only Pinel,[40] the French theoretician and physician at the Bicêtre and the Salpétrière, had genuinely new ideas about the pathology of the diseased imagination, and Englishmen were resistant to these for several decades.

With the promise of an organic marriage of the spiritual (imagination) and the material (animal spirits, fibres, nerves) thwarted, it remained the task of non-scientists to formulate a mythical set of laws uniting them. That poets and philosophers should somehow have felt this their 'task' is significant in itself. More consequential was its effect on literature. Such a 'task' may not have been auspicious for the history of physiology; it was for Romantic poetry and natural philosophy.

Madness was once again returned to sacred grounds, to camps of mysticism and to cults of pantheism which had no basis in scientific fact. Under the influence of powerful minds like Coleridge, Wordsworth, and the German naturalists, neurology regressed in the popular mind, particularly in England, and associationist psychology was declared an invalid. The way was paved for the new phenomenology. The diseased imagination was instead romanticized, endowed with an aura of glory it had never known; so much so that Wordsworth's famous lines may be read almost literally: 'We Poets in our youth begin in gladness,/But thereof come in the end despondency and madness.'[41] The long wars of truth over the supremacy of the body by the mind, or the mind by the body, were finally brought to an end, mind triumphant, as Wordsworth notes subtly, though plainly, at the conclusion of his lyrical ballad of *Goody Blake and Harry Gil* based on a story in Erasmus Darwin's *Zoonomia*.

What appeared in the 1740s and 1750s as the start of a genuine revolution in neurology turned out to be merely an incomplete mechanics, an abortive attempt so far as literature was concerned. Associationist psychologists could not explain the inseparability of matter and force, particularly in the association of secondary ideas. Theoretical physicians instead of keeping their promise to solve the greatest riddle in medicine fell back upon themselves and, in defence, returned to older subjects: redefinitions of madness and the vital substance within the animal spirits. The theory of sensationism lapsed into a kind of vitalistic archaeology. Romantic philosophers like Kant and poets like Coleridge were thus the heirs of an inchoate scientific breakthrough, of what seemed to be an inert materialism, their task of demolition carried out in the name of humanism, the preservation of man as an organic creature: whole, unfragmented, non-disintegrated.

Consequently, it is probably true that the etiology of European romanticism is located as much in the medical researches of the eighteenth century as in the disturbance of the sacred and profane in an increasingly industrial society. Newton's discovery of mathematical laws of light waves contributed directly to concrete imagery in poetry and painting, as Professor Marjorie Nicolson has shown,[42] and his discovery of gravity, to belief in a living center of every moving object, a center forever attracting its parts.[43] But the implications of the neurological 'discovery of the imagination' – as I have been using this phrase – were of equally great consequence. The seat of all creative endeavour was established. Man was irredeemably bifurcated into (1) a physically existential imagination and (2) a complex of fibres, nerves, and animal spirits. Not until these distinctions – now more precise than Descartes' unwieldy pineal gland and a body proper composed of many organs – could be once again united would creative thinkers who valued mental freedom rest.

European Romanticism as we know it was therefore in part a final answer to Cartesian dualism and to the mind–body pathologists who followed Descartes. I know of no moment in the history of modern European culture in which science and literature were more intrinsically interrelated. It cannot be denied that the multifacets of this inter-relationship have warded off students. It is easier, after all, to survey the inter-working of the two in direct linear fashion, as, for example, in the poetry of Donne who explicitly refers to Copernican cosmology, or in the prose satires of Swift who lambastes the ludicrous experiments of the Royal Society in London.[44] More difficult, by far, is it to correlate developments in science and society, such as Robert Merton's thesis – to my knowledge still unchallenged in a demonstrable way by those who maintain that latitudinarian men were equally engaged in scientific experimentation – that puritanism, protestantism, and science developed together, the latter a result of the former two. Still more elusive is the influence of science on modes of human thought and revolutions in aesthetic taste. Any conclusions regarding such influence must be viewed tentatively and skeptically until historians of science catch up with historians generally.

Sociologist Robert Merton has written, 'It has become manifest that in each age there is a system of science which rests upon a set of assumptions, usually implicit and seldom questioned by the scientists of the time.'[45] Would that historians applied such criteria to the Enlightenment, at least as a prolegomenon to understanding. It would then be apparent that one reason among several for the relative wasteland of great imaginative literature, especially in England, was caused by more urgent problem,[46] crucial problems which occupied and deflected men from writing great poetry in abundance and which caused them to turn to the study of 'natural philosophy' – the sciences. The Carlylean tendency to regard the history of human achievement as a succession of inexplicable geniuses arbitrarily bestowing knowledge upon mankind has finally been abandoned as simple-minded and mythical. And Matthew Arnold, in assessing the age of Pope, never understood the reasons that caused men to turn away from verse.[47] Actually, economic conditions in eighteenth-century England were such that a larger portion of the population than in any earlier time could have devoted itself to the writing of creative literature if it wished. To be sure, some portion did, but considered on a per capita and not an absolute basis – and it is an important distinction – the writing of poetry diminished in the eighteenth century. The largest number of educated men became scientists, doctors in particular.[48]

Throughout the eighteenth century, scientists conducted their experiments in the name of morality, or thought they did, and it is amazing to discover how religious most actually were. Scientific empiricism and inductive rationalism were accordingly canonized, beatified, and deified. And yet the fruits of their research seemed underdeveloped. The followers of Descartes in France, who took the physiological half of his dualism (including an early theory of associationism based on nervous traces in the brain) and ignored the rest, were to be answered and repudiated by the German metaphysicians and, slightly later, by English thinkers like Coleridge. Coleridge's theories of the sympathetic and esemplastic imagination are thereby best understood as another chapter in the medical history of the discovery of the imagination. His philosophical thought ultimately rests upon a complete rejection of Locke's associationism and Hartley's sensationism. His flirtation with Hartleyian mechanism is vividly documented, and he may certainly be described as one who sought to heal the Cartesian rift. He is, in a sense, an inquiring physiologist and a skeptical biologist attempting to explain the rift of mind and body in the previous age.[49] There is also a scientific side of Blake, particularly in his demands for an organic union of the senses and the imagination in the non-fragmented man. At no point in Blake's writings – notwithstanding the momentary vision at the end of *Jerusalem* of Locke, Newton, and the bard holding hands – is there hope for the progress of the materialist philosophy or Enlightenment physiology. Indeed, Blake claims to have abhorred Locke and Voltaire from his first reading of them. These and other poets were the posterity – not always consciously, but still the posterity – of Enlightenment science. They accepted associationism as a temporary paradise only to discover shortly thereafter that it was a permanent exile; accordingly, they rejected it and formulated their own theory of imagination. Their thinking had more in common with scientists than with poets of the previous century; their writings were alien to, yet not completely divorced from, the neurologists. John Stuart Mill has written that 'the Germano-Coleridgean doctrine expresses the revolt of the human mind against the philosophy of the eighteenth century.'[50] It was more specifically a revolt against its physiology. As late as 1923 Valéry created in his *Dialogues* a romantic physician Eryximachus who attempts to cure Socrates by combining his nerves and imagination into one, unfragmented, whole framework.

The romantic temperament thrives by indulging in private visionary experience based on moments of transcendental truth; it dwells in a mythical sphere of abundant revelation, however unreal to an outsider, which is similar to the madman's. No earlier thought beatified madness as it did, particularly in its apotheosis of the mad poet, the mad lover, the mad hero, the mad sufferer. I believe it is inaccurate to conceive of the decline of the baroque and the rise of the romantic habit of mind without recourse to physiology and medicine.[51] In their aspirations and ends, the theories of Enlightened medicine and Romantic imagination were similar: Both considered matter and force as inseparable, and both believed that if any ratiocinative process separated them, the result would be disintegration of the organic creature.

The longing to remain whole is not yet inert; it is as ancient as man, and it antedates the beautiful and ridiculous myth of Aristophanes in Plato's *Symposium*. It seems evident that in the period 1650–1800 man's conception of himself suffered an unprecedented trauma, one that fragmented him and left him materially divisible. Not until he could repair the damage of the physiologists would he return to his former selfhood: one man, however sane or demented, but at least one man.

In examining an aspect of the tone and effect of Enlightenment medicine upon artists, I have attempted to make it abundantly evident that the historian's task in this sphere poses insurmountable problems. He knows he is not surveying the impact of discovery A upon person B, and also that as a consequence he must bear the burden of inadequate generality. This burden teaches him that he must proceed cautiously and with gravity since all generality is unsatisfactory, particularly in an age of uncertainty principles; that abstract truths are ultimately fictions, impossible to prove by concrete examples;[52] and that history is myth – the fixing of laws in the realm of untruth.[53] It also teaches him that some generalities are better than others, and that he may ascend closer to the threshold of truth if he stands back and surveys large terrains. It is my contention that the origins of romantic sensibility are incomplete until we survey the cataclysmic shift in physiological thought that occurred in the Enlightenment, a shift that probably contributed more to change man's image of himself than anything since the introduction of gunpowder into Europe in the fourteenth century. Robert Boyle, perhaps the greatest chemist of the period but also a highly original mind, seems to have anticipated this shift when he commented in *The Usefulnesse of Experimental Naturall Philosophy* (1663) that 'those great transactions which make such a noise in the World, and establish Monarchies or ruin Empires, reach not so many Persons with their Influence as do the Theories of Physiology.'[54]

Notes

1. Throughout this essay I use 'imagination' in the strict physiological sense and not in its loose and now common usage, e.g., 'the literary imagination,'

'the artistic imagination,' or imagination as connoting a degree of sensitivity, intuition, or creativity. See Gerhardt von Bonin, *The Evolution of the Human Brain* (University of Chicago Press, 1963) and Harry J. Jerison, 'Interpreting the Evolution of the Brain,' *Human Biology*, XXXV (Sept. 1963), 263–91. Most brain theorists today define the imagination scientifically as the number of associations per unit time. The number of associations depends upon four physical and biochemical processes: (1) the number of neurons in the frontal lobe of the brain; (2) the speed of nervous conduction and synopsis of these neurons; (3) memories which hinder or abet the processes described in (2); and (4) hormones that may or may not affect the whole process. (1) and (2) are best understood, almost nothing is known about (3), and little has as yet been learned about (4). It is possible, and perhaps probable, that the interaction of all four processes may radically change present notions regarding an individual one. Dr. Jerison conclusively shows that there has been no known change in the brain size of the *homo sapiens* since Cro-Magnon man. Brain theorists discuss the healthy and sick imagination in terms of the above four factors. Biochemists surmise, although it is not proved, that (3) and (4) account for most mental illness. It is important to note that the most advanced brain theory may be incorrect: The brain can imagine only that which it is capable of, and it is at least possible that functions (1)–(4) do not permit comprehension of the very process that limits it. E.g., the gorilla knows how to peel bananas, but the limited number of associations per unit time prevent it from formulating theoretical questions. By analogy, the same may be true of man. I am very much indebted to Dr. Saul Zamenhof, Chief, Brain Research Institute, UCLA, for assistance in this note and for discussions in 1968–1969 which permitted me to explore the scientific aspects of my subject.

In the strict sense, 'Enlightened England' is meaningless; I use it merely to designate the period 1660–1800. The most comprehensive scientific definitions for 1660–1800 are found in medical and other dictionaries of the period. See John Harris, *Lexicon Technicum* (1704); Ephraim Chambers, *Cyclopaedia; or, An Universal Dictionary of Arts and Sciences* (1728); Robert James, *A Medicinal Dictionary* (1743–5), *Encyclopaedia Britannica* (1768). The *OED* provides a guide to changing concepts of the word in non-scientific usage. In this period the 'brain' was thought to consist of 'inner' and 'outer' fibres and to correspond to different parts of the body: see William Drage, 'A Tractate of the Diseases of the Head,' *A Physical Nosonomy* (1665); Thomas Willis, *Pathologiae Cerebri* (1667), *An Essay of the Pathology of the Brain*, trans. S. Pordage (1681) and *Two Discourses Concerning the Soul of Brutes* (1683). I have found no agreement regarding definition of the 'fibres' in the brain; most scientists of the time liken them to textile fibres and are ambivalent about their hollowness or solid state. Later on I discuss medical definitions of the imagination, 1660–1800. Unless indicated otherwise, all works cited are published in London.

2. The theories of Artemiodorus (the most important collector of ancient theories of the subject) were transmitted through the Renaissance to the seventeenth and eighteenth centuries, some of which are summarized in *Sir Thomas Browne's Works*, ed. Simon Wilkin (1835). See Philip Goodwin's *The Mystery of Dreams, Historically Discoursed* (1658).

3. Aspects of this revolution have been studied by Alfred North Whitehead, E. A. Burtt, A. O. Lovejoy, R. S. Crane, R. F. Jones, M. H. Nicolson, Herbert

Butterfield, and R. K. Merton. The most conclusive evidence demonstrating that it was a genuine scientific revolution has been gathered by T. S. Kuhn in *The Structure of Scientific Revolutions*.

4. See Aristotle, *De Anima, passim* and *Poetics, passim*; also Plato, trans. Jowett (New York, 1937), *Apology*, 22; *Republic*, 378; *Ion*, 534. An example of anti-Aristotelianism is Akenside's *Pleasures of Imagination* (1744), in which it is significant that Akenside abandoned the definite article of his original title, *Pleasures of The Imagination*.

5. Nerves were defined variously as (1) hollow tubes through which vital liquids flowed; (2) hollow tubes containing undefined but stationary substances; and after the 1740s as (3) solid tubes conducting electric impulses from cell to cell. There was little agreement about all three major explanations and there were several other minor ones. I know of no secondary study that has surveyed eighteenth-century theories of nerves or, as I state in n. 6, of animal spirits. Listed chronologically, some of the works involved are: *Philosophical Transactions* (1661–1780); Ephraim Chambers, 'Nerve,' *Cyclopaedia: or, an Universal Dictionary of Arts and Sciences* (1728), II, unnumbered page; George Cheyne, *The English Malady: or, A Treatise of Nervous Disease* (1733); H. Boerhaave, *Academical Lectures on the Theory of Physic*, 2nd ed (1751–1757), II, 284–5; Thomas Reid, *An Inquiry into the Human Mind* (Edinburgh, 1764).

6. Even more controversial were theories of the 'animal spirits.' Basically, the problem was one of defining the vital substance within the spirits themselves. This problem of definition was further complicated by contemporary theories of acids – in the air, body, and blood – which defied chemical identification. Most of the works cited in n. 5 also discuss animal spirits. Endless debates over the motion of animal spirits sometimes created humility in physiologists. Dr. Robert Whytt, an important neurologist and physician to King George III, wrote of the difficulties of explaining the passions that arise from particular movements of the animal spirits: 'To ascend from small things to great, altho' Sir *Isaac Newton* did not pretend to explain the cause of gravity, yet he made no small improvement in physical Astronomy, when, from this principle alone, he accounted for the various motions of the planets, and banished the imaginary *vortices of Descartes*, which had been contrived, but unsuccessfully, to explain the phaenomena of the solar system' (*Observations on the Nature, Causes, and Cure of...Disorders* [Edinburgh, 1765], vii, preface). The animal spirits controversies have not been surveyed. Important works include: Thomas Willis, 'Treatise of Musculary Motion,' *A Medical-Philosophical Discourse...*, trans. S. Pordage (1681) and *Two Discourses Concerning the Soul of Brutes Which Is That of the Vital and Sensitive Soul of Man*, trans. S. Pordage (1683); *Philosophical Transactions*, 1680–1750 (more than thirty articles on animal spirits); John Harris, 'Nervous Spirit,' *Lexicon Technicum* (1704), II, unnumbered page; Nicholas Robinson, *A New Theory of Physick* (1725) and *A New System of the Spleen, Vapours and Hypochondriack Melancholy: wherein all the decays of the Nerves... are Mechanically Accounted for* (1729); Robert James, 'Animal Spirits,' *Medicinal Dictionary* (1743–1745); *Gentleman's Magazine*, 1740–1765 (seventy-two entries on animal spirits); Malcolm Flemyng, *An Introduction to Physiology; Being a Course of Lectures on the Most Important Parts of the Animal Oeconomy* (1759) and *The Nature of the Nervous Fluid; or, animal spirits demonstrated* (1751); Laurence

Heister, *A Compendium of the Practice of Physic*, trans. E. Barker (1757); 'Spirits,' *A New Medicinal Dictionary*, ed. George Wallis and G. Motherby, 4th edn (1795).

7. See J. D. Boyd, S. J., *The Function of Mimesis and Its Decline* (Harvard University Press, 1968).

8. For commentary on the imitation of the imagination in symbolic painting of the eighteenth century, see Erwin Panofsky, *Iconology* (Princeton, 1955) and Ralph Cohen, *The Art of Discrimination* (1964).

9. See John Yolton, *John Locke and The Way of Ideas* (Oxford, 1956); Kenneth Dewhurst, *John Locke, Physician and Philosopher: A Medical Biography* (1963); Maurice Mandelbaum, 'Locke's Realism,' *Philosophy, Science, and Perception* (Johns Hopkins, 1964); *John Locke: Problems and Perspectives*, ed. J. Yolton (Cambridge University Press, 1969). Richard Ashcraft has shown that Locke was known to his contemporaries primarily as a physician. See 'John Locke's Library: Portrait of An Intellectual,' *Transactions of the Cambridge Bibliographical Society*, IV (Cambridge, 1969), forthcoming. John Harrison and Peter Laslett have written *a propos* of the unusually large number of medical books in Locke's library: 'Whether Locke's ownership of works by Glauber, Gesner, Mayerne, Borelli, Gabelchover, Huyghens, Malpighi, Borrichius, Sachs, Rolfinck, Pisanelli, and the rest make of his catalogue a landmark in the literature of the scientific revolution is for historians of science in our own time to decide,' *The Library of John Locke* (Oxford Bibliographical Society, 1965), pp. 24–5.

10. Eighteenth-century physiologists stressed the role of memory in influencing the process of association; see Locke, 'Of Retention,' *Essay Concerning Human Understanding* (1690), I, 193–201; 'Memory' in Locke's *Medical Journals 1675–1697*, quoted in Dewhurst, *op. cit.*, pp. 100–2; David Hartley, *Observations on Man* (1749), I, 3, 374. Hartley's chapter, Sect. IV, 'Of Memory,' I, 374–82 summarizes the differences of opinion regarding memory held by the physiologists. Much controversy took place over the function of memory in moral ideas that are innate. It was believed by some physiologists that certain of these innate, moral ideas could not be learned by the usual neurological and associationist processes. See also Richard Burthogge, *An Essay upon Reason, and the Nature of Spirits* (1694); Burthogge had studied medicine at Leyden and was a practicing physician.

11. By 1740 there had been so much controversy about the animal spirits that some scientists discredited them altogether, and literary men satirized and described them. See Henry Brooke's exposition in *Universal Beauty, A Poem* (1735), Part IV, ll. 243–52:

> Quick, from the Mind's imperial Mansion shed
> With *lively Tension* spins the *nervous Thread*
> With Flux of animate Effluvia stor'd,
> And Tubes of nicest Perforation bor'd,
> Whose *branching Maze* thro' ev'ry Organ tends,
> And Unity of conscious Action *lends*;
> While Spirits thro' the *wandring Channels* wind,
> And wing the Message of informing Mind;
> Or Objects to th' ideal Seat convey;
> Or dictate Motion with internal Sway.

See also Garth, *The Dispensary* (1699), I, 15–19, III, 82–6; Malcolm Flemyng, *Neuropathia; sive, De Morbis Hypochondriacis et Hystericis* (1740, a Lucretian epic in Latin hexameter), pp. 3–4; Addison, *Spectator*, ed. Bond (Oxford, 1965), I, 367, 471; II, 8, 197, 383, 451, 460, 525, 539, 563; James Thomson, 'Spring,' *The Seasons* (1728), 11. 865–76; Henry Fielding, *Tom Jones*, ed. W. E. Henley (1902), II, 9; V, 9.

12. Locke, *An Essay Concerning Human Understanding*, I, 203.
13. See Nicholas Robinson, *A New System of The Spleen* (1729); P. Frings, *A Treatise on Phrensy* (1746); Jerome Gaub, *De Regimine Mentis* (Leyden, 1747–63), printed with annotations in L. J. Rather, *Mind and Body in Eighteenth Century Medicine* (1965); David Hartley, 'Of the Pleasures and Pains of Imagination,' *Observations on Man* (1749), I, 418–42; Richard Mead, *Medical Precepts and Cautions* (1751), pp. 77–84; see also the index of the *Gentleman's Magazine 1731–1786*, which lists several dozen entries on imagination, many of which are letters to the editor concerning recent books dealing with the imagination. Among the most influential medical definitions of imagination were those of Descartes, who placed its activity entirely in the pineal gland; Thomas Willis, who believed its operations took place in the *corpus callosum* and those of common sense in the *corpora striata*; and La Mettrie (among the associationists), who carried Locke's theory of perception so far as to reduce the entire mind (i.e., judgment, reason, memory, the mechanical parts of the soul) to imagination. Many physicians considered the imagination as the chief cause of all bodily illness: see Thomas Knight, *Transmutation of the Blood* (1725), p. 45. Chambers, in the *Cyclopaedia*, believed the imagination works by mechanical means although he distinguished its operation from that of sensation: 'Whenever there is any motion in that part, to change the order of its fibres, there also happens a new perception in the soul, and she finds something new, either by sensation or *imagination*; neither of which can be without an alteration of the fibres in that part of the brain.' In this note and others I attempt to provide sufficient definitions to imply a 'discrimination of imaginations.' In order that such a list of definitions be complete for 1660–1800 it would be necessary to survey literary analogies as well as scientific ones. Space does not permit that here.
14. Article on 'Imagination.' James's definition is as clear as one can find in medical dictionaries of the age. His *Dictionary* was considered by philosophers and physicians of the age to be the single best summary of contemporary medicine. See Diderot, Eidous, and Touissant, Preface to their translation of James's *Dictionnaire Universal* (Paris, 1746): 'Ce Dictionnaire est précédé d'un Discours historique sur l'origine & les progrès de la Medecine. . . . Il montre enfin Harvey, jettant par sa découverte les fondemens d'une nouvelle theorie sûre & lumineuse & propre à nous faire appercevoir les ressorts cachés qui produisoient des effects dont la cause si long-tems cherchée, avoit jusqu' alors été inconnue.'
15. See n. 1. Chronological associations (a) within a time span $(t^1 - t^2)$ are now thought to depend on the four variables listed in n. 1. Stated mathematically, although such an equation is meaningless since there is as yet no way of calculating (3) and (4): $a(t^1 - t^2) = f(1,2,3,4)$
16. Ephraim Chambers, *Cyclopaedia*, article on 'Imagination,' unnumbered page.

17. 'Entretiens sur la Métaphysique et sur la Religion,' *Oeuvres Complètes*, ed. A. Robinet (Paris, 1965), XII, 30.

18. See William Battie, *A Treatise on Madness* (1758) and John Monro, *Remarks on Dr. Battie's Treatise on Madness* (1758), who comment on this point, and in modern commentary, Michel Foucault, *Madness and Civilization: A History of Insanity in the Age of Reason* (New York, 1965), pp. 101–32. The idea of the freedom of the deranged in the fifteenth and sixteenth centuries is noted in a medical treatise by Jacques Ferrand, *De la maladie de l'amour, ou Mélancholie érotique* (Paris, 1623), translated into English by E. Chilmead in 1640 and issued in many editions by 1700.

19. This belief was repeated in most works discussing the 'diseased imagination' and in medical treatises by Nicholas Robinson, *A New System of the Spleen* (1729), pp. 174–96; Edward Synge, *Sober Thoughts for the Cure of Melancholy* (1742); J. D. T. de Bienville, *La Nymphomanie, ou Traité de la fureur utérine* (Amsterdam, 1758), trans. into English by E. S. Wilmot (1775).

20. Originally published as *Histoire de la Folie* (Paris, 1961).

21. P. 124. See also G. S. Rousseau, introduction to Sir John Hill's *Hypochondriasis: A Practical Treatise* (1766; pub. for the Augustan Reprint Society, Los Angeles, 1969), in which this point is discussed at length.

22. See the controversies over the etiology of the 'diseased imagination' in the 1720s: James Blondel, *The Power of the Mother's Imagination . . . Examin'd* (1729); Daniel Turner, *De Morbis Cutaneis* (1726), *A Discourse Concerning Gleets . . . in respect to the Spots and Marks impressed upon the Skin of the Foetus* (1729), *The Force of the Mother's Imagination upon her Foetus in Utero* (1730). A full study of the medical controversies concerning the diseased imagination is found in G. S. Rousseau, 'Pineapples, Pregnancy, Pica, and *Peregrine Pickle*,' *Tobias Smollett: Bicentennial Essays*, ed. G. S. Rousseau and P. G. Boucé (New York: Oxford University Press), reprinted as chapter 3 in this volume.

23. John Locke, *Journals: 1675–1679*, ed. Dewhurst, p. 89.

24. They were, respectively, answers to Locke's *Essay* and Hartley's *Observations*. See John Yolton, *John Locke* (Oxford, 1956) and Robert Hoeldtke, 'The History of Associationism and British Medical Psychology,' *Medical History*, XI (1967), 46–65.

25. Respectively *Peregrine Pickle* (1751), I, 21–36 and *Tristram Shandy*, ed. James A. Work (New York, 1940), p. 2.

26. See Marjorie Nicolson and G. S. Rousseau, *This Long Disease, My Life: Alexander Pope and the Sciences* (Princeton, 1968), pp. 131–238.

27. I have made a study of the number of such works and find the percentage very high. This may be due, in part, to the large quantities of medical satire written in the eighteenth century and to the many physicians (bona fide doctors, apothecaries, quacks, mountebanks, medical hacks) then writing. The point is emphasized by one writer of the period in an anonymous pamphlet in the Bodleian Library, *An Enquiry into Dr. Ward's Practice of Physick . . . With An Examination into the Origin, and Meaning of the Words Empiricism, Empirick. Quack-Doctor, and Quack. And, An Exact Account of the Present State of Physick* (1749).

28. The Newtonian equation $f = ma$ was actually applied to the animal spirits, although there was no means of calculating a (in Newton, $a = dv/dt$). See the anonymous *Philosophical Conjectures Concerning the Animal Spirits* (1746),

pp. 6–7, in the Royal Society of Medicine, London. There was some opposition to Newtonian equations applied to body motion by anti-Newtonian groups like the Hutchinsonians. Some discussion of developments influencing the slow progress of physiology 1700–1800 is found in Joseph Needham, 'Limiting Factors in the Advancement of Science as Observed in the History of Embryology,' *Yale Journal of Biology and Medicine*, VIII (1935), 1–18.

29. See Albrecht von Haller, *A Dissertation on the Sensible and Irritable Parts of Animals* (1755), trans. by Tissot, *Bulletin of the History of Medicine*, Supp. IV (1936), 651–99, and for Haller's reception in England, *Phil. Trans.*, 1748–57, *Gentleman's Magazine*, XXII (1753), 592–3, and Stephen d'Irsay, *Albrecht von Haller: Eine Studie zur Geistesgeschichte der Aufklärung* (Leipzig, 1930).
30. See John Gay, *A Dissertation Concerning the Fundamental Principle ... of Virtue* (1732), and David Hartley, 'Of the Doctrine of Vibrations and Associations in General,' *Observations on Man* (1749), I, 3–100; see also Hoeldtke, *Medical History*, XI, 48–51, and Joseph Priestley's comments in the introduction to his *Abridgement* (1775).
31. See Miriam Starkman, *Swift's Satire on Learning in 'A Tale of a Tub'* (Princeton, 1950), pp. 3–48.
32. The best evidence of this is found in the correspondence columns of the *Gentleman's Magazine*, 1740–1800. Few subjects received more attention, if letters written by readers (to the editor) are an indication. Studies of mesmerism and hypnotism are also helpful in this connection. See Robert Darnton, *Mesmerism and the End of the Enlightenment in France* (Harvard, 1968). The imagery of nerves, fibres, and animal spirits, in addition to that of the diseased imagination, is common in English and French novels 1740–1800. See some illustrative passages in Smollett's *Ferdinand Count Fathom*, 2 vols (1753), II, 46, 156, 260.
33. See John Livingstone Lowes, *The Road to Xanadu* (Boston, 1927) and G. M. Matthews, 'A Volcano's Voice in Shelley,' *English Literary History*, XXIV (1957), 191–228.
34. I mean in a manner satisfactory to educated men outside the world of experimental science. The response of other scientists alone would not be sufficiently broad to gauge opinion. An analogy in modern science may be current -dissatisfaction with the techniques of psychotherapy. The point is studied in detail by E. H. Carr, 'History, Science, and Morality,' *What Is History?* (Cambridge, 1961), pp. 70–112.
35. See Aram Vartanian, *La Mettrie's l'Homme Machine: A Study in the Origins of an Idea* (Princeton, 1960); F. A. Lange, *History of Materialism* (1877).
36. 'Jerusalem: Plate 15,' in *The Poetry and Prose of William Blake*, ed. by D. Erdman and H. Bloom (New York, 1965), p. 157. See Jean Hagstrum, 'William Blake Rejects the Enlightenment,' *Proceedings of the II Congress on the Enlightenment*, 1964, II, 142–55.
37. 'The Destiny of Nations,' II. 35–97, in *The Complete Poetical Works of S. T. Coleridge*, ed. E. H. Hartley (Oxford, 1912, rev. edn 1966), 132–3.
38. Such is the opinion of many historians of physiology. See Sir Michael Foster, *Lectures on the History of Physiology* (Cambridge University Press, 1901); J. R. Fulton, *Physiology of the Nervous System* (1938); numerous studies of Oswei Temkin's, including 'The Classical Roots of Glisson's Doctrine of Irritation,' *Bulletin of the History of Medicine*, XXXVII (1964), 297–328, and June

Goodfield, *The Growth of Scientific Physiology: Physiological Method and the Mechanist–Vitalist Controversy* (1960).

39. Other less important scientific writers include Drs. William Hunter, John Gregory, William Finch, Hugh Farmer, James Vere, Thomas Arnold, William Cullen, James Makittrick Adair.

40. See Phillipe Pinel, *Recherches sur le traitement moral des aliénés* (Paris, 1800); *Traité médico-philosophique* . . . (Paris, 1801; Eng. trans. by D. D. Davis, 1806); *Nosographie Philosophique* (Paris, 1818). The introductory remarks to D. D. Davis' translation, *A Treatise on Insanity* (1806), are of considerable interest.

41. William Wordsworth, 'Resolution and Independence,' II. 48–9.

42. *Newton Demands The Muse* (Princeton, 1946).

43. It is curious that the considerable influence of Newton's *Principia* on English literature in the eighteenth and nineteenth centuries has not been surveyed.

44. 'The First Anniversary,' ll. 205–6 and *Gulliver's Travels*, Part III.

45. Robert K. Merton, 'Puritanism, Pietism, and Science,' in *Social Theory and Social Structure* (rev. ed; New York, 1957), p. 586, originally in *Sociological Review*, 1938. See also the same author's 'Science, Technology and Society in Seventeenth-Century England,' *Osiri* (1938), IV, 360–632.

46. Here I am evaluating the poetry of England in the eighteenth century with respect to other centuries, and it must be admitted that periods like 1550–1620 or 1798–1850 seem richer in the quality of their poetry. This does not, of course, diminish the achievement of Dryden, Pope, or Johnson.

47. Matthew Arnold, 'Literature and Science,' in *Discourses in America* (1885), pp. 90–94.

48. T. McKeown and R. G. Brown, 'Medical Evidence Related to English Population Changes in the Eighteenth Century,' *Population Studies*, IX (1955–1956), 119–41.

49. See Joseph Needham, 'S. T. Coleridge As a Philosophical Biologist,' *Science Progress*, XX (1926), 692–702; Gordon McKenzie, 'Organic Unity in Coleridge,' *University of California Publications in English* (1939); Basil Willey, 'Coleridge on Imagination and Fancy,' *Proc. Brit. Acad.*, XXXII (1946), 174–87; Elisabeth Schneider, *Coleridge, Opium and Kubla Kahn* (Chicago, 1953); James V. Baker, *The Sacred River: Coleridge's Theory of the Imagination* (Louisiana State University Press, 1957); Althea Hayter, *Opium and the Romantic Imagination* (California, 1968). Coleridge's medical thought as found in *Philosophical Lectures* has not as yet received scholarly study. Some of Coleridge's attack on the physiology of the eighteenth century is found in a lecture entitled 'Materialism, Ancient and Modern,' 15 March 1819: the advertisement in *The London Times* read, 'Mr. Coleridge's Lecture for this Evening is on Dogmatical Materialism – in its relations to Physiology as well as to the religious, moral and common sense of Mankind.'

50. *Mill on Bentham and Coleridge*, ed. F. R. Leavis (1950), p. 108.

51. See the last section of Walter Pagel, 'Religious Motives in the Medical Biology of the Seventeenth Century,' *Bulletin of the Institute of the History of Medicine*, III (1935), 97–128, 213–31, 265–312. Pagel discusses the importance of vitalism in biology 1750–1820 and in the formation of the metaphysical doctrines of some of the Romantics. For discussion of this subject later in the nineteenth and twentieth centuries, see Jean-Paul Sartre, *The Psychology of Imagination* (1950), pp. 141–218. Among the best scientific statements for

and against vitalism are D. Dix, 'A Defence of Vitalism,' *Journal of Theoretical Biology*, II (1963), 338–40 and several books attacking vitalism by F. Crick (the noted biologist), particularly *Of Molecules and Men* (Seattle: Washington, 1966). See also Frank Barron, 'The Psychology of Imagination,' *Scientific American*, CXCIX (Sept. 1958), 151–68.

52. Historians have, of course, done so but not always to the satisfaction of other historians. See the recent works dealing with the origins of Romanticism by R. Wellek, Peter Gay, W. J. Bate, A. Gerard, G. Orsini, N. Frye, M. Abrams. Some aspects of science and the formation of Romanticism are treated in Wylie Sypher, *Literature and Technology: The Alien Vision* (New York, 1968).

53. Ved Mehta's recent interview (*Fly and the Fly-Bottle: Encounters with British Intellectuals*, [Boston, 1962]) with leading contemporary British historians (Trevor-Roper, A. J. P. Taylor, Arnold Toynbee, Pieter Geyl, E. H. Carr, C. V. Wedgwood, Christopher Hill, Herbert Buttlerfield) makes this evident.

54. Second Part, p. 3.

3
Pineapples, Pregnancy, and Pica: Nerves and the 'Mother's Imagination' (1972)

The world of the Georgian wits was populated by articulate physician-authors: Garth, Locke, Mandeville, Armstrong, Cheyne, Hartley, Goldsmith, Smollett and dozens of others among the medically educated who never practiced medicine and now rest in unvisited tombs. Doctors were then the only members of society who understood the body's 'unseen mechanisms', as Willis called them, and (as a consequence of their broadly classical educations) articulated and even versified them. R. D. Laing and Richard Seltzer, Stephen Jay Gould and Oliver Sacks, are the heirs of this tradition; if not contemporary equivalents, then at least equally articulate in communicating with the broad public. The difference now is that today we are quickly able to access much more technical information about the body than Pope's or Smollett's world could.

A discourse of medicine and literature had been developing since the Renaissance that structured knowledge in ways that can seem alien to us several centuries later. Earlier in the 1960s I had written a doctoral dissertation about a physician-author, the much under-appreciated Tobias Smollett (1721–71). Now – in the aftermath of the Pope book and the study of neural imagination (chapter 2) – I asked myself what difference this broad canvas made to the texture of Smollett's writing itself. I found one stunning example in the brilliant opening of Smollett's second novel, The Adventures of Peregrine Pickle *(1751), which deserves to stand next to Sterne's* Tristram Shandy *(1760–67) for the way in which it wittily blends the two kinds of knowledge: medical and narrative.*

The immediate stimulus was a new, American, multi-volume, annotated edition of Smollett's works, which now goes by the shorthand of The Georgia Edition. *I was assigned to annotate and explicate* Peregrine Pickle. *The opening pages engaged me for the way in which Smollett's learning was tapped for the purpose of wit. My reading of primary works of the period 1650–1850 dealing with neural networks in relation to cognition and passion, health and illness had already taken place (1964–69). Smollett himself had been steeped in these discourses of sensibility, as his book reviews and brief essays make clear. But I sought the evidence in narrative forms: proof from the novels themselves that he had used the aesthetic possibilities of 'nervous sensibility' – a term already common in English by the 1740s. I found considerable evidence in his last work,* The Adventures of Humphry Clinker *(1771), posthumously published, in which the main character, Matt Bramble, is a creature of exquisite nervous sensibility: the irascible geriatric squire as sensitive Man of Feeling.*

  The story line in Peregrine Pickle *is linear but its substratum is less evident than meets the eye. Mrs Pickle is pregnant. Midway through her pregnancy she craves a pineapple, which her sister-in-law provides. But Mrs Pickle does not eat it and Peregrine is born without trace – physical or mental – of the pineapple imprint. Had there been, he could have had a birthmark on his body resembling a pineapple or – worse yet – a personality confused about its social class and rank, pineapples then associated with high status while the Pickles were removed from that social niche. Here then is a proto-Tristram Shandy born nine years earlier than Sterne's comic hero (1751 versus 1760). The point is not that Smollett beat Sterne to the front door, or that both authors were steeped in medical writing about the nervous system (which they were), but rather that each recognized the degree to which thinking about neural networks depended upon language itself, especially the metaphoric textures submerged in nervous discourse. This exploitation of theory (neural networks) in the service of linguistic bias (the metaphors embedded in nerves, spirits and fibers) constitutes much of the comic wit of these two novelists and binds them.*

  The mechanistic Newtonian theory involved in these literary applications was straightforward and had been in place for half a century by the 1750s, as I had demonstrated in the 1969 essay (chapter 2 of this volume). Eating the pineapple would have deranged Mrs Pickle's 'imagination' further by making her depressed and hysterical during pregnancy – this in addition to scarring Peregrine's character; just as Walter Shandy's unfortunate question about 'winding up the clock' at the moment of conception damages Tristram for life by irrevocably puncturing his sense of time. Both babies are marred by the abnormal responses of their mothers. But Mrs Shandy does not miscarry despite her husband's ill-timed verbal ejaculation. Likewise, Mrs Pickle's aberrant craving during pregnancy elevates her hysteria to feverish pitch, but she can still produce her sweet little 'pickle' of a boy. The charged language of nerves and spirits was no less ridiculous, Smollett and Sterne thought, than these grotesquely comic women. Imagine construing the animal spirits jaunting along the highway of life as if they were 'little gentlemen', and configuring nerves 'warring in coalitions' as if they were so many living toy soldiers.

Praxis 2: Pineapples, Pregnancy, Pica and *Peregrine Pickle* (1972)

I

'This young lady, who wanted neither slyness nor penetration... replied with seeming unconcern, that for her own part she should never repine, if there was not a pine-apple in the universe, provided she could indulge herself with the fruits of her own country.'[1] Mrs Pickle's remark is calculated and represents the basis of a plan to rid herself of the 'teizing and disagreeable Mrs. Grizzle'. In fact, Smollett tells us that if a certain 'gentleman happening to dine with Mr. Pickle' had not mentioned pineapples, Peregrine might not have seen the light of day: the ridiculous Mrs Grizzle would have remained at the side of her sister-in-law throughout gestation, teasing and torturing her with obsequiousness, dementing her imagination, which at this period seemed to be strangely diseased, and 'marking' her still unborn child.[2] A 'diseased imagination' in the mother, Mrs Grizzle would have argued, produced inferior progeny (Peregrine born with the image of a pineapple clearly defined on his body!), but a 'dish of pineapples' could produce no progeny at all. Mrs Grizzle's search for pineapples for 'three whole days and nights' is thus futile: Peregrine, our hero, must be born. But then, a pregnant woman's desires must not be balked – Mrs Grizzle would have continued – and Mrs Pickle had told her that 'she had eaten a most delicious pine-apple in her sleep'. Unaware of the consequences of these alternatives, Mrs Grizzle makes the wrong choice, furnishing her sister-in-law with two ripe pineapples, 'as fine as ever were seen in England'. Propitiously, however, Mrs Pickle did not partake of the fruit – at least, Smollett never tells us that she did. Her swooning at the dinner table, the ensuing hysteria, the nocturnal pineapple dream – all are devices to encourage Mrs Grizzle to leave the Pickle household.

It is curious that these chapters (v–vi) describing the incubation and birth of Peregrine have received so little attention. This fact is further puzzling when one recalls that they are among the most amusing in the novel. Smollett's eighteenth-century readers would have been amused by Mrs Grizzle's obstetric handbooks and pious medical beliefs, and perhaps entertained by Mrs Pickle's intense dislike of her 'sister'. But they must have been doubly amused by the elder's 'researches within the country' (Smollett's hyperbole) for pineapples: such peregrinations for fruits 'which were altogether unnatural productions, extorted by the force of artificial fire, out of filthy manure!'[3] For her distaste for pineapples and her pretentious medical learning – Aristotle and Nicholas Culpepper[4] – eighteenth-century audiences could have forgiven Mrs Grizzle; they could not forgive her for an inability to put learning into practice. Her immense concern for pineapples, however, must have struck them as particularly topical, for readers of *Peregrine Pickle* would have understood Mrs Grizzle's newfangled theories of the effects of this fruit on pregnant women in the context of eighteenth-century medicine. We, as modern readers, view her actions and statements as part of Smollett's use of learning for the purposes of wit. A clue to the extensiveness of Smollett's knowledge of the subject can be observed in the first full-length book on pineapples in English, John Giles's *Ananas, a Treatise on the Pine Apple* (London, 1767), a book which Smollett may or may not have read. This technical handbook for scientists and expert gardeners confirmed Mrs Grizzle's belief that pineapples were an unlucky sign for pregnant mothers. Much of Smollett's comic irony in these chapters derives from the audience's familiarity with contemporary ideas about pineapples. It is not surprising, however, to discover medical learning used for comic purposes in the novels of an author who was a physician by profession and whose works abound with a variety of scientific references.

For our purposes chapters v and vi of *Peregrine Pickle* contain the key passages in which Mrs Pickle describes her unexplainable desires: for a fricassee of frogs, for a porcelain chamber-pot, for three black hairs from Mr Trunnion's beard – and for pineapples. Then we are told that only in the case of exotic foods did Mrs Grizzle interfere:

> She restricted her [Mrs Pickle] from eating roots, pot-herbs, fruit, and all sort of vegetables; and one day when Mrs. Pickle had plucked a peach with her own hand, and was in the very act of putting it between her teeth, Mrs. Grizzle perceived the rash attempt, and running up to her, fell upon her knees in the garden, intreating her, with tears in

her eyes, to resist such a pernicious appetite. Her request was no sooner complied with, than recollecting that if her sister's longing was baulked, the child might be affected with some disagreeable mark, or deplorable disease, she begged as earnestly that she would swallow the fruit, and in the mean time ran for some cordial water of her own composing, which she forced upon her sister, as an antidote to the poison she had received.[5]

The witty context of this description and the ridiculous actions of the women it involves should not lead us to conclude that Smollett was distorting or ridiculing contemporary theories of embryology. Actually Smollett's accounts of the treatment of pregnant women (such as Mrs Pickle) were based upon experiments with foetuses performed in the third and fourth decades of the eighteenth century and speculations of respected scientists of the Royal College of Physicians such as James Augustus Blondel, Daniel Turner, and William Smellie, a leading obstetrician to whom Smollett was apprenticed and in whose medical library he educated himself. It is essential to note biographically that at the very same time – 1750 – Smollett was composing *Peregrine Pickle*, he was also editing, annotating, and preparing for the press Smellie's *Treatise on the Theory and Practice of Midwifery*, which was published the same year as Smollett's novel. In 1750, then, Smollett was deep in the study of obstetric medicine and, particularly, in abnormal pregnancy, for that subject occupies the largest portion of Smellie's book. Smollett's novel was the first in English (a decade before *Tristram Shandy*)[6] to refer to heated controversies about the role of imagination in abnormal pregnancy, a subject Smellie treated at length and which abundantly stimulated Smollett's own imagination. The humorous episode built upon Mrs Grizzle's extreme distrust and dislike of pineapples also reflects the popularity of this exotic fruit in England and in Scotland in the 1740s an 1750s, its alleged medicinal qualities, and its being a forbidden fruit to pregnant women. Although Smollett did not live long enough to see the appearance of Edward Topham's widely read *Letters from Edinburgh written in the years 1774, and 1775*, he would have taken Scottish pride in Topham's observation that in the 1740s and 1750s Scotland, no less than England, was the garden of Europe – at least as far as exotic *ananas* were concerned:

... if the Scotch are deprived, by the nature of their situation, of enjoyment of natural fruit, they have the opportunity of furnishing themselves with hot-houses ... and, in this respect, have the advantage of the rest of Great Britain. There are few gentlemen of any consequence that are not supplied with fruit by this means; and indeed, melons, pineapples, grapes . . . are produced here with great success.[7]

'Mrs. Pickle's longings', Smollett tells us, 'were not restricted to the demands of the palate and stomach, but also affected all the other organs of sense, and even invaded her imagination, which at this period seemed to be strangely diseased'. The notion that a mother's 'imagination' influenced her foetus and, subsequently, her child, was as old as Aristotle. In his treatise *De Generatione et Corruptione*, the Greek philosopher had discussed the matter at length in 'Rules for the First Two Months of Pregnancy', 'Let none present any strange, or unwholesome thing to her, not so much as name it, lest she should desire it, and not be able to get it, and so either cause her to miscarry, or the child have some deformity on that account.'[8] The sections in Aristotle's book dealing with gestation and pregnancy were extracted in the seventeenth century and bound together, under the title *The Experienced Midwife*. Galenic medicine of the seventeenth century had little to add to Aristotle's precepts. Occasionally an author such as Shakespeare capitalised upon Aristotle's admonishment: thus Pompey, the jester in *Measure for Measure*, comes on stage bellowing that his Mistress Elbow 'came in, great with child, and longing for stewed prunes; Sir, we had but two in the house, which at that very distant time stood in a fruit dish, a dish of some threepence'. Distinguished physicians like Thomas Sydenham and Thomas Willis adhered to the ideas stated in *The Experienced Midwife*, and less illustrious doctors were equally obeisant.

Throughout the seventeenth and early eighteenth centuries this work was a standard textbook for physicians and midwives, and was the major source for Nicholas Culpepper's *Directory for Midwives* (first published in 1651), a book which Smollett must have read at least by the time he prepared Smellie's *Treatise on Midwifery* for the press. Culpepper's *Directory* had become so popular by the 1730s – years during which Smollett was in medical school – that it evoked in 1735 this enthusiastic rhapsody from an anonymous commentator:

And if the [any young doctor] applies himself to the Obstetrical Act, let him turn over Culpepper's *Midwife enlarg'd* night and day. That little Book is worth a whole Library. All that is possible to be known in the Art is there treasur'd up in a small *Duodecimo*. Blessed, yea for ever blessed, be the memory of the inimitable Author, who, and who alone, had the *curious happiness* to mix the profound learning of Aristotle with the facetious Humour of Plautus.[9]

A physician himself, Culpepper repeated Aristotle's advice to pregnant women who, like Mrs Trunnion, wished to give birth to unscarred infants. 'Sometimes there is an extraordinary cause, as imagination, when the Mother is frightened, or imagineth strange things, or longeth vehemently for some meat which if she have not, the child hath a mark of the colour or shape of what she desired, of which there are many examples.'[10]

It was not the pregnant mother's 'strange longings' – a condition known as 'pica'[11] – that disconcerted physicians and midwives so much as the ill effects of these yearnings on the foetus. According to Culpepper even a single instance of bizarre desire would produce 'Hermaphrodites, Dwarfs and Gyants', and this idea was repeated again and again in medical works of the period. Jon Maubray discussed it at length in *The Female Physician, Containing all the Diseases Incident to that Sex, in Virgins, Wives and Widows* (1724). In a chapter entitled 'Of Monsters' he complained that too few authors were 'ready to discuss the proper Causes of *Monstrous* BIRTHS', and continued to give his explanation for the occurrences:[12]

> First then, I take the Imagination to have the most prevalent *Power* in Conception; which I hope may be readily granted, considering how common a Thing it is, for the *Mother* to mark her child with *Pears, Plums, Milk, Wine,* or any *thing else,* upon the least trifling *Accident* happening to her from thence; and *that* even in the latter ripening *Months,* after the Infant is entirely formed, by the *Strength of her Imagination* only, as has been already manifestly set forth at large.

Maubray extended his case to the male as well, noting that 'a Foetus with a *Calf's, Lamb's, Dog's, Cat's-Head,* or the Effigy of any other thing whatsoever,' might be the result of '*a copulating* Man, if he should imprudently set his Mind on such Objects, or employ his perverted *Imagination* that way'. The powerful and lasting effects of the imagination, Maubray contended, were not limited to humankind, but extended also to lower species. 'This absurd *Imagination* takes place even among the very *Brutes,* as Lemnius relates of a Sheep with a Seal's or *Sea* Calf's-Head, having no doubt seen that Animal in the critical Time of *Conjunction* or *Conception.*' Like many doctors in the previous century, Maubray enjoined would-be mothers to suppress their 'absurd

Imaginations', lest they bring into the world no children but 'Monsters formed in the Womb.'

An incident late in 1726 which Smollett probably had heard about, contributed much to the popular fear that women with 'absurd Imaginations' during pregnancy would bring forth monsters: this was the extraordinary case of the pregnant Mary Tofts of Godalming who insisted that she had eventually given birth to at least seventeen rabbits and other curious progeny. Unable to afford rabbits, she nevertheless craved them throughout her pregnancy, and one day, while working in the fields, she actually saw one who may or may not have frightened her. The case itself has been discussed in such detail that I do not pretend to add new discoveries or theories,[13] but will suggest that its widespread fame and the satires it provoked – for example, Hogarth's 'Credulity, Superstition, and Fanaticism' – added further fuel to existing fears. On 5 December 1726, Pope wrote to John Caryll who lived not far from Godalming and might be expected to have heard more details than Londoners: 'I want to know what faith you have in the miracle at Guildford; not doubting but as you past thro' that town, you went as a philosopher to investigate, if not as a curious anatomist to inspect, that wonderful phenomenon. All London is now upon this occasion, as it generally is upon all others divided into factions about it.'[14] If Caryll replied to Pope's inquiry, as he may well have, the letter is not extant, but numerous Scriblerian satires are. One of these was a series of verses by Pope, 'The Discovery: Or, *The* Squire turn'd Ferret. An Excellent New Ballad. To the Tune of *High Boys! up go we; Chevy Chase*; Or what you please', published in December 1726, again in January, and several times thereafter. Here Pope turned his light artillery particularly upon two scientists, Nathaniel St André, a Swiss anatomist and medical attendant on the King, and Samuel Molyneux. Mary Tofts had first been attended by John Howard, a Guildford surgeon and male midwife who had not known her until he was called in on her case. After having devoted most of his time to her for several days, delivering nine rabbits, he moved her to Guildford, to which he invited anyone who doubted the veracity of the reports he had been giving. St André, unfortunately for himself, accepted the invitation and made the trip to Guildford, taking with him Samuel Molyneux, secretary to the Prince of Wales, a scientist of great distinction, particularly important for his work in developing the reflecting telescope. Molyneux was not a medical man and made no pretence to knowledge of midwifery; St André, on the other hand, although he had taken no degree, had been apprenticed to a surgeon and had held the post of local surgeon to the Westminster

Hospital Dispensary. It was the unoffending Moyneux, however, who bore the brunt of Pope's satire – perhaps because Pope knew more about telescopes (he had grown interested in the optical effects of the reflecting telescope) than about midwifery:[15]

> But hold! says Molly, first let's try
> Now that her legs are ope,
> If ought within we may descry
> By help of Telescope.
> The Instrument himself did make,
> He rais'd and level'd right.
> But all about was so opake,
> It could not aid his Sight...
> Why has the Proverb falsely said,
> *Better two Heads than one*;
> Could *Molly* hide this *Rabbit's* Head,
> He still might show his own.

Pope's satire on the 'Rabbit Breeder' was one among many. He himself may or may not have contributed another 'ballad' on the Tofts case to the *Flying Post*, published on 19 December 1726. In the *Flying Post* it was 'Said to be Written by Mr. Pope to Dr. Arbuthnot'; the published title is simple 'Mr P—— to Dr A——t'. Here the satire is chiefly directed at Sir Richard Manningham, son of the Bishop of Chester and godson of Sir Hans Sloane, 'society's most distinguished man-midwife', whose attention to Mary Tofts had been ordered by King George himself. Many others may be found in the Library of the Royal Society of Medicine in an apparently unique scrapbook, 'A Collection of 10 Tracts' on 'Mary Tofts, the celebrated pretended Breeder of Rabbit'. Among these is a one-page set of verses, 'The Rabbit-Man-Midwife', inscribed in an eighteenth-century pencilled hand, 'by John Arbuthnot'. Another is a tract of ten pages, *The Opinion of the Rev'd Mr. William Whiston concerning the Affair of Mary Tofts, ascribing it to the Completion of a Prophecy of Esdras*, written by William Whiston, formerly Lucasian Professor of Astronomy at Cambridge, Isaac Newton's successor. In the apocryphal book of Esdras, said Whiston, 'Tis here foretold that there should by "Signs in the Women" or more particularly that Menstrous [*sic*] Women should bring forth Monsters'.[16] Presumably writing years after the Mary Tofts affair, when the story had been 'long laughed out of Countenance', Whiston insisted that he believed it to be true 'as the fulfilling of this Ancient Prophecy before us'. Still another satire in the Royal Society of

Medicine scrapbook is a pamphlet of 1727, purportedly written by 'Lemuel Gulliver, Surgeon and Anatomist to the Kings of Lilliput and Blefuscu, and Fellow of the Academy of Sciences in Balnibarbi': *The Anatomist Dissected: or the Man-Midwife finely brought to Bed*, 1727.[17] If one of the Scriblerians was responsible for this thirty-five-page pamphlet, it was Dr Arbuthnot, who need not have concealed authorship since *The Anatomist Dissected* was far above the average tract on the rabbit-woman. It professes to be chiefly 'An Examination of the Conduct of Mr. St André, Touching the late pretended Rabbit-bearer', based on St André's defence of himself, *A Short Narrative of an Extraordinary Delivery of Rabbits*. But scrutiny reveals that it is the work of a physician with extensive knowledge of anatomy and experience in childbirth. He points out inconsistencies in St André's account, pausing over matters of the temperature and pulse of a woman in labour, and the influence of her demented imagination on the foetus.

I have strayed afield and treated the Tofts case at length because it stirred a controversy among physicians and other scientists that was to last in England more than forty years. Less than a month after the episode of the 'rabbit breeder', James Augustus Blondel, a distinguished member of the College of Physicians of London, brought out a treatise denying the possibility of such an occurrence: this work was later advertised by Dr Blondel as 'My first Dissertation, *The Strength of Imagination in pregnant Women examin'd*, published upon the Occasion of the Cheat of *God-alming*, hastily, and without Name, as coming from one, who neither designed to be known nor to meddle any more in this Controversy'. Trained in Leyden by Boerhaave, Blondel maintained that deformities in birth were caused by other factors – such as actual delivery – than the mother's imagination. He was challenged by Daniel Turner, another physician and Fellow of the Royal College of Surgeons, who dogmatically asserted the opposite.[18] The two physicians, both distinguished and both fellows of the same society, stirred considerable debate within the College of Physicians. Since Blondel felt he had been attacked personally by Turner, he responded with a 155-page defence of himself: *The Power of the Mother's Imagination Over the Foetus Examin'd. In Answer to Dr. Daniel Turner's Book, Intitled a Defence of the XIIth Chapter of the First Part of a Treatise, De Morbis Cutaneis* (1729). In his preface Blondel stated his purpose: 'My Design is to attack a vulgar Error, which has been prevailing for many Years, in Opposition to Experience, sound Reason, and Anatomy: I mean the common Opinion, that Marks and Deformities, which Children are born with, are the sad Effect of the Mother's irregular Fancy and Imagination.' Without providing any historical survey of the controversy Blondel noted 'that the Doctrine of Imagination, relating to the Foetus, had gone through several Revolutions', and continued to indict the 'Imaginationists'.[19]

'Tis silly and absurd; for what can be more ridiculous, than to make of Imagination a Knife, a Hammer, a Pastry-Cook, a Thief, a Painter, a Jack of all Trades, a Juggler, Doctor Faustus, the Devil and all? 'Tis saucy and scandalous, in supposing that those, whom God Almighty has endowed, not only with so many charms, but also, with an extraordinary Love and Tenderness for their children, instead of answering the End they are made for, do breed Monsters by the Wantonness of their Imagination.

'Tis mischievous and cruel; it disturbs whole Families, distracts the Brains of credulous People, and puts them in continual Fears, and in Danger of their Lives: In short 'tis such a publick Nuisance, that 'tis the Interest of every Body to join together against such a Monster, and to root it entirely out of the World.

In less than six months the 'Imaginationists' led by Turner retorted with a reassertion of the influence of the mother on her foetus. Parodying Blondel's title, Turner called his treatise *The Force of the Mother's Imagination upon her Foetus* (1730). This work was over 200 pages and was a definitive defense of the majority view of the time. One of Turner's arguments was 'the authority of Antiquity': he had culled hundreds of ancient and medieval medical writings and prepared a list of monstrous births in which the mother's demented imagination was the apparent cause. Most common among these, Turner affirmed, was her craving for exotic fruits – plums, cherries, grapes, prunes, and now pineapples. Unlike Blondel, Turner was not concerned with the physiological processes by which the foetus was actually 'marked'. He abstained from proving the truth of his argument by appealing to 'sensation, the nerves, and the circulation of the foetus'. Instead, he cited numerous ancient and modern authorities, known and obscure – Hesiod, Heliodorus, Jacobus Horstius, Ambroise Paré, Johann Schenkius, Thomas Bartholin, Charles Cyprianus, Robert Boyle, Sir Kenelm Digby – who had reported instances of deformed children; moreover, Mary Tofts herself, although 'a cheat', had been frightened by rabbits while sowing in the fields, and in Turner's estimation there was a definite connection between the imagination and the foetus. Learned though his argument seemed, Turner was scurrilously satirised by his opponents in several pamphlets and poems, perhaps the most witty and scrofulous among them a burlesque set of verses written in Butlerian octosyllables entitled 'The Porter Turn'd Physician', and published in 1731.

Although Blondel and Turner after 1730 did not publish works concerning the mother's imagination, the controversy over which they differed continued to be a topic of concern in medical and lay circles for at least three decades. Fellow of the Royal Society, many of whom were Smollett's personal friends, and other scientifically inclined gentlemen, hesitated to drop so controversial an issue. Doctors Hunter and Monro spent considerable time on the topic of their anatomical lectures, and we know that Smollett not only heard these but was in 1748–50 in medical dialogues with these men. As late as 1747, John Henry Mauclerc, an M.D. of no great distinction, published a lengthy treatise entitled *Dr. Blondel confuted: or, the Ladies Vindicated, with Regard to the Power of Imagination in Pregnant Women: Together with a Circular and General Address to the Ladies on this Occasion* (London).[20] Odd as it may appear, Mauclerc's book approved rather than confuted – his own term – Blondel's theory that imagination alone was unable to harm the foetus. 'The Design of the Dissertation', he wrote in the preface of his book intended especially for women, 'is to prove that the Opinion, which has long prevail'd, that the Marks and Deformities, Children bring into the World, are the sad Effect of the Mother's irregular Fancy and Imagination, is nothing else but a vulgar Error, contrary to sound Reason and Anatomy.' Later in his preface Mauclerc writes as if the controversy were still inflaming the hearts of medical men, twenty years after the fact: 'I don't despair of Success: Interest alone should prevail, upon the Party, which is chiefly concerned in this Controversy.' Dr Mauclerc disbelieving in old wives' tales and other odd superstitions, presents – to quote his own words – 'a Sketch of the true Cause of Monsters – I hope, 'tis sufficient for the present, to give a general, and yet a clear Solution of those strange Phenomena [monsters]'. If Blondel had his supporters, so did Turner, although both men had been dead for many years. In 1765, Isaac Bellet, a French physician residing in London, wrote *Letters on the Force of the Imagination in Pregnant Women*, in which he denied the possibility that pregnant women could mark their children, but was compelled to agree rather with Turner's explanations than Blondel's. The result was a second-rate medical work fraught with contradiction, but nevertheless one showing how very much alive the matter still was. Dr Smollett, probably the author of the review in the *Critical Review*,[21] found the book very appealing: 'We declare upon the whole', he wrote, 'that he [Bellet] has fulfilled his scope, and executed his undertaking with great precision, and that he was clearly demonstrated the impossibility of a pregnant woman's marking her child with the figure of any object for which she has longed, or which may have made a deep impression upon the imagination.' It is difficult, if at all possible, to state accurately what Smollett's views on the subject were almost twenty years earlier, when he was writing the early chapters of *Peregrine Pickle*, but there is every reason to believe that even then the probability of a mother marking her child seemed remote to him. Hunter, Monro, and Smellie doubted the possibility, and medically speaking, they exerted much influence on his thought.

Throughout the 1740s case of extraordinary childbirth of every sort continued to interest the English public, especially physicians and scientists. One could compile with ease a long list of works written in the decade about strange childbirths. Nor was the subject treated in books and tracts only. Popular periodicals that enjoyed large amateur audiences devoted much space in their issues to these freaks of nature. *The Gentleman's Magazine*, for example, contained no fewer than ninety-two articles (essays, reviews, and letters) on the question of extraordinary childbirth.[22] Curiosity was especially aroused by a woman who never became pregnant until she drank Bishop Berkeley's tar-war in 1745. More specialised in its reading audience, the *Philosophical Transactions* of the Royal Society (to which Smollett never was elected but many of whose Fellows he knew) was flooded with communications about bizarre births attesting to the interests of its fellow members in the subject. Professor Knapp's biography makes it clear that Smollett was familiar with some of these publications. An idea of the range and diversity of such cases may be gained by listing a few of the titles of these articles:[23] 'Account of a monstrous boy'; 'Account of a monstrous child born of a woman under sentence of transportation'; 'An Account of a monstrous foetus resembling an hooded monkey'; 'Case of a child turned upside down'; 'A remarkable conformation, or lusis naturae in a child'; 'Part of a letter concerning a child of monstrouse size'; 'Account of a child's being taken out of the abdomen after having lain there upwards of 16 years'; 'a letter concerning a child born with an extraordinary tumor near the anus, containing some rudiments of an embryo in it'; 'An account of a praeternatural conjunction of two female children'; 'Part of a letter concerning a child born with the jaundice upon it, received from its father's imagination, and of the mother taking the same distemper from her husband the next time of being with child'; 'An account of a monstrous foetus without any mark of sex'; 'An account of a double child born at Hebus, near Middletown in Lancashire'. So interesting to laymen and amateur scientists were many of these cases that they were abstracted from the *Philosophical Transactions* and reported in abbreviated form as news items in *The Gentleman's Magazine.*

The controversy originally stirred by Blondel and Turner, and about which Smollett must have heard, also provoked considerable commentary in the 1740s in books (essays, novels, and poems). Fielding Ould, among the most famous male-midwives of his age and best known for *A Treatise of Midwifery* (1742)[24] – frequently called by historians of medicine the first important text on midwifery in English – considered the mother's imagination important in the health of the foetus, although he seems to have doubted the validity of Turner's views. Not infrequently the subject appeared in novels. Sir John Hill, an arch-enemy of Smollett in 1751, created a marvellous female character in *The Adventures of George Edwards, A Creole,*[25] who touched a robin redbreast during pregnancy (her fancy having led her to this curious action!) and, thereafter, bore a child with a red breast – or red chest. In the same year that *Peregrine Pickle* was published, 1751, he also wrote a satire on the subject of curious births entitled *Lucina sine concubitu.* Here the reader finds the theories of preformation (according to which organisms are already fully formed in their seeds) and panspermism (minute organisms developed in fluids owing to the presence of germs) satirically treated, as well as those of the influence of the mother's imagination on her child. Although Hill was unable to attain membership in the Royal Society because of his disagreeable personality, his hoax *Lucina* was nevertheless widely read by such professional medical men as Dr Smollett, who had good reason to take note of Hill in 1751.[26] The famous case,[27] two years earlier, of a woman 'who carried with child 16 years', also helped to create the background for Smollett's witty treatment of the state of the mother's imagination.

Peregrine Pickle was published twelve years too early to bear any traces of awareness of George Alexander Stevens's *Dramatic History of Master Edward* (London, 1763). But this collection of extraordinary occurrences in 1730–1760 – written, as the title-page indicates, by the 'Author of the celebrated Lecture upon Heads' – demonstrates how fully formed a type of writing (most accurately described as a *leitmotif*) about abnormal births had emerged by the 1760s. In the opening pages (7–13), Stevens has Thomas recount to David a series of histories, all dealing with abnormalities during pregnancy and shortly after birth. The stories, culled from authors in different countries in Europe, are as bizarre and grotesque as scenes from gothic romances (then coming into vogue). Two histories in particular appeal to David, so much so that Stevens included illustrations of them in his revised edition of 1785. In the first,

'*Aldrovandus* [a seventeenth-century naturalist] relates, that a woman in Sicily observing a lobster taken by a fisherman, and being moved by an ernest longing for it, brought forth a lobster, altogether like what she had seen and longed for'. Stevens's 'lobster woman' is not very different from Mary Tofts, the 'rabbit woman'. The second history richer in complexity and more touching relates, as Stevens writes, 'something singular beyond all these':

> ... [it] is the tale of *Languis*, of a woman longing to bite the naked shoulder of a baker passing by her; which, rather than she should lose her longing, the good-natured husband hired the baker at a certain price. Accordingly, when the big-bellied woman had bit twice, the baker's wife broke away from the people who held her, would not suffer her to bite her husband again; for want of which, she bore one dead child, with two living ones.

Smollett himself had shown interest in cases of extraordinary childbirth five years before the composition of *Peregrine Pickle*. In *Advice: a Satire* (1746), the character Poet refers to the strange conception, or near-conception, of a hermaphrodite:[28]

> But one thing more – how loud must I repeat,
> To rouse th' engag'd attention of the great,
> Amus'd, perhaps, with C——'s prolific bum,
> Or rapt amidst the transports of a drum.

Here Smollett's own note reads: 'This alludes to a phenomenon not more strange than true; the person here meant, having actually laid upwards of forty eggs, as several physicians and fellows of the Royal Society can attest, one of whom, we hear, has undertaken the incubation, and will, no doubt, favour the world with an account of his success. Some virtuosi affirm, that such productions must be the effect of a certain intercourse of organs not fit to be named.' Smollett's source for 'C——'[29] remains a mystery, although his satiric habit of mind was unlikely to fabricate a source. London newspapers in 1745–6 were filled with reports of such odd occurrences and the populace seemed to be diverted, if not instructed, by these accounts. In 1750, Smollett, while preparing for publication Dr William Smellie's *Midwifery*, probably read about cases of unnatural birth in the hours he spent annotating. During that year he may have seen 'Michael Anne Drouvert', the much talked about Parisian hermaphrodite who was displayed in London and written up as a case history in the *Philosophical Transactions*; may have read James Parsons's *Inquiry into the Nature of Hermaphrodites* (1741) or George Arnaud's new book entitled *Dissertation on Hermaphrodites* (1750); or may even have heard that John Hill's forthcoming book, *A History of Animals* (1750), contained a modern epitome of the subject. At any rate, Smollett's extensive reading in obstetrics in the library of Smellie, and earlier in Dr James Douglas's library, would have revealed a wealth of real cases from which to create fictional characters and episodes relating to pregnancy. Smollett's own observations, printed in Smellie's *Treatise* (II, 4–5), 'On the Separation of the Public Joint in Pregnancy', gives ample testimony to and palpable evidence of his interest in cases of abnormal birth.

In fact, Smollett never lost interest in the subject and continued from 1750 to 1764 to edit and revise all Smellie's obstetrical works. Smollett's extensive medical reading coupled with his knowledge of the extra-ordinary case of Sarah Last,[30] who in 1748 underwent normal pregnancy without ever giving birth to her foetus, must have inspired him to draw a parallel case in Mrs Trunnion in *Peregrine Pickle*. Readers will recall the bizarrely constructed chapters in which the pregnant lady is found to have been swelled with air! Smollett was reflecting contemporary fears about strange childbirth when he described the ultimate chagrin of the Trunnions:

> At length she and her husband became the standing joke of the parish; and this infatuated couple could scarce be prevailed upon to part with their hopes, even when she appeared as lank as a greyhound, and they were furnished with other unquestionable proofs of their having been deceived. But they could not forever remain under the influence of this sweet delusion, which at last faded away, and was succeeded by a paroxism of shame and confusion, that kept the husband within doors for the space of a whole fortnight, and confined his lady to her bed for a series of weeks, during which she suffered all the anguish of the most intense mortification.[31]

The first pineapple grown in a hothouse in England may well have been planted during the Restoration. A well-known extant painting, bearing the inscription *Rose, The Royal Gardener, presenting to Charles II the first pine-apple grown in England*, is ascribed to the Dutch artist Danckerts and the gardener is John Rose.[32] Just how long before Monsieur Le Cour, a Frenchman residing in Leyden, Holland, 'hit upon a proper Degree of Heat and Management so as to produce pine-apples equally as good as those which are produced in the West Indies'[33] is not known. Even in Leyden, where winters were less brutal than in England, Le Cour used stoves to grow the tropical fruit. Chambers's compendious *Cyclopaedia* (1728) reports that the gardens of England were supplied with pineapples by Le Cour himself, and John Evelyn wrote in his *Diary* on 9 August 1661: 'I first saw the famous Queene-pine brought from Barbados presented to his *Majestie*, but the first that were ever seen here in England, were those sent to Cromwell, four-years since' (1658).[34] Evelyn continued several days later with a description of the 'rare fruite called the King-Pine', the first he had seen: he tasted it and found it not to his liking.

Whether or not those described by Evelyn were the first grown – it is at least plausible that an occasional fruit had been grown earlier – a more likely possibility is that pineapples were first artificially produced in quantity at any rate, in the hothouses of Sir Matthew Decker's famous garden in Richmond. He was a well-known London merchant (president of the East India Company) of Dutch origin who apparently enjoyed a 'truly Dutch passion for gardening'.[35] Richard Bradley, an authority on gardening in the early decades of the eighteenth century and Professor of Botany at the University of Cambridge, wrote that Sir Matthew's gardener, Henry Tellende, grew the first pineapples for his master 'circa 1723'.[36] Seven years before this, Lady Mary Wortley Montagu had eaten pineapples at the table of the Elector of Hanover. She wrote to Lady Mar about '2 ripe Ananas's, which to my taste are a fruit perfectly delicious', and continued to note surprise that pineapples had not as yet been cultivated in her native land. 'You know they are naturally the Growth of Brasil, and I could not imagine how they could come there but by Enchantment. Upon Enquiry I learnt that they have brought their Stoves to such perfection, they lengthen the Summer as long as they please, giveing to every plant the degree of heat it would receive from the Sun in its native Soil. The Effect is very near the same. I am surpriz'd we do not practise in England so usefull an Invention.'[37] From the lenghty discussion about pineapples that Horatio and Cleomenes have in Mandeville's *Fable of the Bees* (1714),[38] it may be assumed that Tellende and his lord, Decker, had raised the delicious and exotic fruit approximately in 1720–3. Mandeville's characters – not dissimilar to several enthusiasts in *Peregrine Pickle* – comment upon a new and 'fine Invention' as well as on the intrinsic attributes of pineapples: Horatio says, 'I was thinking of the Man, to whom we are in a great measure obliged for the Production and Culture of the *Exotick*, we were speaking of in this Kingdom; Sir Matthew Decker: the first *Ananas*, or Pineapple, that was brought to Perfection in *England*, grew in his Garden at Richmond.'

That garden was still viewed in the 1730s and Smollett, who had come to London in 1739, may have visited it.

As Richard Bradley had explained in his essay 'A particular easy Method of managing Pine-Apples' (1726),[39] the difficulty of cultivation was due to poor hothouses and stoves. Pineapples required a full three years for growth, an exact temperature, ideal moisture conditions, and correctly constructed stoves. As soon this was achieved, the fruit could be grown in domestic gardens, even if at great financial expenditure. Such was the case and it applied not only to pineapples but to other exotic fruits, limes, papayas, guavas, bananas, and even grapes. The cultivation of pineapples in the third and particularly the fourth decades of the eighteenth-century became a hobby – not quite a popular sport – among expert gardeners and aristocrats. Prominent families who could afford the expense sent their head gardeners to Decker's hothouses to observe the new method and educate themselves. Among the first to display home-grown pineapples on their tables were the opulent Earls of Bathurst, Portland, and Gainsborough. The Duke of Chandos, long incorrectly identified as 'Timon' in Pope's *Epistle to Burlington*, not only grew pineapples on his estate at Shaw Hall in Berkshire, but he also sold them 'at a half a guinea a time',[40] a price even Mrs Grizzle would have been willing to pay for her sister-in-law!

Smollett may not have known first-hand the early history of pineapples in England, but he was old enough to be familiar with its more recent peregrinations. Few people were more excited about the new art of growing pineapples than Alexander Pope. Together with his gardener John Serle, whom the poet employed in 1724, Pope was growing 'ananas' by 1734. He had, however, tasted the fruit long before this. On 8 October 1731, Pope wrote Maratha Blount, 'I'm going in haste to plant Jamaica Strawberries, which are to be almost as good as Pineapples.'[41] In the spring of 1735, he wrote to William Fortescue that he was improving and expanding his garden, 'making two new ovens and stoves, and a hothouse for ananas, of which I hope you will taste this year'. Two or three pineapples grown in the Twickenham hothouse and sent to an intimate friend was perhaps the greatest honour Pope could confer. During this period he was continually experimenting for cheaper and better ways to raise the fruit. In August 1738, Pope and Serle 'borrowed' Henry Scott, Lord Burlington's gardener who was an expert in growing pineapples, to consult with him 'about a Stove I am building'. It is possible that Pope was also reading modern handbooks on the subject. Whatever the case, by 1741 Pope thought he had dis-covered with the aid of Scott the long-sought method, and attempted to make it known to his friends. How well-circulated among the London *literati* Pope's 'discovery' was, it is now impossible to tell; but by 1741, the magisterial poet was too conspicuous among men to veil any of his activities, even his pastimes and hobbies. Smollett, then young and still an *ingénue* among the 'wits' in London, kept his ears and eyes open and possibly may have heard about the new pineapple method of Mr Pope, his favourite author among all authors and a poet whose influence was to rub off considerably on his own writings. In any case, Pope soon wrote to Ralph Allen (to whom he occasionally sent a pineapple or two): 'In a Week or two, Mr. Scot will make you a Visit, he is going to Set up for himself in the Art of Gardening, in which he has great Experience, & particularly has a design which I think a very good one to make Pineapples cheaper in a year or two.'[42] Scott and Allen's gardener, Isaac Dodsley, were apparently successful in building the new type of hothouse with new stoves, for Pope wrote next year to Allen: 'I would fain have it succeed, for two particular reasons; one because I saw it was Mrs. Allen's desire to have that fruit, & the other because it is the only piece of Service I have been able to do you, or to help you in.'[43]

Poets and prose writers varied in their response to the fashionable king of fruits, some equating it with luxury and viewing it as a symbol of evil, others seeing in its beautiful colors and exotic shape an expression of the beauty of Nature and God. Pineapples were, as James Thomson wrote in 'Summer' (a poem which Smollett singled out for praise in the preface of *Ferdinand Count Fathom*), the fruits of the Gods in the Primitive Ages of the world:[44]

> ...thou best Anana, thou the pride
> of vegetable life, beyond whate'er
> The poets imaged in the golden age:
> Quick let me strip thee of thy tufty coat,
> Spread thy ambrosial stores, and feast with Jove.

Still other authors wrote about the medical properties of pineapples. In his didactic poem *The Art of Preserving Health* (1744), John Armstong, with whom Smollett was on intimate terms throughout his life and whose theories of imagination her knew, chose the fruit as an example of a product raised that exhibited the differences and the extremes of cold and heat in diet:[45]

> ...in horrid mail
> The crisp ananas wraps its poignant sweets.

> Earth's vaunted progeny: in ruder air
> Too coy to flourish, even too proud to live;
> Or hardly rais'd by artificial fire
> To vapid life. Here with a mother's smile
> Glad Amalthea pours her copious horn.

Smollett knew his friend's poem very well, had read it numerous times, and called it in *The Present State of All Nations* (1768, II, 227), 'an excellent didactic poem.'

Long before the prose encyclopaedists (Ephraim Chambers, John Harris, Robert James) discussed the fruit, medical authors had commented upon it. From the time of Nicholas Culpepper's popular handbook, *A Directory for Midwives*, which Mrs Grizzle had studied so assiduously, pineapples were strictly forbidden to expectant mothers as one of the 'Summer Fruits nought for her and all her Pulse'. In *The English Physitian* (1674) Culpepper had devoted an entire section to the benefits and ill effects of the fruit: 'It marvelously helpeth all the Diseases of the Mother [i.e., hysteria] used inwardly, or applied outwardly, procuring Women's Courses, and expelling the dead Child and After-birth, yea, it is so powerful upon those Feminine parts that it is utterly forbidden for Women with Child, and that it will cause abortment or delivery before the time...Let Women forbear it if they be with Child, for it works violently upon the Feminine Part.'[46] Seventeenth-century herbalists like Thomas Parkinson also warned their readers not to eat the artificial food. But it was not until pineapples were actually grown in English gardens that physicians and obstetricians became alarmed and abandoned superstition for medical science. Observation had revealed that pregnant women who ate this food miscarried again and again for nervous reasons. Dr Robert James, inventor of the famed 'fever powders', writing in the London *Pharmacopaeia Universalis* (1742), commented: 'This Fruit is esteemed cordial, and analeptic; and is said to raise and exhilarate the Spirits, to cure a Nausea, and provoke Urine. But 'tis subject to cause a Miscarriage, for which Reason Women with Child should abstain from it.'[47] One year later James (whose *Medicinal Dictionary* Smollett knew well) was even more precautionary, stating the pineapples definitely caused miscarriage. Similarly, dietitians and other authors on nutrition warned the pregnant woman to refrain from the pineapple. M. L. Lemery, a prolific author on diet whose works were translated into English because of their popularity, wrote in *A Treatise of All Sorts of Goods*:[48] 'Ananas is a delicious fruit, that grows in the West Indies, whose juice the *Indians* extract, and make excellent Wine of it, which will intoxicate. Women with Child dare not drink of it, because they say, it will make them miscarry.' Francis Spilsbury, the author of *Free Thoughts on Quacks* (London, 1777), a treatise explaining the circumstances of Oliver Goldsmith's death, compared pineapples to gout (a strange comparison even for an eighteenth-century apothecary!) since he found a 'universal comprehensiveness' in both. In his words, just as 'the Ananas (vulgarly known under the name of *Pine-Apple*) is considered as containing the taste and flavour of many different fruits, so a great many disorders of the body are, under different appellations to be found in the gout'.[49]

Philosophers as well as medical thinkers pointed to the pineapple as a rare fruit with strange qualities and an exotic taste. Less concerned than physicians with the nervous medicinal aspects of pineapples, they frequently referred to the fruit when discussing the taste. As early as 1690, John Locke singled out pineapples as the best obtainable example of a food whose taste could not be comprehended without actually partaking of it. In a well-known passage in *An Essay Concerning Human Understanding* on 'the Blind Man', to which Fielding referred several times in *Tom Jones*, Locke wrote of the impossibility of words replacing direct sensory experience:[50]

> He that thinks otherwise, let him try if any words can give him a taste of a pine apple, and make him have the true idea of the relish of that celebrated delicious fruit. So far as he is told it has a resemblance with any tastes whereof he has the ideas already in his memory, imprinted there by sensible objects, not strangers to his palate, so far may he approach that resemblance in his mind. But this is not giving us that idea by a definition, but exciting in us other simple ideas by their known names; which will be still very different from the true taste of that fruit itself.

Also speaking of the nervous origin of ideas and the fact that they are grounded in sensory experience (i.e. direct sense experience), Smollett's countryman David Hume noted in the opening paragraph of his *Treatise of Human Nature* (1739) that 'we cannot form to ourselves a just idea of the taste of a pineapple, without actually having taste it'.[51] That is, the rare and uncommon pineapple affords the student of philosophy a splendid opportunity to observe that 'all our simple ideas in their first appearance are derived from simple impressions, which are correspondent to them, and which they exactly represent'. And David Hartley, writing two years before the publication of *Peregrine Pickle*, may not have commented upon pineapples but he increased speculative interest in abnormal pregnancy by the inclusion in his *Observations on Man* of a chapter entitled, 'To Examine How Far the Longings of Pregnant Women are agreeable to the Doctrines of Vibrations and Associations'. Hartley, a physician by profession, underplayed the variety of longings found in pregnant women – an impressive range, as Smollett's female figures in the novels show – and demonstrated instead that abnormal cravings are caused in the first place by means of 'nervous Communications between the Uterus and the Stomach'. Both, Hartley maintained, are in 'a State of great Sensibility and Irritability' during pregnancy, a view Smollett himself had taken in writing about the public joint. Smollett may not have read Locke, Hume, and Hartley – although that is highly unlikely – but he was certainly aware of popular references to pineapples and pregnancies in their works, ideas then so common that they probably required little documentation to a literate eighteenth-century man.

Thus, a decade before the publication of *Peregrine Pickle*, physicians, scientists, gardeners, philosophers, and literary men had all reacted in various ways to the new 'King Fruit' which by 1751 had become much more popular than in John Evelyn's day; all had seen in the body of beliefs and superstitions embracing the fruit something different. If gardeners found it their delight and joy, philosophers were not far behind in using it as an emblem of singular sensory experience. If physicians, especially obstetricians, called it the bane of their pregnant women patients, other scientists (biologists, botanists, physiologists) were equally ominous in their belief that pineapples contained strange and unknown chemical properties.

It was therefore left to a literary man, who was a physician as well as a novelist, to see the comic possibilities in all these prevailing 'nervous' theories. It may also be that Smollett, himself editing the obstetric volumes of William Smellie at the time he was composing his novel, saw that the fantastic (indeed absurd) theories of the mother's nervous imagination together with the many muddles and mysteries that had grown up about pineapples could be wedded into one episode. The early chapters of *Peregrine Pickle* illustrate once again how adeptly Smollett used science, particularly medical learning, for the purposes of wit.[52] His satiric portraitures of characters such as Mrs Pickle, Mrs Grizzle, and Mrs Trunnion place great demands on the modern reader who wishes to comprehend the author's powerful wit. But his contemporary readers would have felt much more at home than we do in viewing his comic spectacle: they would have realised that he was using medical and scientific learning for pure levity and genial farce, and in this sense would have read his works as they would soon be reading those of his great contemporary, Laurence Sterne. We should not be surprised to observe a process of carry-over in Smollett's novels: from his medical writings to the novels and vice-versa. Although he was never a successful physician, if daily practice is a yardstick of measurement, his entire life demonstrates a continuing interest in medical, especially nervous, theory. It is, therefore, to be expected that Smollett's medical works – short essays, unsigned medical tracts written pseudonymously for financial purposes,[53] and virtually all the reviews of medical books in the *Critical Review* 1756–60 and possibly later – would have rubbed off on his fiction. Indeed, the sensibility pervading both worlds, medical and fictive, was one, and Smollett was at their center. Such interaction serves to remind us, that the place occupied by medicine and by the social aspects of that science which daily seemed to take on ever greater consequence in the eighteenth century, is something of which we have yet to take account in our criticism and biography of Smollett.

Notes

1. *Peregrine Pickle*, 1, 32. All references to the first edition of 1751.
2. *Peregrine Pickle*, 1, 31. Passages quoted in this paragraph are from *Peregrine Pickle*, 1, 32–6.
3. *Peregrine Pickle*, 1, 32. Grown in elephantine stoves in specially-built hothouses, pineapples were considered an exotic and artifical fruit in the eighteenth century. I discuss the fruit more fully below.
4. *Peregrine Pickle*, 1, 31: 'She purchased Culpepper's midwifery, which, with that sagacious performance dignified with Aristotle's name, she studied with indefatigable care, and diliegently perused the Compleat House-wife, together with Quincy's dispensatory, culling every jelly, marmalade and conserve which these authors recommend as either salutory or toothsome, for the benefit and comfort of her sister-in-law, during her gestation.'
5. *Peregrine Pickle*, 1, 31.
6. Some attention to embryological theory of the eighteenth century and *Tristram Shandy* is given in Louis A. Landa, 'The Shandean Homunculus: The Background of Sterne's "Little Gentleman"', in Caroll Camden (ed.), *Restoration and Eighteenth Century Literature* (Chiacago, 1963), 49–68. I have discussed Smollett's novels and medicine in 'Doctors and Medicine in the Novels of Tobias Smollett' (Princeton University dissertation, 1966).
7. *Letters from Edinburgh* (Edinburgh, 1776), 'On . . . Gardening', 229.
8. *Aristotle's Compleat Master Piece: Displaying the Secrets of Nature in the Generation of Man* (32nd edn, 1782), 33–4. See Fielding H. Garrison, *An Introduction to the History of Medicine* (4th edn rev., 1929), 101ff.

9. *An Essay for Abridging the Study of Physick* (1735), 17.
10. *A Directory for Midwives* (London, 1684), p. 145.
11. The name given to the condition by Ancient Greek physicians, and also called 'citta' or 'malatia'. See Hermann Heinrich Ploss, Max and Paul Bartel, 'The Longings of Pregnancy' in *Woman: an Historical and Anthropological Compendium* (1935), II, 455–60. A recent study of the medical aspects of *pica* is by M. Cooper, *Pica* (Springfield, 1957).
12. *The Female Physician*, 368. See also part ii, chap. 7, which discusses numerous cases of foetuses that had been marked. Maubray was one of the first physicians in London to offer private instruction for midwives. See F. H. Garrison, *History of Medicine*, 399. A great believer in monsters, he earned notoriety in 1723 by assisting in the delivery of a Dutch woman, who produced a monstrous manikin called *de Suyger*, with 'a hooked snout, fiery sparkling eyes, a long neck and an acuminated, sharp tail'. Maubray called it a moldy-warp (mole) or sooterkin.
13. The most complete account is by S. A. Seligman, 'Mary Tofts: the Rabit Breeder', *Medical History*, v (1961), 349–60, which is based upon extant contemporary accounts. Another less extensive treatment is by K. Bryn Thomas, *James Douglas of the Pouch and his pupil William Hunter* (1964), 60–8.
14. George Sherburn (ed.), *Correspondence of Alexander Pope* (Oxford, 1956), II, 418–19.
15. I follow the text given by Norman Ault in *Minor Poems of Alexander Pope, The Twickenham Pope: vol. vi* (1964), 259–64. St André was a Fellow of the Royal Society and contributed papers to the *Philosophical Transactions*. His appointment as Surgeon and Anatomist to the Court seems to have been made rather for his linguistic ability than his medical ability. Mary Tofts's confession put an end to his Court position, and he was never again to attain medical recognition.
16. According to the title-page of the tract, these pages were copied from the second edition of Whiston's *Memoirs*, published in London in 1753. The interpretation of the Tofts case does not appear in the first edition of 1749, and, so far as I can determine, was not published separately. See K. Bryn Thomas, *James Douglas* (1964), 65.
17. See Marjorie Nicolson and G. S. Rousseau, *'This Long Disease, My Life': Alexander Pope and the Sciences* (Princeton, 1968), 114: 'Early in 1727 when the small talk of London seems to have been divided between Mary Tofts and Lemuel Gulliver – *Gulliver's Travels* had appeared the preceding autumn and provoked almost universal applause – it was inevitable that at least one pamphlet on the rabbit woman should be attributed to Jonathan Swift. Those who have done him that dubious honor have failed to notice that Swift had returned to Ireland a month before the Tofts affair, and while he probably heard of it in letters, he had no such background for parody as Scriblerians in London.'
18. The chronology of works in the controversy was as follows: Turner, *De Morbis Cutaneis* (1726; first published in 1714); Blondel, *The Strength of Imagination in Pregnant Women Examin'd* (1727); Turner, *A Discourse concerning Gleets ... to which is added A Defence of ... the 12th Chapter of ... De Morbis Cutaneis, in respect of the Spots and Marks impress'd upon the Skin of the Foetus* (1729); Blondel, *The Power of the Mother's Imagination Over the Foetus Examin'd* (1729); Turner, *The Force of the Mother's Imagination upon her Foetus in*

Utero... in the Way of A Reply to Dr. Blondel's Last Bok (1730). Turner's first work, *De Morbis Cutaneis*, was written in part as a defense of Malebranche's theory that the mother marks her child. See *Father Malebranche's Treatise concerning the Search after Truth*, trans. T. Taylor (Oxford, 1694), 'Book the Second Concerning the Imagination'. As a conclusion to this book, Malebranche wrote: 'When the Imagination of the Mother is disordered and some tempestuous passion changes the Disposition of her Brain... then... this Communication alters the natural Formation of the Infant's Body, and the Mother proves Abortive sometimes of her foetus' (60). When the Tofts case revived the issue among medical men, Turner turned from Malebranche to a then real-life example. In this and other footnotes, I have dealt at length with these medical tracts because they were clearly read by the masses in their time and now are so little known.

19. *The Power of the Mother's Imagination*, XI. Blondel singled out from medical literature the six most common causes of spotted children: '1. A strong Longing for something particular, in which Desire the Mother is either gratified, or disappointed. 2. A sudden Surprise. 3. The Sight and Abhorrence of an ugly and frightful Object. 4. The Pleasure of Looking on, and Contemplating, even for a long Time, a Picture or Whatever is delightful to the Fancy. 5. Fear, and Consternation, and great Apprehension of Dangers. 6. And lastly, an Excess of Anger, of Grief, or of Joy' (2). Later (4) Blondel notes that item (1) of the list above was the most common of the six especially in the case of certain fruits, 'the strong Desire of *Peaches*, or Cherries'. Presumably he would have included in this list pineapples!

20. An earlier version of this work appeared in 1740 with the title-page, *The Power of Imagination in Pregnant Women Discussed: with an Address to the Ladies in Reply to J. A. Blondel*. So far as I can learn Mauclerc published no other works.

21. *Critical Review*, XX (July 1765), 63–5. On pp. 125–33, Bellet provides a fair estimate – in his opinion – and a history of the controversy from the time of Malebranche. There is reason to believe that Smollett and Bellet had met and that Smollett was impressed by his knowledge of medical history.

22. A list of references is too long to be given here. The two cases that attracted the most attention were 'A Foetus of Thirteen Years', *The Gentleman's Magazine*, XIX (1749), 415, and in the same publication, 'Fatal Accident: Woman carry'd a child sixteen years', XIX (1749), 211.

23. Respectively *Philosophical Transactions*, XLI (1740), 137; XLI (1741), 341; XLI (1741), 764; XLI (1741), 776; XLII (1742), 152; XLII (1743), 627; XLIV (1747), 617; XLV (1748), 325; XLV (1748), 526; XLVI (1749), 205; XLVII (1750), 360. Some of these cases were abridged and printed in popular monthlies such as *The Gentleman's Magazine* and *The Monthly Review*.

24. First published in Dublin and numerous times thereafter in London. For Ould's contributions to midwifery see John R. Brown, 'A Chronology of Major Events in Obstetrics and Gynaecology', *The Journal of Obstetrics and Gynaecology*, LXXI (1964), 303; and Fielding H. Garrison, *An Introduction to the History of Medicine* (4th edn rev., 1929), 338–40.

25. First published in 1751 and reprinted in *The Novelists' Magazine*, XXIII (1788). One episode is discussed by William Scott, 'Smollett, John Hill, and *Peregrine Pickle*', *Notes and Queries*, CC (1955), 389–92.

26. Two weeks before the publication of *Peregrine Pickle*, Dr Hill anticipated Smollett's novel by bringing out *The History of a Woman of Quality: or the Adventures of Lady Frail*. While Smollett believed the presence in his novel of Lady Vane's 'Memoirs' would enhance sales, Hill's earlier publication greatly diminished sales.

27. *The Gentleman's Magazine*, XIX (1749), 211.

28. James P. Browne (ed.), *The Works of Tobias Smollett, M. D.* (1872), 1, 294.

29. There is no mention of this case in the *Philosophical Transactions* or other scientific literature I have examined. Perhaps 'C——' was the famous 'Charing Cross hermaphrodite', about whom Dr William Cheselden had written in the *Anatomy of the Humane Body* (1713) and about whom Dr James Douglas wrote many medical fragments in the 1720s. Smollett was too young to have seen that curious organism. By 1751 this hermaphrodite may have become too stale a subject for satire, although K. Bryn Thomas, *James Douglas and his pupil William Hunter* (1964), 190, does not think so. I am inclined to believe it was a more recent occurrence, about which Smollett was informed, as the tone of his note indicates.

30. Among numerous accounts of her case the most interesting I have found is in the *The Gentleman's Magazine*, XXI (1751), 214–15; 'About the beginning of August 1748, Sarah Last, a poor woman in Suffolk, had the usual Symptoms of pregnancy, which succeeded each other pretty regularly thro' the usual period, at times she as seiz'd with pains ... the child did not advance in birth ... after the pains were gone off, the woman grew better ... her menses return'd at proper seasons as if she had been deliver'd of a child, and continued to do so for several months ... the poor woman recover'd, and is now perfectly well.' The editor commented that 'the foregoing case is not singular; we see two of the same kind recorded in the Memorirs of the Royal Academy of Surgery at Paris for the last year ... and [one] communicated to the French academy'.

31. *Peregrine Pickle*, 1, 72. Smollett's description of Mrs Trunnion's expectant state tallies well with observations made in John Pechey's *Complete Midwife's Practice Enlarged* (5th edn, 1698), especially the section 'Of False Conception' (57–62). According to a manuscript in the Hunterian Museum, Pechey was included among required authors to be read by students at the Glasgow Medical School, which Smollett attended 1736–9.

32. Without pretending to summarise the vast literature dealing with the date and author of this painting, I mention the following: George W. Johnson, *A History of English Gardening* (1829), 72–81; Alicia Amherst, *A History of Gardening in England* (1910), 238ff.; Miles Hadfield, *Gardening in Britain* (1960), 126; J. L. Collins, *The Pineapple* (Honolulu, 1960), 70–86; William Gardener, 'Botany and the Americas', *History Today*, XVI (Dec. 1966), 849–55, where the picture is reprinted. In his *DNB* life of John Tradescant the younger, gardener to Charles I, G. S. Boulger writes: 'There is a tradition that the younger Tradescant first planted the pineapple in England in the garden of Sir James Palmer at Dorney House, Windsor, where a large stone cut in the shape of a pineapple by way of extant ... The pineapple pits were therefore pre-Charles II. Surely then John Tradescant the younger grew pineapples here for Charles 1. The fact that there is no painting of John the elder or John the younger presenting a home-grown pineapple to Charles I

does not disprove the possibility. The Tradescants would have had ready access to pineapples thanks to Sir William Courteen who was one of their principal benefactors. Sir William took out the first settlers to Barbadoes in 1625. The West Indies were one of his regular trade routes.' This theory is supported by M. Allan in *The Tradescants: Their Plants, Gardens and Museum 1570–1662* (1964), 143–5. Tradescant was known for his exotic fruits, as is seen in Tom Brown's *Amusements Serious and Comical*, particularly the section entitled 'The Philosophical or Virtuosi Country'.

33. Robert James, M. D., *A Medicinal Dictionary* (1743–5), article entitled 'Ananas'.

34. E. S. de Beer (ed.), *The Diary of John Evelyn* (Oxford 1955), III, 293, 513.

35. Hadfield, *Gardening in Britain*, 126.

36. *Dictionarium Botanicum* . . . (1728), article entitled 'Ananas', and 'A Particular Easy Method of Managing Pine-Apples' in *New Improvements of Planting and Gardening* (1726), 605. Bradley's assertion was challenged in 1780 by Horace Walpole, who wrote to the Reverend William Cole: 'There is another assertion in Gough [*British Topography* 1768], which I can authentically contradict. He says Sir Matthew Decker first introduced ananas. My curious picture of Rose, the royal gardener, presenting the first ananas to Charles II proves the culture here earlier by several years' (W. S. Lewis (ed.), *Letters to the Reverend William Cole* (New Haven, 1937), II, 239). Walpole had acquired the painting from William Pennicott in 1780. In his popular handbook, *The Gardener's Dictionary* (1724), Philip Miller attributed the first pineapple to Tellende. See also E. S. Rohde, *The Story of the Garden* (1932), 178.

37. R. Halsband (ed.), *The Complete Letters of Mary Wortley Montagu* (Oxford, 1965–7), 1, 290.

38. Edited by F. B. Kaye (Oxford, 1924), II, 193–5.

39. Pages 605–6. Bradley, among other authors, notes that a forty-foot stove was necessary to ripen one hundred pineapples. An average pineapple took three years to ripen and *c*. 1726 its total cost from the time of purchasing seeds was £80.

40. Hadfield, *Gardening*, 166, and C. H. Collins Baker and M. I. Baker, *The Life of James Brydges First Duke of Chandos* (1949), 103. See also George Sherburn, '"Timon's Villa" and Cannons', *Huntington Library Bulletin*, VIII (1935), 143.

41. *Correspondence of Alexander Pope*, III, 233 and 453.

42. Ibid., IV, 360. See also IV, 405, 420; and Benjamin Boyce, *The Benevolent Man: a Life of Ralph Allen* (Cambridge, 1967), 114. On 25 March 1746, Pope wrote to Swift about the new fruits in his garden: 'I have good Melons and Pine-apples of my own growth. I am as much a better Gardiner, as I'm a worse Poet, than when you saw me: But gardening is near a-kin to Philosophy, for Tully says *Agricultura proxima sapientiae*' (*Correspondence*, IV, 6). Without documentation Hadfield, *History of Gardening*, 187, states that 'a year later [1742] Allen was advised not to take Scott's advice'. But Pope could not have been the unmentioned person since he fully approved of Scott's method. For Pope's activities as a gardener and ideas about gardening during Smollett's mature years, see Edward Malins, *English Landscaping and Literature 1660–1840* (1966), 26–51.

43. *Correspondence*, IV, 429.

44. J. L. Robertson (ed.), *The Poetical Works of James Thomson* (Oxford, 1908), 'Summer', II. 253–4.

45. *The Art of Preserving Health* (1796), including a *Critical Essay* by Dr John Aikin, II. 334–40. Aikin commented on foods like pineapples as an example

of a 'too luxurious diet' (p. 14). The medical aspects of pineapples were also discussed in scientific publications. For example, see William Bastard, 'On the Cultivation of Pine-Apples', *Philosophical Transactions*, LXVII (1777), 649–52, in which the author describes his hothouse in Devonshire and the effects of the fruit on the body. Armstrong, a Scotsman who practised medicine in London, probably did not taste pineapples in Scotland. Dr John Hope, the Regius Professor of Botany at Edinburgh University and a populariser of Linnaeus in Scotland, allegedly grew in 1762 the first pineapples in Scotland, although I can discover no certain means of verifying this allegation.

46. *The English Physitian* (1684), 189–90. Pineapples were not mentioned in the edition of 1651, presumably because they were then unknown in England. Smollett referred to Culpepper's medical handbooks in several novels, and Joseph Addison listed the *Directory for Midwives* among essential books in an eighteenth-century 'Lady's Library'. See Donald Bond (ed.), *The Spectator* (Oxford, 1965), 1, 155.

47. *Pharmacopaeia Universalis* (1742), 118. John Quincy made the same point in his *Complete Englrish Dispensatory* (rev. edn, 1742), 194.

48. *A Treatise of All Sarts of Foods, Both Animal and Vegetable*, trans. by D. Hay, M. D. (1745), 350 and 75–6. The signatures of several distinguished physicians of the Royal College of Physicians appear on the frontispage as approving the medical aspects of the book: among them are Edward Brown, Walter Charleton, and John Woodward, whom the Scriblerians satirised. For other comments by dietitians about the medicinal aspect of pineapples, see A. Cocchi, *The Pythagorean Diet, of Vegetables Only, Conducive to the Preservation of Health . . .* , trans. from Italian (1745), 74–6 and Sir Jack Drummond and Anne Wilbraham, *The Englishman's Food* (1939), pp. 228–9. Numerous comments about the danger of pineapples for pregnant women may also be found in Ephraim Chambers's *Cyclopaedia: Or, An Universal Dictionary of Arts and Sciences* (1728), article entitled 'Ananas', and in George Cheyne's *An Essay on Regimen* (2nd edn, 1740), pp. 76–7. Chap. xix of *Roderick Random*, in which the hero meets the French apothecary Lavement, makes it clear that Smollett was thoroughly familiar with the medical effects of different diets.

49. *Free Thoughts on Quacks* (1749), 164–5.

50. Alexander Campbell Fraser (ed.), *An Essay Concerning Human Understanding* (Oxford, 1894), II, 37–8. Although Locke was an expert botanist and did a great deal of plant research in the Oxford Botanical Garden in 1650–60, there is no evidence that he himself ever grew pineapples. See Kenneth Dewhurst, *John Locke (1632–1704) Physician and Philosopher: a Medical Biography* (1963), 8–9.

51. L. A. Selby-Bigge (ed.), *A Treatise of Human Nature* (2nd edn rev., Oxford, 1928), 5.

52. I have borrowed this phrase from the excellent article of D. W. Jefferson 'Tristram Shandy and the Tradition of Learned Wit, *Essays in Criticism*, 1 (1951), 225–48.

53. Some of these have been studied and attributed to Smollett by G. S. Rousseau, 'Matt Bramble and the Sulphur Controversy in the XVIIIth Century: Medical Background of *Humphry Clinker*', *Journal of the History of Ideas*, XXVIII (1967), 577–90.

4
Nerves and Racism: Le Cat's Neurology of Racism (1973)

This essay followed closely on the heels of its predecessor, moving from the concerns of 'diseased imagination' to those of the neuroanatomy of skin when viewed in the context of race. I was bewildered by the connections in Enlightenment scientific thought, especially in the light of its developing anthropology. Besides, presentist concerns in the early 1970s drove me there at a moment when you could still use the word 'Negro' in America without being called a racist, yet – paradoxically – it was acutely evident that something was profoundly troubled in our race relations. I found one clue in the writings of the obscure French physician Nicolas Le Cat who had won a prize for his radical beliefs about the consequences for race of nervous mechanistic physiology. Note 1 below provides a brief summary of his work. But Le Cat was also representative of the growing Continental attraction to the concept of a thoroughly 'nervous body' and became wholly invested in it. His writings, soon known in cities from Edinburgh to Copenhagen and St Petersburg, extended the already large domain of the nerves.

Yet Le Cat was quickly written out of the history of science, or never written in. So, in 1971–73 there was little to consult, nothing to read, except a few primary documents in Latin, French and German. It seemed odd then that no one in the growing camps of historians of science had noticed how he was esteemed in his own time in the British Isles: not merely in England and in the Royal Society but in the 'hotbed of genius' – as Smollett called it – in Edinburgh. He was intuitively searching for an explanation to explain race through the neuropores, or external orifices: the neural canals before a theory about them had developed. The reception of my essay on publication was no more enthusiastic than Le Cat's general niche had been after his death. Richard Popkin, the American historian of skepticism, referred to it, noting its importance for eighteenth-century rational thought, especially for Enlightenment theories of Jewish national and racial types. Its application could even be seen to account, in part, for the stereotype of the 'wandering Jew'. Perhaps the subject – race in relation to anatomy and physiology – was too explosive to be discussed even in historical contexts.

Now, thirty years later, the biology of racism remains an underdeveloped historical topic in North American scholarship. Perhaps it is still too explosive. And Le Cat himself remains unknown: nine out of ten Enlightenment scholars could not tell you who he was or what he wrote. Yet the mighty Albrecht von Haller had formally congratulated Le Cat (n. 1) when his book won a prize and he seemed to be a rising star in the firmament of neuroanatomy – the same Haller who so securely straddled Enlightenment physiology and nervous sensibility and lent weight to the growing Sensibility Movement in Europe. Haller recognized a member of his own camp extending the domain of neurophysiology, no less than a Gerald Edelman or Oliver Sacks would today in the ranks of young neuroscientists seeking to understand consciousness.

Le Cat and the Physiology of Negroes (1973)

'The origin of *Negroes*,' Ephraim Chambers wrote in the 1728 *Cyclopaedia*, 'and the cause of that remarkable difference in complexion from the rest of mankind, has much perplexed the naturalists; nor has anything satisfactory been yet offered on that hand.' A generation later, in the 1750s, this was still true, although Claude Nicolas Le Cat was to influence the picture considerably. It is hard to know if Chambers, no scientist or medical man, would have been at all impressed by Le Cat's theories. But if he had heard or read them, he might have modified somewhat his statement in the *Cyclopaedia*.

From the vantage of the history of science, Le Cat's entire career, quite unsurveyed, incidentally, is as exciting as that part of it represented by his contribution to the age-old debate about the colour of Negro skin, its origins and history, from the beginning of man to the eighteenth century. Born in 1700 and dead by 1768, Le Cat was the chief physician and surgeon of the Hôtel-Dieu, the leading hospital in Rouen, a member of many French and foreign scientific societies, and the author of over a dozen medical treatises. In 1762 he retired from his hospital post, and during his remaining seven years wrote most of the books that utilise his researches, observations, and reading of over fifty years.[1]

His scientific contribution to the race argument has either been neglected or thought so insignificant until now that one looks in vain for his name in most modern reference books in the history of science and medicine as well as in encyclopaedias and dictionaries of biography. And yet, careful scrutiny of his works reveals that he played a role in advancing biological understanding of skin colour. He himself was apparently aware of this role, and he accordingly devoted his greatest scientific energies to what we today must regard as his most significant medical work, *Traité de la couleur de la peau humaine en général & de celle des Nègres en particulier*, published in Amsterdam in 1765.[2]

143

Le Cat's treatise contradicts previous theories maintaining that bile is responsible for the colour of human skin; this argument had been advanced as indisputable scientific fact in the earliest writings of Egyptian medicine, later appeared in Homer, Strabo, Ovid, and Pliny, and was advanced throughout the Renaissance and for much of the eighteenth century. The *Teatro critico* of Father Feijoo is typical of the impressionistic manner in which the bile argument was set forward: succinctly, without experimental support, and as an *ipse dixit* argument.[3] Other eighteenth-century naturalists, including Raymond de Vieussens, Buffon, La Mettrie, D'Holbach, and numerous travel writers, also repeated the argument as if it were gospel truer than truth.

In Italy, Albinus and Sanctorini supported a bile theory (although these men recanted and at several junctures even displayed scepticism about the belief), and in France, where it seems to have been extremely popular, it attracted numerous advocates, and none more vocal than Pierre Barrère, a Perpignanese physician and medical author who strenuously championed it in 1741 in a dissertation on the cause of skin colour, *Dissertation sur la cause physique de la couleur, des Nègres, de la qualité de leurs cheveux, & de la génération de l'un & de l'autre*. Germans, Scandinavians, and Englishmen also gave the belief their stamp and seal, and it is accurate to say that by 1750 the belief was prevalent – truly as popular as the 'monster-mongering' sport, to use the phrase of Professor Jordan in his edition of Samuel Stanhope Smith's *Essay on the Causes of Complexion*[4] – that blacks were another species of man, *sans* the ordinary human organs, tissues, and heart, and (of course) *sans* soul.[5] Le Cat's theory, in contrast, introduces a black substance, 'ethiops' (in other words, melanin and its cell the melanocyte), which, he maintained, is present to some extent in all creatures, white and dark, but to a greater degree in blacks; and it is this that distinguishes them. This theory had been Malpighi's,[6] and as I shall show in the paragraphs below, Le Cat, who had read and studied Malpighi's works, developed it. Establishment of the precise connection between the theories of the two men is important because one cannot understand the significance or implications of Le Cat's theory of ethiops without first understanding Malpighi's.

Both Malpighi and Le Cat believed that ethiops is contained in the nerve tips, where it permanently resides. But whereas this idea is merely suggested, without detailed development, in Malpighi's writings, Le Cat made it the central focus of his argument. Furthermore, he tried to show that ethiops is not governed by the liver, pancreas, or gall bladder, but is indigenous to the membrane surrounding the tips of nerve cells. Le Cat based this assumption on microscopic experiments he had done with frogs and other animals. In the frog, for example, 'ethiops' (i.e. melanin) *is* in fact present anatomically in nerve cells, but not in human beings. In our anatomy, pigment is exclusively located in epidermal tissue, which is apart from the nerve cells. Le Cat would not have known this; microscopes in the 1730s and 1740s were not powerful enough to distinguish sharply within human dermal tissue. Nerve tips, under weak miscroscopes of the type Le Cat is likely to have used, would appear to extend as far up as the epidermis, whereas, in fact, they do not; they are subdermal. It was not until the nineteenth century that microscopy enabled medical men to see that an epidermal-subdermal barrier (basement membrane) exists and that nerve cells do not penetrate this barrier.

Le Cat, who was logical and reasonable in his inference that human anatomy is almost identical with animal anatomy (frogs, chameleons), was so much convinced of the presence of ethiops in nerve cells that he directed his energies to other questions about the physiological nature of ethiops. For example, he asked how blacks originally acquire this ethiops – a question we might think would have interested Malpighi but which apparently never did. Le Cat tried to formulate an answer, but it was not as clear as we would hope: ethiops, he maintained, comes not from the sun, climate, or torrid zones alone, but from these climatic conditions in conjunction with the peculiar physiological traits Negroes developed over longs periods of time. Not a perfectly clear formulation, to be sure, but in 1765 there was no Darwinian evolutionary theory of selection. Yet Le Cat's staunch belief that ethiops is somehow indigenous to blacks reveals a colour argument scientifically more sophisticated than the theories of his contemporaries or near-contemporaries Malpighi, Feijoo, Sanctorini, and Barrère.

Like all scientific hypotheses, Le Cat's must be judged for its ultimate accuracy. In this regard it fails, as I shall show in detail below. But it ought also to be viewed in the context of his basic assumptions concerning physiology and the common assumptions of his age. In this regard, Le Cat's theory shows up rather well on several counts, not merely one. First, he believed that the nervous system controls the organism – not a revolutionary assumption in the 1760s, but one that was in constant need of focusing and that required application to the racist debates in medicine. In assuming this view, he was in line with the most progressive mainstream of current European medicine and physiology. It was a view demonstrating that he had read and understood Willis's brain theory and Haller's radical but nevertheless accurate thesis about nervous action in relation to muscular contraction. Second, he was right to assume that ethiops is somehow controlled in its action by the nervous system. We today know that the pituitary, an integral part of the nervous system, regulates many of the functions of melanin; Le Cat could not have known this, but was not very far from the truth in assuming that nerve tips extending into the epidermis regulate pigment cells.

If his conception of the nervous system is lacking in certain areas, we ought to be tolerant within reasonable limits. For example, Le Cat believed that the animal spirits, not subject to the laws of physics and chemistry, pervade the hollow tubes of the nerves. This is untrue, but most scientists – good scientists – of his epoch also believed it. Moreover, Le Cat held that a mucous sheath (*corps muqueux*) wraps the entire nerve cell.[7] Although this idea is not entirely true, it is closer to the truth than the notion of many of his contemporaries, and it is certainly a more advanced concept of the anatomy of this part of the nervous system than Malpighi's. Controversies about the precise physiological structure of the 'outsides' of nerve cells had vigorously been carried on throughout the eighteenth century in England, where the question was debated in the Royal College of Physicians and in the Royal Society, as it was too on the Continent by Boerhaave, Hoffman, and lesser-known figures. It is true that Le Cat could not add substantially to these debates or radically change the theories of these men. But he did spend more time than they examining ganglions, the ends of nerves, under the microscope, and eventually he developed a fairly sophisticated conception of nerve tufts (*papilles nerveuses*) which he likened to the nipple-like structures of the tongue. Furthermore, he demonstrated that they expand and contract mechanistically, especially when regenerating themselves.

Considering the assumptions of physiologists in his age, therefore, Le Cat did not fare badly. In fact, he did exceptionally well, erring only in the points described above and, importantly, in his mistaken idea that the nervous system of humans is exactly, or almost exactly, the same as in frogs. To summarise his anatomical reasoning, he built his theory on some of the best physiology of his day and buttressed his assumptions with microscopic observations of several decades.

But even so, he was unsatisfied about the precise nature, histologically, of ethiops. And as a result of his dissatisfaction he reconsidered the matter, he says in his *Traité*, many times before satisfying himself. The most puzzling question, he believed, related to the *origin* of ethiops. He had seen this substance expand and contract under the microscope, so there could be no question of its physical nature: it could not be non-material, as were animal spirits. He was also certain, although it is hard for us to know why, of its presence at birth in blacks, and that there was no possibility of its being acquired after maturation. It was transmitted from generation to generation by the sun's rays, he thought, but these rays alone could not *produce* the substance. Heat could expand it, he believed, in the same way that heat causes other types of physical expansion.

Since Le Cat's experiments with various animals played an important role in his theory in the *Traité*, something, however brief, must be said about these, as well as about the significance of these experiments for modern medicine. Le Cat was convinced of the necessity of microscopic investigation, unlike many of his contemporaries, rationalists at heart who placed little faith in the microscope. He had seen ethiops in many animals and fish, but especially in the cuttlefish or squid (*sèche*). For two decades (1740–60) he observed their large black cells under various kinds of microscopes and deduced that human skin tissue must be similar. What he actually saw under the lens were melanocytes, microscopically quite prominent and very large in squid; but he was ultimately incorrect to assume that melanocytes in black men were structured similarly to those he observed in cuttlefish. Such reasoning by analogy was far from outlandish (scientists today, for example, experiment on mice and then extrapolate all their findings to humans); nor was his thoroughly logical assumption that Negroes have some sort of greater melanocyte production than do whites. Time has proved him correct, although his reasons were different from ours. But he had no conception of the melanocyte cell itself, its nature, anatomical structure, boundaries within the basal layer, accumulation at the base of the epidermis, chemical composition, and evolution throughout the life of a normal human being.[8]

If Le Cat's theory is 'translated' into modern medical terminology (and extreme caution must be employed in such a translation), these approximate statements obtain. Melanocytes are scattered throughout the epidermis but do not appear, whether in whites or blacks, in the basement membrane or dermis. These two layers, dermis and epidermis, are separated by a boundary (the basement membrane) through which nerve tissue does not penetrate. Therefore, it is quite impossible, by the standards of modern anatomy, to imagine melanocytes in the dermis, or, conversely, nerve cells in the epidermis. Moreover, these melanocytes do not differ significantly, if at all, in chemical composition in whites and blacks, although their number does. Blacks are known to have many more melanocytes per epidermal area than whites, but present-day knowledge of the hormonal activity of melanocytes is not sufficient to indicate if this disparity influences bodily functions. But it does influence skin pigmentation, thereby accounting in part for the difference between fair and dark peoples. There are of course other factors, mostly genetic, that influence this coloration, but they need not be explained in detail here.

To turn now to Le Cat within this brief 'translation': as I have already indicated, Le Cat was wrong on several counts, especially in his notion that nerve cells penetrate through the basement membrane. But he must be given credit for his intuitive leap in suspecting that bile cannot influence pigment, and thus for changing the whole course of physiological theory about skin. He must, it seems to me, also be given credit for his suspicion that the nerves play a more extensive role in the body than was thought at this time. Haller, Whytt, and other neurologists demonstrated in his own age that the brain required further examination, but it was Le Cat who suggested, however primitively, that the blood channel and nervous system were connected more intimately than most medical men thought.[9] Le Cat, viewed in this light, clearly emerges as a more important physiologist than Malpighi, especially if his contribution to the racist debates is the yardstick of measurement.

Malpighi, who died in 1694 (only six years before Le Cat was born), believed in an altogether different theory, one much less scientific and sophisticated: that all men were originally white, but that sinners among them had degenerated into black. In putting forward this remote divine cause of black skin, Malpighi impeded rather than advanced arguments regarding race among scientific men. It is true that he later abandoned his divine cause and substituted a proximate physiological cause: namely, a mucous sheath separating the dermis from the epidermis, recognition of which solved the physiological riddle puzzling anatomists for centuries. But he was wrong here, as wrong as Le Cat, although in a different way: the basement membrane, Malpighi's mucous sheath, does not contain melanin. Malpighi also theorised about a 'mucous liquor' determining skin colour, but he never stated where this liquor is located or how it operates, and Le Cat sensed this gap early in his researches. He dedicated his experiments, in part, to a refinement of this theory, but never could convince himself that ethiops was confined to a single sheath within the dermis. In other words, Le Cat argued for more area within skin tissue, for the whole basal cell and its surroundings as a zone wherein ethiops was contained. Malpighi, on the other hand, was persuaded that a localised substance must necessarily be the cause of differences in skin colour.[10] Having established to his own satisfaction that the cutis as well as cuticle of blacks is white, he reasoned that blacks differ anatomically only in this mucous liquor. In this regard he was certainly more advanced than all his seventeenth-century colleagues, but not so advanced as Le Cat, who consciously tried to *show* the connections between the 'mucous liquor' and the nervous system – a connection that we are now just beginning to learn does exist.[11] Le Cat, in his own way, was saying that the nervous system (brain, nerves, etc.) has some control of pigment activity (we know that the pituitary controls the hormonal activity of melanocytes). No one would wish to argue that this discovery in anatomy should bear Le Cat's name, but he was closer to the truth than his colleagues in France, and certainly those in England. And it is precisely in the bold imaginative leap of this connection, however primitively made, that Le Cat demonstrated his sound scientific intuition.

His contemporaries failed to understand him. Most never deemed his ideas worthy of the labour of serious comprehension: they continued the racial debate, usually asserting once again all the inadequate previous theories – but no one veered from the age-old lure of the bile theory. Riolan, Littre, and Morgagni, for example, were perplexed by the origin of black skin, and hypothesised that since most Negro skin had white patches, black men must originally have been white: a curious argument possessing little anatomic veracity. Later on, the sun turned their bile black (so the argument continued) and also their skin. For these scientists the relation of sun and bile was cause and effect: too much sun caused bile to blacken, and bile determined skin colour. QED. Albinus, an eighteenth-century Italian scientist (whose name, incidentally, has no connection to 'albino'), proved to his own satisfaction that Negro bile, both hepatic and cystic, was black.[12] Sanctorini concurred with Albinus in considering bile the *only* substance in the body capable of influencing skin colour.[13] These men, oppressed by the tyranny of the ancient theory of bile, with centuries of weight behind it, had either not read Malpighi or did not comprehend him. (It is naturally possible that they read and rejected him, but this seems unlikely in view of the zeal with which Sanctorini and Albinus approached the theories of others; one wonders, moreover, why they would not have refuted him in print if they had renounced his theories.) Elsewhere than in Italy, the situation was not different. Winslow in Denmark was undecided,[14] and Grossard, a Le Cat student who later became a professor at the medical school in Montpellier and who also happened to have undertaken important research into the lymph system, impressionistically speculated that *lymph* was more important than bile in determining colour; but he was surprised to discover in autopsies that Negro lymph is every bit as white as the white man's.[15]

Then in 1741, a momentous episode in the eighteenth-century history of this medical debate occurred. Barrère, in France, published experiments asserting that Negro bile is black, and that it alone causes the black

pigment in Negro skin.[16] Not the theory but the experiments won him
attention. The bile theory was centuries old; but Barrère now endowed it
with an authority it had never had. His book stated that his conclusions
were based entirely on laboratory studies, thus creating the impression
that black bile and its effects, long suspected but never seen, were as
verifiable as the second law of motion. But the careful reader would have
found that Barrère gave himself away. Blacks acquired the black bile, he
postulated, by dwelling in hot jungles. He himself had not, of course,
seen black bile in Negroes, nor could he account for the fact that
generations of white men living in Africa never turned black. He somehow
took black bile on 'faith', having viewed something abnormal resembling
it, perhaps, in a few diseased bodies.

It is therefore greatly to Le Cat's credit that, only twenty years after
Barrère's theory won universal acclaim, especially in France, he intuited
and then demonstrated that it was specious. Historians of science may in
the future show that certain French and English medical men anticipated
Le Cat in this regard, but even so, some credit, however little, must go
to him; at least, he must be rescued from the total oblivion in which he
has until now remained. This is all the truer when it is remembered
that towns like Rouen were somewhat isolated. If Le Cat had done his
experiments in Paris, with the aid of many exceptional colleagues, we
might feel more wary of granting him much honour; but he swam
against the tide alone, in a small northern French city that had never been
a medical centre. An idea of his courage in rejecting the dominant belief
in bile as the single and sole determinant of skin colour is glimpsed by
examining reviews of Barrère's theory in comparison to those of Le Cat.
If Barrère was recognised and praised, Le Cat was disparaged as a shallow
rationalist, even by English scientists who ought to have known better.
Monsieur Eloy, author of the four-volume *Dictionnaire historique de la
médecine*, published in 1778, commented favourably upon Barrère's bile
theory but criticised Le Cat's nerve–ethiops theory as a wild hypothesis:
'Il explique ensuite le sentiment qu'il a adopté, mais comme il n'est fondé,
ni sur l'observation, ni sur l'expérience, on est en droit de le renvoyer dans
la classe des hypotheses qui sont plus ingénieuses que concluantes.'[17]

Two years after Le Cat's systematic demolition of the bile theory in
the *Traité*, the Abbé Demanet published a *Dissertation physique et historique
sur l'origine des Négres et la cause de leur couleur* (1767),[18] wherein he repeated
the old bile arguments without mentioning Le Cat. Such an omission in
itself is insignificant, but it reveals the typical neglect of Le Cat before
the beginning of the nineteenth century. While it is true that his
research on skin was occasionally mentioned during the last quarter of

the eighteenth century – for example, in Jean Paul Marat's *Philosophical Essay on Man* (London, 1773) – his theory, however inchoate, of the interactions of ethiops and the nerve system was either too advanced or physiologically too radical, or appeared too clouded by physiological details, to admit of acceptance or recognition in his own time. Or it may have been left unregarded altogether, though this possibility is hard to understand in view of Le Cat's reputation. This was the man, after all, who had won the esteemed Berlin prize in physiology and about whom the editors of the *Gentleman's Magazine* said in 1753 he 'ought to be universally read'.

Throughout the last quarter of the eighteenth century, the scientific-medical community debated questions regarding the origin of Negroes and their black skin. As revolution approached and man's thoughts were deflected, it abated; but until then the question consumed them, though it seemed to arrive nowhere. Although some writers pointed out the loopholes inherent in the hot sun–black bile argument, none gave quite such specific reasons as Le Cat. Samuel Stanhope Smith professed to have read much literature before turning up any tangible conclusions in his *Essay on the Causes of Complexion...*, and ultimately admitted that not much new could be said on the subject. Essentially a synthesiser, he was satisfied to relegate Le Cat to a single mention in a voluminous footnote in which the Rouen surgeon is, of course, lost.[19] Whether Smith actually read Le Cat is doubtful, but his estimate cannot be misconstrued under any circumstances. (He had at least heard of Le Cat and his theories, which is more than can be said for other writers; most authors simply disregarded Le Cat altogether, and I have already suggested that this is not likely to have been prompted by his obscurity.) In 1768, three years after Le Cat's *Traité* was published, the first edition of the *Encyclopaedia Britannica* appeared. In the article entitled 'Negroes', the anonymous author commented: 'Dr Barrère alleges that the gall of Negroes is black, and being mixed with their blood is deposited between their skin and scarf-skin.' But no mention appears of Le Cat, or of his magisterial, though cautious, challenge to the theory of bile as the cause of colour. Though this anonymous author took notice of 'Dr. Mitchell of Virginia' (John Mitchell, author of 'an Essay upon the Causes of the Different Colours of People in Different Climates'), he apparently had never heard of Le Cat, or if he had, could not see the difference between the sophistica-tion of Le Cat's theory and the primitiveness (as well as repetitiveness) of Barrère's.[20] Like Barrère, he confused himself in this article by citing bizarre cases of colour change; yet he never paused to ask what the physiological basis of skin colour was.

Le Cat himself took time out to study such fantastic cases of blacks turning white, or whites turning black, but in each instance he attributed the change to severe illness, body change during pregnancy, or wild growth of the 'ethiops'. That is, he perceived these were exceptional cases, and drew no paradigms from them. His balance of induction and deduction was intelligently managed, and one observes few cases in the *Traité* of his going out on a limb or forcing a conclusion from an isolated example. But when it was time to generalise, he surrendered prejudice and tradition to his empirical findings. He ruled out climate as a primary cause: a Norwegian clan migrating to the Sudan could never become black, at least not in the course of a few centuries. He thereby discarded adaptive conditions and concentrated on physiological processes. If he could have known the approximate age of the world, he might have been able to anticipate Charles Darwin in *The Descent of Man and Selection in Relation to Sex*, and might also have reasoned that there must have been selection for lightly pigmented individuals in higher latitudes since they could better utilise sunshine. But the chronology of the world in 1765 was still in doubt, and so it remained until the nineteenth century; the age of man, indeed, is still to be determined. And Le Cat, who really cannot be criticised for this lack of knowledge, demonstrated his abilities as a model-maker by refuting the bile theory and turning to the nervous system's interaction with other systems.[21]

The significance of this essay for a symposium on racism in the eighteenth century is not easily grasped. For the men it treats, Le Cat, Malpighi, and to a lesser extent Albinus, Santorini, and Grossard, were never involved in the debates about race. Philosophers like Voltaire and Diderot held their personal opinions about the real status of black men, and especially about their physiological similarity or dissimilarity to white men. But Voltaire and Diderot never engaged, to my knowledge, in medical experiments, as did Le Cat; besides, they allowed other concerns – nationalistic, economic, religious, philosophic – to influence their final decision about the species to which black men belonged.[22]

Le Cat, so far as I know, had no such complicated concerns. He was not a 'philosophical' scientist in the way his English colleague Dr Robert Whytt was; he was content to experiment and report his observations. This is not to imply that his scientific assumptions were simple or lacking in any way, but Le Cat, unlike the French *philosophes* discussed elsewhere in this volume, had less ambitious plans for himself. He desired, understandably, to rise as high as possible in medical research, and for this aspiration he was respected in his own age. Yet he remained content to leave it to others to comment on the social implications of his discoveries.

Perhaps, then, the significance of Le Cat's work is that there is no significance. I am personally persuaded that thinkers who debate a topic like racism without understanding something about the physiological bases of skin colour cannot be sophisticated thinkers. They may have a great deal that is important to say about other topics, e.g. the nature of nervous man, God, the life process, the human condition, and so forth. But this is a different matter from making a significant statement, one worthy of recording in the annals of history, about racism. Too many examples of my point abound in this volume for me to provide detail; and no one is going to think less of a Montesquieu or a Voltaire because either thought blacks were a different species of man, apparently without taking the trouble to read contemporary scientists like Le Cat. But we may be certain that if more people in the eighteenth century had read scientists like Le Cat, the nature of the debates discussed now would be different. Laymen cannot be expected in any age to comprehend the technical writings of medical men, but the ideas of a Le Cat, for example, were explained in popular magazines like the *Gentleman's*, and were epitomised in everyday language for the common man.

If I may conclude on a modern note, it seems to me that the situation today is not altogether different from that in Le Cat's age. Thinkers from the common man to professional philosophers have their personal views about the black man, his capabilities, limitations, potential.[23] Yet not very many of these thinkers have taken the trouble to read the recent radical theories of Dr Jensen and his team.[24] The content of these theories is not in question: they may be right, they may be wildly wrong. But they have been put forward by scientists of very high calibre, with credentials beyond question, who hopefully have scientific truth as their first concern. Who knows if historians of science two centuries in the future will prove Jensen and company correct? Who knows what changes in the social structure of American life will be effected by Jensen's theories, if they are accurate? Or has a monumental change occurred, and do we now live in an age when certain theories are simply too dangerous to be put forward regardless of veracity? There are big questions, but must be left for another occasion.

Notes

1. There is no biography. Information, and precious little exists, is scattered: see N. F. Eloy, *Dictionnaire historique de la médecine ancienne et moderne* 4 vols (Paris, 1778), 1, 565–71, for the only brief sketch. Nothing at all is said about Le Cat in the standard histories of medicine by Arturo Castiglione, Fielding

H. Garrison, Sir William Osler, Theodor Puschmann, Henry F. Sigerist, Charles J. Singer, René Taton, and E. A. Underwood. René Taton's *Enseignement et diffusion des sciences en France au xviii^c siècle* (Paris, 1964), briefly discusses Le Cat's anatomy courses at the Hôtel-Dieu in Rouen. Robert Darnton, *Mesmerism and the Enlightenment* (Cambridge, 1967), mentions Le Cat in relation to hypnotism. Le Cat's private papers survive and are available in the Archives of the City, Rouen, France. On 31 January 1739, Le Cat was elected a foreign member of the Royal Society, London (Thomas Thomson, *The History of the Royal Society* (London, 1912), appendix xli). After this time his anatomical works were regularly translated into English and reviewed in English journals. His interactions with Dr James Parsons, F.R.S., are described by John Nichols, *Literary Anecdotes of the Eighteenth Century*, 6 vols. (1812), v, 475–6. By 1753 Le Cat, many of whose communications were now published in the *Philosophical Transactions of the Royal Society of London*, was sufficiently well known to be referred to by a columnist in the *Gentleman's Magazine*, XXIII (1753), 403 as 'the ingenious writer... who ought to be universally read'. In 1765 Le Cat won the prize of the Berlin Academy by answering their set of physiological questions on the structure of nerves. Offered by the Academy since 1753 but without a candidate, the prize answers were published as Le Cat's *Dissertation sur l'existence & la nature du fluide des nerfs & son action* (Berlin, 1765). The actual questions and an account of Le Cat's achievement in answering them are found in A. von Harnack, *Geschichte der Königlich Preussischen Akademie der Wissenschaften zu Berlin* (Berlin, 1901), 1400. Le Cat was congratulated by the acclaimed scientist Haller for his attainment.

2. Extending to almost two hundred pages, it was not translated into English or reviewed in English periodicals as almost all Le Cat's other works had been. The reasons are not clear: perhaps the subject matter was too controversial for the more sedate reviews and too conservative for others.

3. Benito Feijoo, *Teatro Critico Universal* (Madrid, 1736), VII, third discourse, 69–94.

4. Published in the John Harvard Library Series (Cambridge, 1965), viii.

5. A long list of comments and works could be compiled for the period 1700–80. Without making a search, I have found no fewer than two dozen comments in travel books alone. See, for example, Edward Long, 'Negroes', in *The History of Jamaica* (1774), chap. i, the third book. Some of these works appear in Winthrop Jordan's 'Guide to Smith's References', in Smith, *Essay*, pp. 253–68.

6. See *Opera Omnia* (1686), II, 221.

7. *Traité de la couleur*, 30 ff.

8. This concept arose in the mid-nineteenth century and required at least Schwann's theory of the cell. See Bobbie Williams, 'Human Pigmentation', *General Anthropology* (1973), 487–523, the best scientific treatment of skin colour I have seen. I am grateful to Professor Williams, Department of Anthropology, University of California, Los Angeles, for making this unpublished material available to me.

9. This is a chapter of the history of science not yet surveyed.

10. *Opera Omnia*, II, 215–38. This notion was transmitted to the eighteenth century as is made clear by dictionaries and encyclopedias. See, for example, Abraham Rees's article 'Complexion' in *The Cyclopaedia; or, Universal Dictionary* (Philadelphia, 1810–24), IX, no pages.

11. I.e., the pituitary regulating functions of melanocytes. Precisely why this is the case is unknown as yet, as most histology textbooks explain. In general, little is known about the influence of the nervous system on hormonal activity. As an example of another area in which the influence of the nervous system is not well understood there is the glyal cell, separating the blood system and the brain. Before the 1920s it was not known that medicines could pass through this barrier, i.e., penetrate from the blood into the brain, and consequently affect the nervous system.

12. See Bernardus Albinus, *De Sede et Causa Coloris Aethiopum* (Leiden, 1737), 267–78.

13. *De Statica Medicina* (The Hague, 1664), 221. Sanctorius, an influential writer of aphorisms, never developed his theory.

14. Jacob Winslow, *An Anatomical Exposition of the Structure of the Human Body*, trans. G. Douglas, 2 vols (1733), and *A Description of the Integuments of the Vessels* (1784).

15. Grossard never published books, but his ideas and writings were circulated in France among interested doctors. He consulted frequently with Le Cat in Rouen and at scholarly meetings.

16. *Dissertation sur la cause physique de la couleur des Nègres, de la qualité de leurs cheveaux, & de la génération de l'un & de l'autre* (Paris, 1741). A brief survey of Barrère's life appears in Elroy's *Dictionnaire historique de la médecine*, 1, 265. Although Barrère's books never attained the same important in England as Le Cat's, his name and theory (i.e., as champion of the bile theory) carried great weight there. Nothing in France contributed more to the prestige of Barrère's theory than the extensive review and serious treatment he received in the *Journal des Sçavans* (February 1742), 97–107. See, for example, Edward Long's discussion of the scientific origins of Negro skin in his important treatise *A History of Jamaica*, 3 vols. (1774), II, 351–2, which reckons with Barrère but has never heard of Le Cat: 'Anatomists say, that this *reticular membrane*, which is found between the *Epidermis* and the skin, being soaked in water for a long time, does not change its colour. Monsieure Barrère, who appears to have examined this circumstance with particular attention, as well as Mr. Winslow, says, that the *Epidermis* itself is black, and that if it has appeared white to some that have examined it, it is owing to its extreme fineness and transparency; but that it is really as dark as a piece of blackhorn, reduced to the same gracility [*sic*]. That this color of the *Epidermis*, and of the skin, is caused by the bile, which in Negroes is not yellow, but always as black as ink. The bile in white men tinges their yellow skin; and if their bile was black, it would doubtless communicate the same black tint. Mr. Barrère affirms, that the Negroe bile naturally secrets itself upon the *Epidermis* in a quantity sufficient to impregnate it with the dark colour for which it is so remarkable. These observations naturally lead to the further question, "why the bile in Negroes is black?"' The tone and weight of Long's prose in this passage makes it clear that Barrère's authority is beyond question and that he represents the most valid school of thought. Only Buffon receives an equal amount of esteem in Long's chapter.

17. 1, 571. An earlier version of the dictionary appeared in 1755.

18. Published in Paris. I have found no biographical information about Demanet of any note.

19. *Essay*, 53.
20. The author throughout refers to Barrère as an authority. The truly amazing thing, from my vantage at least, is the attention Barrère's treatise received and the almost complete neglect of Le Cat's.
21. A study in depth of Le Cat's scientific writings would, of course, have to explain why Le Cat was able to posit connection. His books on the physiology of the nervous system made him eminently qualified. See, especially, his prize-winning volume, *Dissertation sur l'existence & la nature de fluide des nerfs* (Berlin, 1765), in many ways one of the genuinely radical theories of the age. Also of help were his medical treatises on the anatomy of the passions, such as *Traité des sensations & des passions en général* (Paris, 1767).
22. Much 'history' can of course be accumulated documenting virtually every aspect of the racism debates of the eighteenth century. But every age adheres, whether it knows it or not, to primitive assumptions about what constitutes 'scientific belief'; and we cannot penetrate to the core of racism in the Enlightenment unless we know precisely what it believed about the scientific bases of the skin question.
23. This is an important point. What a vast sense of the monumental changes of belief and emphasis, as well as theories of cause and effect, one derives by approaching the problem vertically rather than horizontally, starting, let us say, with Robert Boyle's analysis in Peter Shaw (ed.), *The Works of the Hon. Robert Boyle* (1772; ed., 1699), 1, 714–19, 'Of Colours: Experiment-xi'. A brief reading list after 1780 might include: E. G. Bosé, *De Mutato per morbum colore corporis humani* (Leipzig, 1785); Robert Know (ethnologist), *The Races of Man* (1850–62) and *Man: His Structure and Physiology* (1857); Franz Pruner-Bey, 'Notions preliminaires sur la coloration de la peau chez l'homme', *Bulletins de la Société d' Anthropologie*, V (1864), 65–135; C. H. G. Pouchet, *Des colorations de l'épiderme* (Paris, 1864); L. Dunbar, *Ueber Pigmenterungen der Haut* (Berlin, 1884); Ashley Montagu, *Man's Most Dangerous Myth: the Fallacy of Race* (New York, 1952, 3rd edn rev.); Richard Bernheimer, *Wild Men in the Middle Ages* (Cambridge, 1952); J. S. Slotkin, 'Eighteenth Century Social Anthropology', in *Readings in Early Anthropology* (1965), 244–356; John S. Haller, *Outcasts from Evolution: Scientific Attitudes of Racial Inferiority, 1859–1900* (Urbana, 1971).
24. Arthur R. Jensen has argued that genetic rather than environmental factors account for differences in IQ for the most part. He further claims that those environmental factors that do operate are likely to be nutritional, dating to the prenatal period: 'How Much Can We Boost IQ and Scholastic Achievement?', *Harvard Educational Review*, XXXIX (winter 1969), 1–123. Those who replied to Jensen in the next issue of the journal seem more bent on airing their own views than in considering his: 'Discussion', *HER*, XXXIX (spring 1969), 273–356.

5
Nerves, Spirits and Fibres: Toward the Origins of Sensibility (1975)

The work of a decade was converging, but it was specifically the juggernaut of Michel Foucault and Thomas Kuhn that inspired this essay. By 1973 I had read and mulled over both philosophers, but only the latter was then being debated in any intensity in America, Foucault not yet having exploded in the American consciousness in the way he would in just a few years after some of his works had been translated into English. I was mesmerized by his Nietzschean ideas, especially those he so adroitly and rhetorically applied to the long eighteenth century, or – what he facilely called – 'l'age classique'. At home, closer to the grindstone, I had been nervously trying to understand where all the sensibility he delineated, especially its deranged forms in madness, led. The word itself – sensibility – should have provided a clue. But I could not believe the path from nerves to sensibility to be direct or linear and worried about the byways. As I implored in note 12, in the context of literary critic Northrop Frye's aim to define this Age of Sensibility, 'what is sensibility and where does it come from?'

Foucault and Kuhn had both loosely suggested, however differently, that socioeconomic developments were in themselves insufficient conditions to bring about revolutions in thought; for this there must also be a concurrent shift in the basic paradigm, or unit, of knowledge (as Kuhn called it) or a break in the episteme *(Foucault's coinage for the frame that holds up abstract thinking in the first place). The internal logic of the Foucaldian epistemes and Kuhnian paradigms held out hope for someone like me in the mid-1970s who sought to understand what the pan-European Sensibility Movement was and why it produced so much distinguished literature and philosophy, especially in Scotland but also in Europe. Therefore, it seemed reasonable to test the hypothesis that the Newtonian revolution applied to the body had been responsible for causing man to think differently about himself. Historically these Newtonian applications to the body had been made foremost by the*

Dutch physician Hermann Boerhaave who, together with his Dutch students, was influential in the Scottish Enlightenment, so many of whose leading physicians he had taught. It was a tall order to bring together all these strands in 1974–75.

The essay also faced the challenge of satisfying diverse literary camps eager to answer Frye's question about the origins of an Age of Sensibility. Chronologically, the origins could have been around 1780 or 1750 or 1730. I located them earlier, as the essay demonstrates, and one of my reasons was the lexical clues found in the 's words': sense, sensitivity, sensible, sensibility and many others whose root was 'sen-' for the senses. The other end of the chronological tunnel c. 1800–1820 also shed light through its revolutionary Romanticism. If American literary critic Harold Bloom was right in a series of books he had then begun to publish about 'the anxiety of influence' (his phrase), then Romanticism had been part of a response to earlier Sensibility: almost as if man's changed sense of himself, and the new verbal expressions describing this shift, had produced a new post-Newtonian, or – more accurately – anti-Newtonian aesthetics of disbelief. So both the origins and their much later outcome was being disputed.

But Sensibility is not a contemporary label: Frye, Foucault, Kuhn and company. It was generated in the seventeenth century, as I show below, by diverse types of thinkers, often clerical but also philosophical and medical, who referred to themselves as creatures of sensibility living in an epoch of nerves. Read no further than Cheyne for this. So it was nonsense to pretend that 'sensibility' was merely a modern label affixed to an old epoch. Besides, Haller himself would alter the course of physiology after Newton's death in 1727 by maintaining that humans were sustained by an 'irritability and sensibility' (his terms) permitting them, almost in the Darwinian sense, to survive and be fit. In this context Haller was just the type of logical paradigm builder Kuhn had described. I asked myself how to situate Haller in the flow from Willis to Cullen, and puzzled over whether his irritability and sensibility represented the beginning or end of something. Where had Haller been educated, what was he reading? What were his intellectual origins, his most fundamental beliefs? All roads led back to Leiden and Boerhaave, and – before Boerhaave – to Thomas Willis, for it was just as necessary to ask what Boerhaave read. Here – in Willis' revolutionary brain theory – I found the momentary oasis of nervous thought uniting the various strands that later coagulated into an ethic of nervous sensibility as it would flourish in the high Enlightenment. Before the 1960s Willis was a name constricted to the annals of neurophysiology. Now, after 1975, he gradually became known to historians and humanists of diverse backgrounds. Today, in the early twenty-first century, no university course on the rise of sensibility can omit his name. Where were the scientific origins of sensibility? The brief answer universally given to students is 'in Willis' brain theory.'

My essay describes Willis' career and influence in some detail, as well as that of his student John Locke, but Willis' ideas could not have been disseminated and made popular without an intermediary. This man was George Cheyne who extrapolated and described Willisian neuroanatomy in popular works like The English Malady *(1733), a book so well tuned to the needs of ordinary readers then that it was practically paradigmatic (in Kuhn's sense). However, Cheyne was barely known in 1970–75: no one was thinking or writing anything about him. You could not even find a modern reprint of his best-selling* The English Malady, *a puzzling state of affairs when one recalls the Cheynean industry that later developed in the 1980s and 1990s.*

Various pathways now began to clear: not merely from Haller to Boerhaave via Leiden and Edinburgh, but from Cheyne to novelist Samuel Richardson, his constant correspondent, crossing the borders from medicine to fiction, realism to fiction, and then from Richardson – loosely and often unpredictably – down through the edifice of the developing novel, and then from the Victorian novel into the modern neuroscience of consciousness. A Kuhnian-style paradigm about 'nervous sensibility' began to suggest itself: that sensibility came into its own at the moment when man became demonstrably 'nervous'. This was important because it augured the birth of sensitive mankind, alert to the environment and emotional landscape in unprecedented ways.

The body, after all, told the truth about sensibility through its neurophysiological mechanisms. Then, the generation of Coleridge and Wordsworth took up these ideas for the philosophical problems embedded in nervous sensibility, and construed their 'nervous words' as understood without the need to unpack them. A generation of sustained Enlightenment mechanistic thought had emptied into an ocean of Romantic rebellion, resisting its basic Newtonian tenets and somewhat auguring the way modern neuroscience has distanced itself from the positivism of late nineteenth-century science. To me the parallels in 1975 seemed extraordinary. For no decent modern neuroscientist would maintain that the neurophysiology of consciousness is sufficient in itself to describe the whole process of consciousness. So too the Wits of the Age of Sensibility, in literature and philosophy, who idiosyncratically tested the heritage of Willis and Locke by altering the types of consciousness represented in their literary works. Alas, poor Yorick still tells the truth! – there is more in the animal spirits than even his readers in 1760 recognized.

This essay was frequently cited during the first five years after its publication. However, it came into its own in the 1990s. One of my students has discovered that since the year 2000 there have been over one hundred references to it.

Nerves, Spirits, and Fibres: Towards Defining the Origins of Sensibility (1975)

We have all heard a great deal in the last decade about Kuhn's 'paradigms'. His definition in *The Structure of Scientific Revolutions* has itself become something of a classic:

> Aristotle's *Physica*, Ptolemy's *Almagest*, Newton's *Principia* and *Opticks*, Franklin's *Electricity*, Lavoisier's *Chemistry*, and Lyell's *Geology* – these and many other works served for a time implicitly to define the legitimate problems and methods of a research field for succeeding generations of practitioners. They were able to do so because they shared two essential characteristics. Their achievement was sufficiently *unprecedented* to attract an enduring group of adherents away from competing modes of scientific activity. Simultaneously, it was sufficiently *open-ended* to leave all sorts of problems for the redefined group of practitioners to resolve.
>
> Achievements that share these two characteristics I shall henceforth refer to as 'paradigms', a term that relates closely to 'normal science'.[1]

During the last decade we have also read and heard that large segments of the scientific community are not happy with Kuhn's definition. They argue that the deflection of human energy by unprecedented, open-ended theories is inadequate to describe the origin of scientific revolutions.[2] Nevertheless, Kuhn's paradigms have considerable worth; if nothing else, Kuhn's definition typifies and describes his own achievement. No other single concept in the last ten years has deflected thinkers so much from their own pursuits, nor is any in the recent history and philosophy of science so open-ended to have caused students of every background to scrutinize it and even to imitate it, as in Michel Foucault's theory of the *episteme* in *Les Mots et les Choses*, first published in 1966.³ Even now and at the risk of belaboring a now well-known theory, it is worth repeating that Kuhn's 'paradigm' refers to *books*, and that his concept of paradigms was formulated by examining the way that science textbooks charted the route to 'normal science'.⁴

What does such a theory do for us students of the eighteenth century? We might extend Kuhn's list by adding many works, for example Locke's *Essay Concerning Human Understanding*, clearly an 'open-ended' work that deflected men through its use as a scientific textbook. But would we add Hume's *Treatise of Human Nature* or Adam Smith's *Theory of Moral Sentiments*? It all depends on the level at which we decode, and on our understanding of Kuhn's definition. Application of the theory, as we can see, has already presented a problem; but in fairness to Kuhn we ought to remember that he reserved the term paradigm for unprecedented works demonstrating open-ended theories and deflection in the highest possible degree. Thus, while it probably can be shown – and I say *probably* because it has not yet been shown – that Locke's *Essay* deflected all sorts of men in addition to ethical philosophers and soon established itself as a scientific textbook leading to understanding of the 'new science,' the science of man, the same – and here I want to be somewhat the loose Humean – perhaps cannot be said of the treatises by Hume and Smith. If nothing else, we can probably prove beyond a shadow of doubt, that these later works were far less tentative than Locke's; furthermore, that in the case of Hume, the point made was too precise to leave ensuing practitioners hovering in open-ended doubt; and moreover, that in Smith's case the contents summarized the theories of others and made them available to everyone in a new form rather than put forward a radically new and open-ended theory itself. A similar case can be made for certain works by Diderot, Rousseau and other *philosophes*, for La Mettrie, Le Cat, Marat. In paradigmatic terms, if I may take the liberty of expanding on Kuhn's original term, the Rousseauistic doctrine *je sens, donc je suis*, is the end rather than the beginning of a revolution.

We are therefore left with Locke, a condition that will surprise or horrify some and that others will call reductionistic or even foolish. But if we accept Kuhn's theory (and despite its difficulties it is still the best available in 1973) and follow it to its logical conclusion, Locke's *Essay* alone among textbooks about the 'new science' of man satisfies Kuhn's two extraordinary conditions. Is this in itself not extraordinary? Not at all extraordinary in the fact that Locke's is a seventeenth-century work (published in 1690), nor in the further fact that no other scientific works than Newton's, Franklin's, and Lavoisier's are mentioned for the

eighteenth century, but rather in the fact that Locke's *Essay* is the first
to deal with a *science* (as all the others do: physics, astronomy, optics,
chemistry, etc.) that had not as yet developed: the science of man. Here
Kuhn's theory about Franklin and electricity is equally instructive. '*Only*
through the work of Franklin and his immediate successors did a theory
[of electricity] arise that could account with something like equal
facility for very nearly *all* these effects and that therefore could and did
provide a subsequent generation of 'electricians' with a common paradigm for its research'.[5] In other words, by the time Locke published his
Essay a *theory* of the new science of man had evolved that was sufficiently unprecedented and open-ended to deflect at least three subsequent
generations of moral scientists: Mandeville, Shaftesbury, Hutcheson,
Hume, Adam Smith, La Mettrie, the *philosophes*, and dozens of others.
Call this science what you will: social science, the science of morals, or,
as Peter Gay has called it, the 'science of man,'[6] and give it any label
you fancy, a crisis, a revolution, an epoch of transition. One thing
however is clear: without Locke and his immediate successors, the theory
could not have developed.

What then precisely was it about Locke's *Essay* that allows for, indeed
insists on, this paradigmatic treatment? Surely it was his application of
crucial aspects of the physical sciences to a realm – ethics and politics –
that was not previously imagined to yield to any scientific types of
explanation. Locke's integration of ethics and physiology has little, if
anything, to do with the fact that he himself was a physician. For every
physician in 1690 who was integrating seemingly non-allied terrains,
there were dozens, perhaps hundreds, who saw no connection at all.
And even among the integrationists, only one thinker's genius inclined
him, for whatever mysterious reason now lost to time, to grasp at the
one realm – ethics – most requiring an unprecedented theory arrived at
by radical integration of disparate areas of study. Physico-theologists
like Ray and Derham almost annually were integrating the physical
sciences into the study of religion. The difference between their endeavors, which certainly led to no revolution in knowledge, and Newton's is
evident: if there is one thing the *Principia* and *Opticks* are *not* it is physico-
theologies. Newton, the man whose open-ended theories deflected men
for over a century by replacing the old textbooks they used to study
with his new ones, also kept science and religion apart in the deepest
assumptions of his books. We all know about his protestations to the
effect that he could not tell anyone '*why* is gravity, only *what* is gravity'.[7] Paradigmatic achievement does not depend upon integration as
an efficient cause. In Locke's case it happens to function as such, but

this is partly owing to the rapid acceleration of scientific research after the Restoration, and partly to Locke's own monumental genius in recognizing that integration beneath the level of overt surface statement of these disparate realms – ethics and physiology – would result in the open-ended effect about which Kuhn speaks. Stated otherwise, Locke intuitively reasoned, as Descartes had not, that the whole argument about knowledge pivots around the concept and definition of 'sensation'.

Now my isolation of Locke's *Essay* as paradigmatic is in itself of no great interest except that we have tended to think of 'sensation', hence the ensuing cults of 'sensibility' and 'sentiment', as mid eighteenth-century phenomena. We speak of a 'sensibility movement' *commencing* with Richardson in the 1740s and persisting until something called 'romanticism' eclipsed it. Here I wish unequivocally to profess my temporary suspension of belief in nominalism; for the moment I am uninterested in semantic labels and tags; it is paradigms and paradigmatic works related to sensibility that I'm after. I trust everyone knows that Northrop Frye has called English literature between Richardson and Wordsworth the product of an 'Age of Sensibility',[8] and those still in doubt may soon be convinced by Robert Brissenden's forthcoming book, *Virtue in Distress: Studies in the Novel of Sentiment from Richardson to Sade.* And yet the half-century gap between 1690 and 1740, between the appearance of Locke's *Essay* and Richardson's *Pamela*, an epoch separating men two generations apart, has eluded us. If we follow through on Kuhn's argument and my subsequent reasoning, should Frye's 'Age of Sensibility' not have occurred fifty years earlier? Have we been dangerously promoting an historical fallacy by alleging that its appearance was a mid- and late eighteenth-century phenomenon?

Not really. For such reasoning erroneously assumes that imaginative literature – and by this I mean poetry, fiction, the drama – is influenced by science *at once*, and we know this is not true of our eighteenth century merely by noticing that it took Newtonian science at least one generation, and longer, to 'demand the muse'. It is no less dangerous at this point and consequential for the future of eighteenth-century studies, for us to confuse *imaginative literature* and *speculative science*.

What I am therefore suggesting is first that the origins of the eighteenth-century revolution in intellectual thinking regarding the 'science of man' owes its superlative debt to John Locke. Secondly – and this is the important point – that sensibility, not merely sentimentalism,[9] is at the very heart, the rock bottom, of this revolution *precisely for the two reasons* given in Kuhn's definition of paradigms and subsequent revolutions.

It was unprecedented and open-ended. But sensibility was not a mid eighteenth-century phenomenon, certainly not in philosophy or the natural sciences, it was a late seventeenth-century development owing its superlative paradigmatic debt to books – and here, again, I adopt Kuhn's language – like Thomas Willis' *Pathology of the Brain*,[10] but also to one unprecedented, integrative work, Locke's *Essay*. Mid eighteenth-century neurological treatises of Haller and Robert Whytt and many others who entered the arena of debate were not the paradigmatic works that led to a revolution in the scientific approach to the study of man, or to the sensibility movement in literature. These were not the works that paved the way for *Clarissa Harlowe*, *A Sentimental Journey*, and *Justine*; nor were the earlier treatises of Dr. George Cheyne, national spokesman for the English malady, with whom Richardson corresponded so prolifically. At the deepest level of decoding, the level at which I believe Kuhn has decoded, the revolution in sentiment occurred in the last quarter of the seventeenth century. It took imaginative writers like Richardson and Sterne a half century to 'catch up', as it were, and more importantly, it also took most *scientific* thinkers like Cheyne, Haller and Whytt almost an equal length of time to understand what had transpired in the interim.

These observations should not surprise us. Almost fifty years ago, R. S. Crane warned that if we wish to understand the origins of sensibility 'we must look to a period considerably earlier than that in which Shaftesbury wrote'.[11] Today I am suggesting that it is dangerous to think that the revolution in sensibility was a mid- or late eighteenth-century phenomenon. To trace its origin to Shaftesbury, or even to Locke, is to indulge in sheer mysticism and to have no sound philosophy of history. I realize that I have been partly responsible for some of the confusion in *This Long Disease My Life*, but I hope I am not so culpable as either Mr. Crane or Mr. Frye, who was satisfied to repeat what every Victorian and Edwardian schoolteacher taught by rote, and to garnish the point with niggardly consolation to the effect that English literature between Gray and the Romantics 'is not *altogether* dull'.[12]

I must now demonstrate that at least in scientific thought the revolution in sensibility, that is in self-consciousness, was not an eighteenth-century phenomenon; in other words show that unless one decodes at Kuhn's level one leaves the largest questions unanswered, and moreover that decoding at the level Kuhn's paradigms imply is a minimum, not a maximum, for students of the eighteenth century. If this is done and favorably viewed, one significant consequence is that all propositions of the form 'the social sciences were born in the eighteenth century' must be thrown out of court on grounds of false aetiology. They flowered then; they certainly were not born then. The social sciences of man, about which mid-eighteenth century Frenchmen had much to say, may not have had an influence on the manifold aspects of daily life until the mideighteenth century, but it is absurd to suggest 'birth' at that time if we accept the rigors of explanation demanded by thinkers like Wittgenstein, Popper, Kuhn, Foucault, Lévi-Strauss, and other recent analytical minds. What then were the 'paradigms' of sensibility, and the consequent revolution created by these paradigms? Crudely speaking, they were sets of physiological texts like Willis' *Anatomy of the Brain* and *Pathology of the Brain*, published after the Restoration that were sufficiently 'open-ended', to deflect all types of scientists, not merely other anatomists and physiologists. This is not to suggest that the physiological dimensions of these texts is the crucial aspect for the paradigm. It is not: physiology textbooks had been written since Egyptian and Greek days and certainly by the second century A. D. when Galen published his paradigmatic physiological work *On the Natural Faculties*.[13] After 1660, however, their numbers increased owing to regental and university support of scientific research. Before naming these texts, it is essential to note that these books were attempts (however cognizant or not their authors) to answer Cartesian science, especially the mind/body problem. That is – and here again I follow certain of Kuhn's philosophical theories – the history of science, like the history of ideas, is best conceived as a continuum in which paradigmatic works periodically deflect 'groups of practitioners'.[14] Until we discover which are the *paradigms* and which the responses, we cannot meaningfully understand revolutions in intellectual thought: the rise of sensibility as a self-consciousness is a perfect example of this process and continuum.

Before the Restoration, Descartes' *Discourses* and treatise on *The Passions of the Soul* were paradigmatic works, especially in anatomy, and deflected all types of natural scientists and directed them to the study of physiology. But for various political reasons, the Interregnum for one, deflection in England was abortive. The next such paradigmatic works were Thomas Willis' texts on the brain published in the 1660s and 1670s, and translated almost immediately in the early 1680s. Although there were other paradigmatic texts before the nineteenth century (e.g. Boerhaave, Whytt, Haller, Cullen), these were not of the same parity as Willis'. Willis' special genius, like Descartes' before him, lay not in the scientific veracity of his theory but in his ability to deflect men. The theory itself was of course unprecedented: he was the first scientist unassailably to posit that the seat of the soul is strictly limited to the *brain*, nowhere else. Fleeting shadows and brief anticipations of this revolutionary theory can be found before 1660, but nothing clear and loud and plain.[15] In the sense of cause and effect, it was this theory of Willis' that inspired a revolution in intellectual thought concerning the nature of man, and that greatly enhanced the doctrines of anti-Stoics and anti-Puritan divines of the Latitudinarian school about which R. S. Crane has written in his genealogy of the man of feeling. Every competent anatomist of the late seventeenth century knew that nerves, morphologically speaking, carry out the tasks set by the brain. But not every physiologist or anatomist knew, or if he did know would have agreed, that the soul is located in the brain. Without this knowledge it is impossible to account for the intense interest after the Restoration (but not before) in nerve research, and consequently the preparation for sensibility as an emotional state of mind.

Here it is delightful and perhaps amusing, but no more, to recount that Willis was Locke's tutor at Oxford, that Locke is known to have sat at Willis' feet and enthusiastically copied into notebooks everything he (Locke) thought he might use later on, and that it is no forcing of the known facts to say that Willis had a profound influence on Locke in his most formative years. It would be nothing less than treacherous, however, to argue that it was Willis' theory of the brain that 'deflected' Locke into writing the *Essay*. In rehearsing the influence of Willis on Locke my intention is not to minimize other factors (e.g. religious, political) in the development of Locke's imagination, but to question whether the deepest substratum of the *Essay* is not more intelligible when viewed in the light of Locke's education.

If we continue this line of inquiry regarding the revolution in anatomy and physiology, it becomes evident why nerves, and their subsidiaries – fibres and animal spirits – could not be accounted as the basis of knowledge, and consequently of human behavior, until the seat of the soul was limited (not moved) to the brain. For this organ alone, the brain, depends upon the nerves for all its functions. Once the soul was limited to the brain, scientists could debate precisely how the nerves carry out its voluntary and involuntary intentions. And the history of science reveals that they did this: no topic in physiology between the Restoration and the turn of the nineteenth century was more important than the precise workings of the nerves, their intricate morphology and histological arrangement, their anatomic function. It is true, this collective scientific endeavor could not have been under-taken without Harvey's previous discovery in the 1620s of circulation of the blood expounded in another paradigmatic work, *de Motu Cordis*.

But neither would it have been possible without Willis' revolutionary theory of the brain.

These admittedly sweeping abstractions about an aspect of the history of science have now been minutely documented by Edwin Clarke, our most distinguished historian of physiology alive. In an important article entitled 'The Doctrine of the Hollow Nerve in the Seventeenth and Eighteenth Centuries', Clarke concludes:[16] 'Despite the welter of speculation and observation concerning the supposed hollow or porous nerve which had accumulated in the seventeenth and eighteenth centuries, little advance beyond Galen's original suppositions had in fact been made'.[17] Why did this welter exist in the first place, and what difference does it make to a history of sensibility? If indeed the soul is limited to the brain, as Willis and his followers in the 1660s contended, then nerves alone can be held responsible for sensory impressions, and consequently for knowledge; and more consequential, the nerves must necessarily be hollow tubes rather than solid fibres so that the brain's unique secretion, animal spirits, can freely flow through them to the body's vital organs. It was essential to the deepest and probably most unconscious assumptions of these anatomists that the old model of nerves as hollow tubes be sustained. But the rapid and marked acceleration of Clarke's 'welter of speculation' *after* Willis' paradigmatic books on the brain, is equally imperative to notice. Once Willis' paradigms are understood, a context for physiology manifests itself, and we can begin to understand why the war between the mechanists and vitalists developed at the end of the seventeenth century.

The *mechanists*, unlike their vitalist or animist opponents, were dualists. Followers of Descartes, they accepted his mechanistic explanation of all bodily functions except that of the soul, which, again like Descartes, they located everywhere in the body but whose activities, they asserted, do not act in any mechanistic fashion. When asked by vitalists how the soul *does* act, mechanists from the time of Descartes and Newton to that of La Mettrie and Haller more than a century later, answered it does not matter because: (1) the soul has little power in and of itself – virtually everything depends on the clockwork movements of the body, a perfectly constructed machine whose basic motions would be enacted whether or not the soul willed them voluntarily. After Willis brilliantly limited the soul to the immediate area of the cerebrum and cerebellum and its surrounding network of nerves, the mechanists avidly set about to prove, although they did not succeed, that all nerves were in fact hollow tubes through which the quasi-magical fluid secreted by the brain flowed. Unless the mechanists could prove that nerves were porous,

cavity-like structures, they would have to surrender their most fundamental assumption about the dualism of body and soul (or mind).

But precisely this attempt to prove that nerves were porous cavities
gave animists like Stahl and his many followers in the eighteenth century
their greatest impetus. Monists of varying degrees, the Stahlians – Stahl,
Whytt, Cullen, to mention just the most celebrated – had never accepted
Descartes' dualism of body and soul, although they had been deflected
by his theories from the very start. Instead they preferred to adhere to
an animate, functioning soul whose mechanical operations throughout
the body were maximized to the greatest possible degree. In other
words, every part of the body is chemically and physically governed by
this soul, which does not function predictably, rationally, or mechanistically, but is influenced by non-mechanical, unconscious phenomena.
It is hard *not* to notice how the whole dispute between Cartesian mechanists and Stahlian animists is radically displaced by Willis' limitation
of the soul to the brain. After 1680 mechanists and animists alike, both
dualists and monists, had no choice but to refute Willis' unprecedented
contention by demonstrating that nerves are solid, or to agree with
him. For if the nerves *are* solid fibres rather than porous hollow tubes,
no argument exists by which to explain the brain's control over the rest
of the body – not at least before the discovery of electricity in the mid
eighteenth century. That is, no means exists otherwise by which to
account for knowledge gained by experience, for non-innate knowledge.
But no one in the eighteenth century not even Boerhaave, could prove
the *solidity* of the nerves; i.e. no one could disprove Willis' theory by
adducing concrete microscopic evidence. The only remaining alternative was to work away at proving the one condition that would in
turn prove Willis' theory, the hollowness of the nerves, and this is
precisely why there was such a 'welter of speculation,' for 150 years.

If we stop at this point, surely we scatter to the wind the most essential
thread and consequence of the argument, the manner in which the idea
that nerves control human consciousness gradually took hold of Europe.
We also lose sight of the fundamental concept of 'sensation' around
which the entire debate had centered from the start.

If Willis had not appeared on the scientific scene with his striking
theory about the autonomous brain, the nerves could never have held
the dominant sway they did. For by the 1660s anatomy was sufficiently
well developed as a subject for serious study, especially with regard to
circulatory, tissue and lymph systems of the body, to insist that the
nerves are the slaves of the brain and conversely, that the brain is
thoroughly enslaved to the nerves and unable to function without

them. This had been unequivocally demonstrated by Vesalius, Van Helmont, and their contemporaries. Physiologists and other scientists in that case would have continued to debate the problem of how to prove the hollowness of nerves, the precise morphology of its fibres (which no one had seen microscopically), and the chemical composition of animal spirits. But organs other than the brain, such as the heart, stomach, bowels, may have commanded superior positions as subjects for investigation by philosophers as well as anatomists.

Willis' paradigmatic leap, if we continue in this line of decoding, was regionalization of the soul to the brain in a series of *experiments and books* possessing just the right balance between observed fact and unprecedented hypothesis to deflect bewildered scientists for over a century, to the time of Boerhaave and Haller, Whytt and Cullen, Brown, Novalis – that is to say to the very end of the eighteenth century.[18] Unless the consequences of this complicated imaginative leap are fully understood, we can never comprehend the origins of ideas resulting in Crane's 'cults of sensibility' so clearly disseminated into the air of mid-eighteenth century culture. A new assumption about the anatomy of man arose through Willis' deflection of scientists, including mechanists, vitalists and animists of every variety and persuasion. The unspoken assumption was hardly a 'paradigm' in Kuhn's sense, but still a radically new assumption arose about man's essentially nervous nature. From pure anatomy, it was just one step to an integrated physiology of man and another to a theory of sensory perception, learning, and the further association of ideas. Locke, in the course of time Willis' best student, took these steps, perhaps not visibly in the written *Essay* but in the stages that may be construed as the preformation of the *Essay*, and the schools of moral thinkers he in turn deflected – Shaftesbury, Hutcheson, Hume, Adam Smith, and many others – carried his brilliant act of integration to its fullest possible conclusion. Collectively they developed a scientific approach to every aspect of the study of man *by means* of a theory of sensory perception and a theory of knowledge that directly followed from their understanding of the physiology of perception. Today, we are still the heirs of that revolution. Witness our specialized scientific approaches to the study of man: psychology, sociology, anthropology, psycho-history, psycho-linguistics, and so forth.[19]

If we understand the revolution set in motion by Willis *and* Locke – the theories of the former without the latter would not have had the terrific effect as quickly as they did – then we can at last come to terms with sensibility. We still require narration of the whole story of this development; for I have outlined the crudest sketch and essential features only. Even so, the outline demonstrates some salient facts about European intellectual history in the seventeenth and eighteenth centuries: first that no adequate theory of perception arose, or could arise, until physiological questions pertaining to anatomy were at least partially solved, not by actually answering the deepest questions – we know they were *not* answered – but by endowing the answers with enough authority to permit men seriously to study them, i.e. the serious study of physiology. Second, that a scientific approach to the study of man, such as the one we see flourishing in the eighteenth-century schools of Dutch anatomy, Scottish morality, English empirical philosophy, and even French ethical thought (especially as it has been so persuasively presented by Lester Crocker in his books on the subject), required as a prerequisite a developed science of physiology. Call this science anatomy or morphology of the nervous system if you will; in either case it was new. Speculation about it had existed for centuries as a marginal aspect of more general science, but at the end of the seventeenth century it came into its own and permitted, as it were, the new science of man to begin to practice. To decode further at this level, the 'revolution' in anatomical thinking was not an eighteenth-century phenomenon but a late seventeenth. Mechanism, animism and vitalism were responses to previous radical ideas and not radical new ideas themselves. All three depended for their life-blood on the institutionalization of physiology as a serious endeavor in itself, and there is good reason that all three philosophical positions were not hotly debated before the end of the seventeenth century. While Willis and his contemporaries[20] can hardly be credited with making the study of physiology respectable by teaching, studying and restudying it, texts like his and those of his student John Locke directly contributed to the 'revolution' we now call the scientific study of man. Whether these men also made it possible in the first place depends on one's theory of cause and effect.

We can now begin to understand all sorts of connections not evident earlier. By comprehending precisely how 'sensation' was at the heart of the revolution, in physiology, we can observe how it was also the parent of a child called the science of man. We can also see why theological systems, even dissenting ones, based on a theory of the soul that was more or less anatomically grounded, were ultimately asked to account for the phenomenon of sensation. But we can do much more. We can now understand realms that hitherto have seemed disparate: the cults of melancholy, hypochondria, the 'English Malady', as Cheyne called it, Richardson's novel of sentiment, later on the well-formed and mature 'man of feeling', Sterne's bizarre variations on this theme, the eighteenth-century insistence, indeed obsession, with the relation of mind (soul) to body, and, still later, Romanticism with a capital R. We can understand why Mrs. Donnellan, no scientist or learned lady, could directly link (in the sense of outright cause and effect) Richardson's wretched health with his atypical sensibility as a writer:

> ... the misfortune is, those who are fit to write delicately, must think so; those who can form a distress must be able to feel it; and as the mind and body are so united as to influence one another, the delicacy is communicated, and one too often finds softness and tenderness of mind in a body equally remarkable for those qualities. Tom Jones could get drunk, and do all sorts of bad things in the height of his joy for his uncle's discovery. I dare say Fielding is a robust, strong man.[21]

This is no 'attempt to console Richardson for his perpetual ill-health', as Ian Watt has suggested;[22] rather than consolation this is clear indication, at the deepest and most unconscious level, of a revolution in thinking that had been set into motion in the late seventeenth century. Mrs. Donnellan's unstated premises had been scientifically worked out by Willis, Locke, and many others. Crudely stated in the form of a syllogism, (A) the soul is limited to the brain, (B) the brain performs the entirety of its work through the nerves, (C) the more 'exquisite' and 'delicate' one's nerves are, morphologically speaking, the greater the ensuing degree of sensibility and imagination, (D) refined people and other persons of fashion are born with more 'exquisite' anatomies, the tone and texture of their nervous systems more 'delicate' than those of the lower classes, (E) the greater one's nervous sensibility the more he is capable of delicate writing. The ordering of the unspoken assumptions here could not be clearer if it tried. They – the assumptions – may conceal a mythology only partly grounded in physiological research; they doubtlessly thrive on an innate and steadfast distinction between persons of different origins and backgrounds. But these assumptions nevertheless formed part of the substratum of thought of an epoch extending over several generations and are not easy to reconstruct at this removal of time. Richardson symbolized to contemporaries like Mrs. Donnellan the man *par excellence*, of exquisite and truly delicate sensibility, and other women as well knew *why* he was able to write so delicately, even if we do not now.

We can now begin to understand that the novel of sentiment, especially as it developed in the 1740s under Richardson's influence, ultimately owed nothing to the notorious neurological debates between Haller and Whytt, or even to Richardson's earlier debt to Dr. George Cheyne with whom he was on intimate terms and from whom he learned so much about his perverse bodily constitution. Nor did it owe much to Hutcheson's *Passions and Affections* (1728) or to his *Ideas of Beauty and Virtue* (1725), or to Shaftesbury's *Characteristics* (1711). The debate between Haller and Whytt did not erupt until 1751,[23] four years after *Clarissa Harlowe* was published. Even if it had broken earlier, before Richardson composed *Clarissa*, and even if it could be proved beyond a shadow of doubt that Richardson were being fed every detail, blow-by-blow, of the Haller–Whytt controversy; even if Richardson himself revealed to us in the preface to *Clarissa* or elsewhere that his knowledge about sensibility and science, sentiment and the heart, derived from his intensive reading about the controversy; that would prove nothing more than token influence.

But we, like Kuhn early in the 1960s, have not been decoding at a level of mere surfaces and linear or necessarily one-to-one direct influences. And it is consequently of no more concern to us whether Richardson, for example, read Haller *and* Whytt, or Haller *or* Whytt, than whether he read Boerhaave and Cheyne or Hutcheson or Shaftesbury before them. His reading is of immense concern to Richardsonians only. What counts to those among us who would understand the deepest levels, the most original ideas, which made the cults of sensibility possible in the first place – whether in the novel or elsewhere in imaginative literature – is the simple fact (and it is so simple that we have never bothered to notice it) that no novel of sensibility could appear until a revolution in knowledge concerning the brain, and consequently its slaves, the nerves, occurred. If Sterne or Smollett or even Jane Austen of *Sense and Sensibility* fame, had chronologically preempted Richardson by writing for the first time about the delights of sensibility, it would make no substantive or even impressionable difference to us. For Mrs. Donnellan has already told us why Richardson could perform so well in this species of writing, and, presumably, she would have found similar explanations for an ailing Sterne or Smollett or Austen, or for that matter a suitable Hogarth, Reynolds, or Blake. Her explanation for exquisite refinement in painting or music would not have substantially differed from the one she gives for Richardson; and her testimony is valid because, like dozens of other similar passages in eighteenth-century letters, it was uttered without any premeditation and in a moment of total sincerity. There is no elaborate reasoning given by her, because her assumptions are those of the age. Nor does it matter in the least whether she was right in any absolute sense: each age is entitled to believe what it wishes, to create the revolutions in knowledge it wishes; and even if we wish it otherwise, generations of men will continue to fabricate their own mythologies despite our later protestations. It is *our* task – and I hope I will be forgiven for such heavy-handed moralizing – neither to falsify the ideas of previous ages, nor to give emphasis or credit where it is not due. But an even greater task for intellectual historians is a steadfast refusal to reduce highly complex contents to embarrassingly simple strains that neither do justice to reality nor ask or answer the 'big' questions. The fact, for example, that it took almost thirty or forty years for English writers to grasp the full extent of the brain–nerve revolution is of no more interest, except to literary specialists, than what Richardson read. Recent writers in this century have taken that long, if not longer, to understand at the level of unspoken assumption paradigmatic works by Darwin, Einstein, Freud, Heissenberg, and others. The exact chronological distance between writers of sensibility (Richardson, Sterne, MacKenzie, Sade) and the revolutionaries themselves – in this case Willis and his Oxford and London colleagues – must remain an academic sport to engage the attention of certain specialists of the interrelations of science and literature. So, also, must the manner in which the brain–nerve revolution influenced the sum of medical research from 1680 onwards concern medical historians primarily.

We can also understand why it is virtually unnecessary for us to demonstrate the influence of particular thinkers on these writers of sensibility when we decode at this substratum of thought. It would almost be improper. Given that a physiological theory of perception was a necessary condition to explain feelings of every sort, especially the diversity of simple and complex passions, it is of little interest to us, and certainly no cause for celebration, if we discover an identical passage or perfectly clear analogue in a scientific work known to have influenced the writer in question. We must consider as arbitrary the scientific author who wrote the following: 'Feeling *is nothing but* the Impulse, Motion or Action of Bodies, gently or violently impressing the Extremities or Sides of the Nerves, of the Skin, or other parts of the Body, which ... convey Motion to the Sentient Principle in the Brain'.[24] It is immaterial to us, and moreover to those who would understand the true origins of sensibility, if the author of this passage is – moving backwards – Haller, Whytt, Hartley, La Mettrie, Hume, Cheyne, William Hunter, Nicholas Robinson, Ephraim Chambers, Boerhaave, Hutcheson, Shaftesbury, or any one of a dozen others. It happens to be Cheyne, but any of these men would have written it. Only if it appeared in a work written before the paradigmatic books of Willis, Locke and their colleagues would we be concerned.[25] My contention all along is that it could *not* have appeared earlier; that it was impossible before the revolution in brain theory to expect all feeling to be nothing but motion in the nerves. Even more important, we can now begin to understand why all diseases, not merely those hysterical and hypochondriacal, were eventually considered 'nervous' and after a reasonable amount of time were internalized by persons of fashion as visible emblems of refinement and delicacy – thereby as tangible proof of distinct upper-crust difference from the lower and middle classes. It slowly but surely becomes clear that Richardson, Sterne, Diderot, Rousseau, MacKenzie, and even Sade were the posterity of two generations of thinkers who had increasingly 'internalized' – that is the important word – the new science of man, leading thought about him from his eyes and his face to his nerves and brain, from what he looks like to what he feels, and from what he feels to what he knows. Internalization means that man is no longer satisfied to understand himself as a doer of deeds and a thinker of thoughts. He – man – wants to know precisely how his feelings have shaped his knowledge; for the first time he is unable to keep them separate. Richardson penetrates his own fictive creation Clarissa, as does Sade his Justine, by turning inwards and internalizing the relation between her anatomy, feelings, actions, and finally knowledge. And internalization is impossible without an analogue, whether stated or implied, of body and mind. But we would falsify history if we continued to believe that this analogue owes its birth to the mid-eighteenth century; in fact it was already fairly mature in Shaftesbury's formative years in the first decade of the eighteenth century.[26] Smollett's analysis of his last and greatest hero, Matt Bramble, couched in words well meditated only at the level of unconscious assumption, could be the epigraph of all writers from Richardson to the Marquis de Sade: 'I think his peevishness arises partly from bodily pain, and partly from a natural excess of mental sensibility'.[27] And Goldsmith's account of Sir William Thornhill in *The Vicar of Wakefield*, perhaps his most genuinely benevolent character, loses no opportunity to ground itself in a body–mind analogy that was old by 1766: 'Physicians tell us of a disorder in which the whole body is so exquisitely sensible that the slightest touch gives pain; what some have thus suffered in their persons this gentleman found in his mind. The slightest distress, whether real or fictitious, touched him to the quick, and his soul laboured under a sickly sensibility of the miseries of others'.[28] Body–mind analogies could not have become conventions in sentimental literature without an antecedent theory of nervous diseases, and this theory owes little to Burton, Bacon, and Descartes, and other anatomists of melancholy and a great deal to Willis, Boerhaave and Cheyne. It is as if the infinite universe, upon which Pope and Addison had dwelled at such length, had to close up again, this time involuting itself on man's inner nervous universe.

But we can also begin to understand why that most puzzling of modern enigmas, Romanticism, was in turn the heir to a heritage of the cults of sensibility, thereby going beyond the best all–encompassing definition we thus far have, that of Harold Bloom. For if one accepts his persuasive idea that it is only proper to speak of Romanticism in literature at the moment when conventional motifs of 'the quest' are internalized,[29] then we can start to see why the intricate process of internalization itself required a specific neurological legacy. It is not true that Romanticism, understood in this way, could have occurred at any time. First a revolution in knowledge about nervous man, set in motion by certain paradigmatic works, had to occur and then the cults of sensibility, religious, social, moral, literary, merely fashionable, had to play themselves out. While they did, theories about man became increasingly internalized and it was no longer important to pretend, as Swift had, that man could merely be a thinking creature. Imaginative writers could now return to 'the quest', centuries old, and internalize it as readily, as naturally, as scientific thinkers of every persuasion had been internalizing philosophical theories about man.

Postscript of 1975

I wrote this essay about five years ago, in 1970, after its ideas had rumbled about in my mind for about an equal period of time. As I reread it now and imagine myself deconstructing it, in J. Derrida's sense, I am most pleased by, although most troubled by, its version of the integration of philosophy, the history of science and medicine, metaphysics, and literature.

Itself it is not one but all of these; and it consciously strives to account for a new way of thinking about 'nervous man' at the end of the seventeenth century by looking at the most disparate intellectual territories. The literary-critic reader will probably not recognize Willis' name even though he was the most important brain theorist between the Renaissance and modern times. The reader who is an historian of science or philosophy, moreover, will not be so familiar as he may wish with the imaginative writings of Richardson, Swift and their contemporaries, especially if he plans to take the trouble to decipher the essay proper. Yet a keen vision based on some knowledge of all these realms is necessary if one is to trace the birth, in Kuhn's language, of a new paradigm, the genuine origins and effect of Locke's *Essay Concerning Human Understanding*, and the ways in which it deflected all types of students, scientific and humanistic, to drop their own work and assimilate Locke's.

The imaginary argument that learning then, ca. 1680, was not so compartmentalized into 'disciplines' as it is now will ultimately not stand up as a defence of or an attack on integration carried to my extreme; nor will the absurd notion that interdisciplinary investigation more closely approximates an absolute truth. For sometimes the 'truth' is simple and requires no complex apparatus to explain it. My brief survey of the new consciousness about nervous man as a creature of nerves, spirits and fibres bears these caveats in mind and shows again how forceful science is in the modern world, at least since 1600. For if there is any axiom that emerges from the foregoing material, it is the apparent fact that Locke could not have conceived of or written the *Essay* without a scientific model – Willis' new theory about the brain – incorporating a new theory of perception. On this argument everything else stands or falls. To disprove it is at least to carve out the heart of nervous man, and perhaps even to reduce the power of science. But I am not aware that it has been disproved, and I have yet to discover an adequate study of the *Essay* or of the origins of Locke's theory of perception. Had it been Locke's *Essay* and its ideas that permitted Willis the genius to leap to his new theory of the brain, then everything would be different. Nervous man may still have emerged but the rest of the intellectual universe of the eighteenth and nineteenth centuries, and possibly even later, would have been turned upside down, inside out, as a consequence of such bizarre influence.

Alternatively, if Locke could have 'imagined' Willis' brain theory, and in turn could have 'imagined' a new theory of perception allowing him to formulate his notion of primary versus secondary association, then a readily available scientific model would not have been necessary for Locke's paradigmatic book. But Locke did not 'imagine' Willis' theory; and no amount of poetic philosophizing can banish Willis as a central intellectual force in Locke's development before 1690. Even though Locke himself was trained in medicine and for a while practiced as a physician, I do not believe that he was capable of 'imagining' Willis' brain theory, any more than I believe that Descartes would have been capable of 'imagining' Harvey's circulation of the blood, or Milton Galileo's telescope if he had not actually read or heard about it, or as we know, gone to Arceti in 1638 to see it for himself. Peter Medawar has given us a clue in *The Art of the Soluble*: scientists and philosophers and poets all imagine vividly, depending on their temperaments and talents, but philosophers and poets rarely conjure up or stumble upon great scientific theories *that later hold up in professional laboratories under the duress of the need for proof.* I will not generalize for every case; but so far as the origins of nervous man are concerned one can rest assured that science (Willis) stimulated philosophy (Locke), not the other way round, to evolve a new configuration that would eventually create a revolution in literature (Richardson). I cannot go on in this essay to show how the new theory of nervous man gathered barnacles – how it eventually assumed mythic proportions by the time of Byron and Valéry, and how it has dominated other notions about man in the twentieth century. Here I show my own bias when I suggest that probably seven or eight out of every ten new theories develop in precisely this way, i.e., by science changing philosophy and literature, not vice-versa. I am perfectly well aware that the reverse – literature causing a radical change in science – has occurred from time to time in European civilization but surely not so often as the norm, its opposite.

Such cocksure pronouncement, especially proffered as the sturdy foundation of a methodology in the history of ideas, cannot fall on welcoming ears among literary critics, the same contemporary critics who have watched their predecessors – Sidney, Dryden, Pope, Johnson, Wordsworth, Arnold, T. S. Eliot – historically profess the superiority of poets to scientists, superior in the sense of gazing at the 'truth,' the universals, and superior by virtue of forming active, virtuous minds within a 'culture.' One especially thinks of the celebrated paragraph about the 'Man of Science' and the 'Man of Poetry' in Wordsworth's Preface to his and Coleridge's *Lyrical Ballads*. But Wordsworth, like so many other critics, was explicitly thinking about the poet versus the scientist here, not the scientist versus the *philosopher* as I have been in this essay. Literature (Richardson, the writers of sensibility) in almost every case shows the effects last of the three, almost as a series of ripples after the great wave. Even certain literary critics like Matthew Arnold suspected what a powerful foe, or at least force, science would be in the future. Rarely, though, did they speculate about its ability to change human consciousness of itself by influencing philosophy and psychology – sciences manqué. Arnold and others like himself always thought literature 'grasped' something. Science created the 'something,' crude as this may seem.

All this, it will be argued, will hardly do as a proper deconstruction. It has not told us precisely how one goes 'from Willis to Locke,' reading closely into both their relevant sets of texts and showing how the one implies the other. It has not rigorously proved its case, nor provided nearly enough facts as evidence. Nor has it defended, more particularly, the argument it so boldly announced as the heart of its thesis in note 11:

> Benevolence and related ideas of 'doing good' almost certainly *could* have developed before 1750, or 1660, or (for that matter) 1640, for they are everywhere present in the Bible, and in medieval and Renaissance Christian ethical teaching. But a theory to explain the self-conscious personality could *not have derived from earlier times* (earlier, that is, than the Restoration) *because there was no scientific model for it.*

A proper commentary will demonstrate, to be sure, that abstract concepts requiring a certain type of context – e.g., the new broadened concept of *energy* after Einstein and other relativists transformed it[30] – cannot develop *before* a scientific model paves the way. But any adequate deconstruction will also consider every other possibility: will consider the nature of those abstract ideas that developed without scientific models, will ask why certain ideas ('nervous man') require *a priori* scientific models and others (love, friendship, freedom – less scientific concepts though nonetheless abstract) do not, and will of course search for evidences of 'nervous man' before the Restoration.

All this is more than can be done in the space permitted me here, but it must be done in the name of methodological rigor and hopefully as a proof of some type validity. Without it 'sensibility' will never be understood as having anything other than a philological or metaphorical history. Certainly it will not be clear, or even wildly suspected, why it developed at the precise moment and in the very particular way it did.

Notes

1. Thomas S. Kuhn, *The Structure of Scientific Revolutions* (Chicago, 1962), p. 10.
2. The literature of Kuhnian criticism is enormous and cannot be reduced to a few bibliographical references. Perhaps the single best criticism is one not directly attacking Kuhn but substituting for his 'paradigm' a different but not unrelated theory of the '*episteme*'; see Michel Foucault, *Les Mots et les Choses* (Paris, 1966); English version, *The Order of Things: An Archaeology of the Human Sciences* (New York, 1970). Although I vigorously disagree with Foucault about the simple and crude facts of European scientific history 1600–1800 I have

been enormously influenced by his way of doing intellectual history, i.e. decoding beneath visible surfaces, as I have by Robert K. Merton's theory 'that in each age there is a system of science which rests upon a set of assumptions, usually implicit and seldom questioned by the scientists of the time', in *The Sociology of Science*, ed. B. Barber and W. Hirsch (New York, 1962), 41.

3. I have attempted to show some of the differences between Kuhn and Foucault in 'Whose Enlightenment? Not Man's: The Case of Michel Foucault', *Eighteenth-Century Studies*, VI (1973), 238–56. No evidence I know suggests that Foucault has actually read or heard of Professor Kuhn's 'paradigms'.

4. Kuhn, *Scientific Revolutions*, 10: 'In this essay, 'normal science' means research firmly based upon one or more past scientific achievements, achievements that some particular scientific community acknowledges for a time as supplying the foundation for its further practice.'

5. Kuhn's exact sentence (*Scientific Revolutions*, 15) is this: 'Only through the work of Franklin and his immediate successors did a theory arise that could account with something like equal facility for very nearly all these effects and that therefore could and did provide a subsequent generation of electricians with a common paradigm for its research.'

6. Chap. iv of Peter Gay's *The Enlightenment: An Interpretation, Vol. II: The Science of Freedom* (New York, 1969), 167–215.

7. See Alexander Koyré, *Newtonian Studies* (Cambridge, MA, 1965), 63–7, and *From the Closed World to the Infinite Universe* (Baltimore, 1957), 131–4, for analysis and discussion of Newton's reasons.

8. Northrop Frye, 'Towards Defining an Age of Sensibility', *English Literary History*, XXIII (June 1956), 144–52.

9. I do not consider the two identical, although they are obviously related in dozens of aspects. Historically and generally speaking *sensibility* was the larger of the two, touching almost every aspect of life; *sentimentalism* came later, especially in imaginative literature like fiction and poetry, was the more religious, moral, literary, and far less aristocratic of the two, and was also the one that lent itself more readily to radical modifications and variations from an already blurred original base. In every case the distinction is gray, never black or white. Some excellent philological explorations into these labels have already been undertaken: see E. Erametsa, *A Study of the Word 'Sentimental'* (Helsinki, 1951); R. F. Brissenden, '"Sentiment": Some Uses of the Word in the Writings of David Hume', in *Studies in the Eighteenth Century: Papers Presented at the David Nicol Smith Memorial Seminar Canberra 1966* (ed. R. F. Brissenden, Canberra, 1968), 89–106.

10. *Pathologiae Cerebri* (1667), trans. S. Pordage as *An Essay of the Pathology of the Brain* (1681); Willis' two other most important works are *Cerebri Anatome* (1664) and *De Anima Brutorum* (1672), the last also trans. by Pordage in 1683 as *Two Discourses Concerning the Soul of Brutes Which is That of the Vital and Sensitive Soul of Man*.

11. R. S. Crane, 'Suggestions Toward a Genealogy of the "Man of Feeling"', in *The Idea of the Humanities* (2 vols., Chicago, 1967), I 188–213, originally published in *ELH*, I (1934), 205–30. Crane's exact words are (I 190): 'If we wish to

understand the origins and the widespread diffusion in the eighteenth century of the ideas which issued in the cults of sensibility, we must look, I believe, to a period considerably earlier than that in which Shaftesbury wrote and take into account the propaganda of a group of persons whose opportunities for moulding the thoughts of ordinary Englishmen were much greater than those of even the most aristocratic of deists'. Crane's intuition about chronology 'earlier than . . . wrote' is sound, but his reasons are altogether unacceptable. He maintains that sensibility was 'not a philosophy which the eighteenth century *could have* derived full fledged from ancient or Renaissance tradition. It was something *new* in the world – a doctrine, or rather a complex of doctrines, which a hundred years before 1750 would have been frowned upon, had it ever been presented to them, by representatives of every school of ethical or religious thought' (I 189–90, italics mine). Benevolence, and related ideas of doing good almost certainly could have developed before 1750, or 1660, or even 1600, for that matter: they could have derived from ideas present in the Bible and medieval scripture. What could *not* have developed, because there was no scientific explanation for it, was a theory to explain away the self-conscious personality. *Sensibility* used more *narrowly*, as a term to connote self-consciousness and self-awareness, has a different history from Crane's term, and although it is not altogether unrelated to his usage, it is the sense in which many 18th century writers employed it, as Brissenden and Erametsa have both shown in their philological treatises.

12. Mr. Frye wrote in 'Towards Defining An Age of Sensibility': 'I do not care about terminology, only about appreciation for an extraordinarily interesting period of English literature, and the first stage in renewing that appreciation seems to me the gaining of a clear sense of what it is in itself', and Frye's concluding statement: 'Contemporary poetry is still deeply concerned with the problems and techniques of the age of sensibility, and while the latter's resemblance to our time is not a merit in it, it is a logical enough reason for reexamining it with fresh eyes'. But what is sensibility and where does it come from? Frye gives *no* definition of the term at all. In fact he uses it merely to denote a chronological time period, roughly 1740 to 1800, or the death of Pope to the *Preface* to the *Lyrical Ballads*.

13. See the Loeb edition translated and edited by A. J. Brock (1916). Galen's enormous influence on the history of medicine is a subject in itself, especially the manner by which some (but not all) of his physiological doctrines remained virtually intact during the seventeenth and eighteenth centuries. Of unusual interest to my thesis are the following: R. B. Onians, *The Origins of European Thought About The Body* (Cambridge, 1954, 2nd edn); E. Voeglin, *Anamnesis* (Munich, 1966); K. E. Rothschuh, *Physiologie: Der Wandel ihrer Konzepte* (Munich, 1968); Peter H. Niebyl, 'Galen, Van Helmont, and Blood Letting', in *Science, Medicine and Society* (ed. Allen G. Debus, 2 vols, New York, 1972), II 13–23; *The History and Philosophy of Knowledge of the Brain and Its Functions: An Anglo-American Symposium* (Oxford, 1958; rev. ed. 1972); F. Solmsen's study of physiological theories prevalent in the time of Plato makes it evident at least by implication that Galenic concepts would have been sufficiently 'open-ended' to create interest; but whether they deflected enough men to be 'paradigmatic' in Kuhn's sense I cannot say. See F. Solmsen, 'Tissues and the Soul', *Philosophical Review*, LIX (1950), 435–68.

14. Kuhn's phrase for the scientific community that becomes deflected after a paradigmatic work, p. 10.
15. See *The History and Philosophy of Knowledge of the Brain and Its Functions: An Anglo-American Symposium* (Oxford, 1958; rev. edn 1972), especially three papers in the Third Session: Walter Pagel, 'Medieval and Renaissance Contributions to Knowledge of the Brain and Its Functions', 95–114; Walther Riese, 'Descartes's Ideas of Brain Function', 115–34; W.P.D. Wightman, 'Wars of Ideas in Neurological Science – from Willis to Bichat and from Locke to Condillac', 135–48.
16. In *Medicine, Science, and Culture* (ed. L. G. Stevenson and R. P. Multhauf, Baltimore, 1968), 135.
17. Clarke, *Medicine, Science, and Culture*, 135, rightly notes two exceptions: 'But there were two investigations in the eighteenth century, the results of which were readily available to all, which pointed to the future that nerves are solid. In 1717 Leeuwenhoek saw and illustrated the single myelinated nerve fiber, the center of which (the axis cylinder or axon) he took to be hollow... Of greater significance, however, was the second discovery, made by Fontana in 1779... Again, this was the myelinated axon of today, but Fontana's work seems to have had little immediate effect, probably because of the suspicion engendered by most eighteenth-century microscopic investigations'. My own researches on nerves and animal spirits corroborate Clarke's findings: I have found no evidence that the discoveries of Leeuwenhoek and Fontana were acknowledged, understood, or digested. This development is not surprising in view of the fact that Leeuwenhoek himself never realized what he had observed.
18. Clarke, *Medicine, Science, and Culture*, 123–41, has performed the research and settled the matter once and for all. His statement (p. 124) about scientific models in physiology is revealing and germane to the rise of 'sensibility' as a serious subject for scientific concern: 'In general, the customary sequence of events during the accumulation of knowledge regarding a part of the animal or human body is that its morphology is established first of all; *thereafter its physiology can be investigated*. This has been true with structures like the heart, but in the case of nerves the advancement has been more complicated because of the greater complexity of nervous tissue and organs. Here, during the seventeenth and eighteenth centuries, speculation predominated in respect to both form and function. It is probably true that the ancients, having accepted the suggestion that the nerve acted by means of a substance passing through it, *also had to* postulate a hollowness or porosity so that this *would be possible*. Structure was therefore *determined* by the demands of function' (italics mine). But Willis' paradigmatic works created a revolution in science in that he made it *possible* – in Clarke's sense – to explore the physiology of nerves in the first place. Until the seat of voluntary and involuntary motion was limited to the cerebrum and cerebellum and their network of surrounding nerves, speculation about Clarke's 'form' (i.e. morphology) was necessarily erratic and uncontrollable.
19. Especially in *Pathologiae Cerebri* (London, 1667) and *De Anima Brutorum* (Oxford, 1672). Almost every modern historian of physiology has spoken about Willis with wonder and awe, e.g. Sir Michael Foster, *Lectures on the History of Physiology During the Sixteenth, Seventeenth and Eighteenth Centuries*

(Cambridge, 1901; rep. with an intro. by C. D. O'Malley, New York, 1970), 269: 'Though Malpighi as we have seen devoted much attention to the histology of the nervous system, we find in his writings very little concerning its functions...One man alone perhaps during this century stands out prominently for his labours on the structure and functions of the brain, namely Thomas Willis'. One of Willis' most thorough biographers, Dr. Hansruedi Isler (*Thomas Willis*, 1965; trans. 1968, x–xi), maintains that 'Willis' achievements in neurophysiology comprise the first useful theory of brain localization of psychic and vegetative functions as well as the first interpretation of nerve action as an energetic process. His new concept of nerve action led him to the idea – and the term – of reflex action, whereas his localization theory gave rise to the development of experimental physiology of the central nervous system. In order to complete his account of the nervous system Willis described the bulk of the nervous and psychic diseases: the three books he published from 1667 to 1672 contain *the most complete text of neuropsychiatry since Greek antiquity*. Most later interpretations have been influenced by his ideas, either directly or indirectly' (italics mine). John R. Fulton, surely the most distinguished twentieth-century historian of neurophysiology, considers the cornerstones of modern neurology to be based on six books by 'Willis, Whytt, Magendie, Hitzig, Ferrier, and Sherrington' (*Physiology of the Nervous System*, New York, 1949, 177). Fulton, while recognizing some of the important discoveries of Robert Whytt, considers him relatively unimportant in the line of revolutionary theories about brain localization like those of Willis: 'In his memorable *Cerebri anatome*, published in 1664, Thomas Willis, the Oxford neurologist, suggested that the cerebrum presided over voluntary motions and that the cerebellum governed involuntary movements. Willis had noted that when the cerebellum was manipulated in a living animal the heart stopped, and that if the cerebellum was removed the animal died. Suggestive was the idea that the cerebellum facilitated involuntary action... There was little further advance until 1809...' (*Physiology of the Nervous System*, 506). Kenneth Dewhurst, another biographer of Willis', has also stressed Willis' revolutionary role in the development of modern science: see *Thomas Willis as a Physician* (University of California: Los Angeles, 1964). In two other important works, he demonstrates Willis' profound influence on Locke: *John Locke Physician and Philosopher: A Medical Biography* (London: The Wellcome Institute for the History of Medicine, 1963) and 'An Oxford Medical Quartet – Sydenham, Willis, Locke, and Lower', *British Medical Journal*, II (1963), 857–80. R. K. French (*Robert Whytt, The Soul, and Medicine*, London, 1969, 134) is right to note that 'Towards the end of the century opinion inclined away from placing mental functions in structures within the brain, and many, agreeing with Steno that Willis had been too speculative, favoured a more general placing of the soul in the substance of the Brain'. Steno's *Dissertation on the Anatomy of the Brain* (1669) and Ridley's *The Anatomy of the Brain* (1695) were among these works. But *all* these books were answers to Willis and merely attest to his 'paradigmatic' ability to deflect, as Kuhn says, 'succeeding generations of practitioners'.

20. While there is a great deal of evidence pointing to Willis' influence on Locke in the *Essay* (see, e.g. Isler, *Thomas Willis*, 176–81, in which Isler traces the

source of many Lockean passages to Willis), there is less known about his influence on later physiologists. To rehearse that Whytt and Haller were perfectly well aware of his theories about voluntary and involuntary motions, and all the replies, rebuttals and disagreements regarding his all-important 'intercostal' nerve, is to indulge in simplicity about the history of science: one might as well set out to prove that Pope had heard of Milton (see R. F. French, *Robert Whytt*, 32ff). The response of vitalists and animists equally demonstrates clear knowledge (even if it is not always unequivocally stated) of every aspect of Willis' brain theory. See H. Driesch, 'G. E. Stahl', in *The History and Theory of Vitalism* (trans. C. K. Ogden, London, 1914), 30–6; G. Canguilhem, *La Formation du Concept de Reflexe* (Paris, 1955); L. J. Rather, 'Stahl's Psychological Physiology', *Bulletin of the History of Medicine*, XXXV (1961), 37–49.

21. *The Correspondence of Samuel Richardson* (ed. Anna L. Barbauld, 4 vols., London, 1804), IV 30.

22. Ian Watt, *The Rise of The Novel* (Berkeley, 1957), 184.

23. For the most exhaustive survey of the controversy, see R. K. French, 'The Controversy with Haller: Sense and Sensibility', in *Robert Whytt*, 63–76.

24. The passage is found in George Cheyne, M. D., *The English Malady* (2 vols.,London, 1733), I 71. No one was more important to nerves, spirits and fibres than Cheyne.

25. One can test the hypothesis by consulting scientific works written in the 1650s, especially by Hobbes and some of the early Cambridge Platonists; nowhere is the brain invoked in this manner before 1664, the date of publication of Willis' *Cerebri Anatome*.

26. A list like this perhaps raises questions about Defoe and Fielding: 'were they not exposed to the same ideas as Richardson?' 'Were their nerves any less sensible?' Yes, Fielding received similar exposure – everyone did – but his indigenous physiology, the era of Mrs. Donnellan would have argued, was much less exquisite than Richardson's. The truth is of course more elaborate than this but, however crude, her answer is an approximation, and we see it splendidly mirrored in Johnson's estimate of Fielding as a 'barren rascal'.

27. Tobias Smollett, *Humphry Clinker* (London, 1771), J. Melford to Sir Watkin Phillips, 18 April.

28. Goldsmith, *The Vicar of Wakefield* (ed. Arthur Friedman, 5 vols, Oxford, 1966), IV 29.

29. Such is Harold Bloom's theory in *The Visionary Company: A Reading of English Romantic Poetry* (New York, 1961). Bloom, taking his cue from Northrop Frye, titles his first chapter 'The Heritage of Sensibility'.

30. Energy as an idea with a history and life of its own needs to be viewed in the curve of time from Burke and Godwin to Freud and Joyce. Burke, subscribing to a traditional view of energy current from the time of the Greeks, felt threatened by it when present without *property*; and as a consequence suggested in the *Reflections on the French Revolution* that he could be more comfortable when repressing it. Godwin, whose politics and private religion were as different from Burke's as can be imagined, also subscribed to a traditional view of energy as an absolute force, but unlike Burke considered it 'the most useful quality of all' (William Godwin, *Caleb Williams*, Everyman Library Edition, p. 254). So traditional a view, though disparate and often

imaginative, persisted until the early twentieth century when Einstein's theory of relativity destroyed the older, and physically inadequate, notion of energy. Whether the new *scientific model* of energy (a paradigm in Kuhn's sense) has in turn precipitated a new consciousness parallel to that related to 'nervous man' I am not able to say, for I have not studied the versions of 'energy' in this context.

6
Nerves and Nymphomania: Bienville and Female Sensibility (1982)

The vast Enlightenment discourses of nerves, spirits and fibers feeding into the cults of sensibility had been plural rather than single. That is, they had been polyphonic rather than monophonic, in our postdisciplinary sense for the diversity of their narrators' voices; and they touched on all aspects of human life, not merely expressive and representational gestures; and they also shaped the concepts of belief that infiltrated manners and morals. My cluster of topoi in the 1960s and 1970s – nerves, spirits and fibers – had also been embedded in narrative discourse and ordinary speech-acts. But it would have been unthinkable that mood and emotion, even consciousness itself, could have been framed along these lines two centuries earlier, in the 1550s rather than the 1750s. By the eighteenth century the 'I' was fully capable of describing itself as a 'nervous subject' and expounding precisely what that entailed. The more that animal spirits became taken for granted as a type of Enlightenment DNA, and lorded over by commanders in the nervous system, the more they themselves – these anatomical captains – became hard-wired into cultural discourse. By the 1780s 'delicate nerves' became the fully-formed emblems of rank and class. They touched all aspects of life: for example, medical education and sectarian religion, where sermonists routinely referred to them, as well as the developing science of sex that would be so important in the century leading to Freud and psychoanalysis.

In this essay of 1982 the paradigm was sex as sensation. The logical sequence was nerves, responding to stimulus and forming sensation, which in turn provided pleasure (orgasm) and pain (as far as the weirdest Sadean types). But what was the difference between 'impression' and 'sensation'? What, indeed, were the 'impressions' one had in imagining, and then in the experience itself, of sexual activity? Imagination played a crucial role at the start of the chain process at the precise moment where its images in the brain impinged upon the animal spirits flowing through the nerves. This act was

185

anatomically determined and enabled by mechanistic, physiological processes. I also searched for paradigmatic and theoretical physicians other than Le Cat who would expound the interstices of these chains of thought. The classic 'philosophical physicians' – Boerhaaye, Stahl, Hoffman – had innovated in logic and application, but not at the margins of knowledge. Stahl's animism was crucial for nerves, but he eschewed sustained discussions of sex, perhaps the result of his Germanic pietism. Likewise Hoffmann, if for different reasons. More contemporarily, Foucault had sensitized his readers in the late 1970s to labels and to the loaded activity of naming itself: where the label, or name, takes on the function of the thing itself. He had implied that a homology is established before it can interrogated. All science proceeds, to a certain degree, by acts of naming and accumulations of glossaries: this double-act constitutes the basic fabric of the process of collecting and classifying. But whereas Foucault was obsessed by the predominantly male *categories of sodomy and homosexuality he had pronounced little about their female counterparts. He discusses Bienville in the* History of Madness, *but Bienville's theory of insanity interested me less than his more basic investment in nervous sensibility, especially its various coinages to delimit the grades of nervous affliction: hypomania, erotomania, metromania, and, now – according to Bienville – the neologism of nymphomania. For Bienville invented the word, if not the concept, depending upon one's beliefs about the formation of words and things.*

Circa 1980 there were no histories of nymphomania – it may now seem hard to believe but histories of sexuality themselves barely existed then. So rapid has been the dissemination of new discourses with the assistance of technology. Sexual minority discourse in America, framed ideologically, was then fundamentally explosive, morally charged and even incendiary, often a weapon to brand, castigate and exclude. One entered its labyrinth at peril because it was also customarily seen as politically incorrect (what were the scholar's ulterior motives?); specifically interpreted in the case of 'nymphomania' as insulting to women and quasi-pornographic when a male discussed female genitalia. If Bienville's contemporaries were suspicious about his own motives, as I show, mine, two centuries later, were not much less. I could have written about nymphomania narrowly within the history of medicine, narrating as if Bienville were merely another eighteenth-century French doctor, but this approach would have overlooked his radical sexual applications of nervous sensibility. It was therefore predictable that the essay would arouse suspicion at first. It generated more interest after Foucault's writings had an opportunity to make the history of sexuality in America respectable by the mid-1980s. Soon historians like Jeffrey Merrick were compiling works about the sexual underworlds of Bienville's France, and scholar-physicians such as Vernon Rosario books about a European 'science of homosexuality' *(1997) and*

The Erotic Imagination: French Histories of Perversity *(1997), in which Bienville figured as a paradigmatic figure. By the millennium he had come into his own, except that almost nothing is known about the life of the man himself.*

The Invention of Nymphomania (1982)

Michelet tells us that God changed his sex in the middle of the thirteenth century. If this is true, it is nevertheless inconceivable that Chaucer's grandparents could have told us why. Similarly although less obviously, if prurience is common to all age, as Rémy de Gourmont contends in *The Natural Philosophy of Love* (1922; translated with a postscript by Ezra Pound), erotic sensibility is not. For erotic sensuality is a removed zone from erotic sensibility: some Greek friezes display enchanting bacchanalian scenes (Keats's fair 'Attic shape'); Boccaccio's nymphs tease their victims into unrestrained sense; Rowlandson's erotic watercolours superimpose a grotesque mould on the human anatomy with the aim of elevating the cock or clitoris of even decent gazers; and the Franco-Swedish Nils Lafrenson's sensual drawings of the eighteenth century lure the viewer, by a process of calculated sympathy, to imitate the frenzied state of carnal pleasure served up to him. Every epoch has its hedonists, its sensualists, as well as possessing an invisible dichotomy between those who actually 'live the life of sense' as opposed to those who merely day-dream or write about it. Even the iconography of the mythological figure Eros renders this dichotomy patent, as Joseph Kunstmann's book *Ewige Kinder*, translated as *The Transformation of Eros* (1964), abundantly testifies. Nevertheless, what distinguishes one age from another in this regard is not the quantum of its sensual erotic activity – as if such a quantum could ever be measured – but artistic and theoretical speculation about it. I want to explore the latter in a very particular context: others have provided an overview of sexual beliefs and practices in the period; my aim is to distinguish the epoch from other ages by showing how its deep-seated cults of sensibility helped to establish the first scientific approach to sex.

Here philology is instructive. Let us consider the word *nymphomania*, so common today. According to the large *Oxford English Dictionary* the first printed appearance of the word is in 1802 in an English translation of Dr William Cullen's *Nosology*, originally published in 1769. A half century earlier Samuel Johnson did not include the term in his *Dictionary of the English Language* (1755); he may not have heard of it, although this would be surprising in view of the fact that the *Dictionary* is rather complete for common medical words. But Cullen, the prominent Scottish professor of medicine, certainly knew a good deal about nymphomania by 1755 and could have informed Johnson what it was. Cullen had written about 'a condition of nymphomania', technically known as *furor uterinus* or 'mania of the uterus' in his *Synopsis methodicae* (Edinburgh, 1769), an early taxonomy of nervous diseases. According to Cullen this mania was a common female disorder, yet he does not speculate about his use of the term. Indeed Cullen may not have been aware that he was the first writer to refer to nymphomania. Decorum possibly played some role in Cullen's silence about the history and genesis of the word; more likely, doctors had been invoking the term for several decades without committing it to print and Cullen believed it required no gloss as his main readers would be medically trained.[1]

Two years after the appearance of Cullen's *Synopsis* a little-known European doctor, Bienville, published a remarkable work entitled *La Nymphomanie, ou Traité de la fureur utérine*. Printed in Amsterdam in 1771 by the daring printer of the works of Jean-Jacques Rousseau – Marc-Michel Rey – it appeared in an octavo of 168 pages.[2] Soon after publication it was reissued in 1772 in another French edition and was shortly thereafter translated into several foreign languages. It was also published as from 'Padua, January 13, 1775' in an English translation by Dr Edward Sloane Wilmot, an obscure young Englishman resident in Italy who must not be confused with the far-better-known Edward Wilmot, M.D., King George III's Physician-in-Waiting. The title-page of this translation begins with the words *Nymphomania, or a Dissertation concerning the Furor Uterinus*... For almost a decade I have been using a well-marked copy of the 1771 Amsterdam edition that belonged to C. K. Ogden, the twentieth-century aesthetician who collaborated with I. A. Richards in *The Foundations of Aesthetics* (1935).

Almost nothing is known about Bienville except that he lived in Holland for most of his adult life, and even this fact is cited by Michaud and others without evidence.[3] Bienville's life is actually shrouded in such obscurity that there is little point in speculating about it since there is no evidence to confirm anything, neither letters, diaries nor private papers. He wrote several other scientific works – including two treatises defending inoculation for the smallpox – but no biographical materials have been found. As the contents of the *Traité* are technical and thereby suggest that he was medically trained, he may also have practised medicine. But thus far a biographical search has failed to produce evidence of any medical degree or any mention of Bienville's name among the annals of late eighteenth-century French social history. Bienville seems to have played no role in the Revolution, neither before nor after heads were decapitated at the Bastille. There is a brief reference to him in Richard Hunter's and Ida Macalpine's *Three Hundred Years of Psychiatry 1535–1860* (Oxford, 1963, p 349), but nowhere else in recent psychiatric or medical-historical literature do I find him discussed. This is a curious gap for someone who ought to have attracted more attention in two hundred years. Considering this dearth of biographical material, the student of Bienville must study the *Traité* itself.

The book opens with a discussion of a subject called *metromania*, a word capable of various etymologies – including one dealing with poetic verse and numerical measure – but whose correct etymology in this instance is related to the Greek *metra*, meaning womb. This development arises because the concept *metromania* (womb fury) has been confused with *metermania* (a rage for reciting verses). The confusion is even found in as reliable a source as Dr John Quincy's *Lexicon Physico-Medicum: or, a New Physical Dictionary* (1719; many editions by 1811) which defines *metromania* as 'a rage for reciting verse' while failing to indicate that it is a synonym for *nymphomania*.

Whatever etymological confusions exist, Bienville maintains that *metromania* or *nymphomania* – and he uses the terms interchangeably – 'begins with a melancholy delirium, the cause of which is found in a defective matrix' (i.e., defective uterus). He argues, however, that scrutiny shows a deeper cause than 'melancholy delirium', namely 'a mental derangement caused by the imagination'. The amplification is significant for a number of reasons, primarily because this is the first treatise on nymphomania and it sets an example ascribing great powers to the 'imagination'. Bienville considers the imagination so crucial to the development of this sexual disorder that he devotes his final section, the longest and most substantial of all the parts, to a discussion of it.[4] He even considers it necessary to set down his own philosophy of imagination. It was of course an age in which the imagination had come into its own; had come under the fine lens of a microscope, as it were. Philosophers from Locke and Hartley onwards had devoted part of virtually every learned treatise to this topic, and medical and scientific writers did not lag behind.[5] Whether the topic under discussion was time, space, infinity, or memory, genius, invention, sooner or later imagination was invoked. Bienville followed in the tradition by alleging that a medical treatise, such as his purportedly was, must take its stand on 'the vexed question of imagination'.[6]

'The imagination', he writes, 'is a mirror that reflects the things which interest man and which cause him to take action.' What definition could be broader? The imagination is at once everything and nothing. The analogy of mirror and reflection was old by the 1770s, as is his subsequent contention that 'it is the imagination that is almost always the mother of the greater part of the passions and their overflow'. The imagination, as a result of this authority, must be regarded, Bienville contends, 'as the bailiff of self-respect'. In this relationship of master and slave, the passions are also enslaved to it, not the other way around. For example, masturbation – an activity that Tissot had recently considered at great length in *L'Onanisme; ou dissertation physique sur les maladies produites par la masturbation* (Lausanne, 1760; translated by A. Hume as *Onanism; or a Treatise upon the Disorders produced by Masturbation* in 1766, it had gone through five English editions by 1781) – masturbation is also enslaved to the imagination, even though it is not literally a 'passion'. In Bienville's *Traité* it is enslaved to the imagination to a much greater degree than in previous writers: 'this harmful mania of masturbation, of which the imagination is the *sole* contriver...'[7] Whereas earlier medical authors speculate about the possibility of religious causes (i.e., souls possessed of the devil), Bienville will have none of this: he is a thoroughgoing mechanist who views the operations of the imagination as strictly controlling those of the passions by means of the nervous system. He therefore recommends 'still continuing the other bodily remedies' as a cure for nymphomania; and presumably he means pills and potions. But he concludes that ultimately treatment must be of 'the imagination'.

At this point perspective is supplied by standing back somewhat from the treatise. There is nothing radical or new about Bienville's medical assumptions: only a myopic medical historian possessing little familiarity with the epoch and its literature would ascribe terrific importance to these assumptions. Even the designation of the imagination as 'sole contriver' of masturbation is not so important as all that. The place where Bienveille ought to stir ears and raise eyebrows today is in his choice of subject: his isolation of and emphasis on nymphomania *for the first time* and his courage in making it the subject of a whole treatise. Just this novelty of subject matter may account for the shabby reception given to the *Traité* on its publication.[8] After all, it was one thing to write, as Tissot had, about masturbation and quite another to expend so much energy on a mere female disorder. Men – especially young men – were the backbone of every great nation. Of what use could a lengthy medical work about a radical female malady be?

On the Continent the *Traité* was barely noticed, perhaps because the title was so explicit about a medical area not yet respectable; alternatively, perhaps because of the reasons I have already suggested. When it was translated into English in 1775, it was consistently savaged by a few reviewers. The comments are worth scrutinising for the history of nymphomania as well as for the history of the reception of new medical theories, and for prejudices that arise in medicine when a sexual issue is involved. Edward Bancroft (1744–1821), the American-born naturalist and spy for Franklin, attacked the treatise mercilessly. 'Neither the theory nor the practice delivered in this performance have any share of merit; and we must therefore regret that Dr Edward Sloane Wilmot (if a man of this name and description really exists) should have been so regardless of his own reputation, and of public decency, as to promote an English impression of this worthless production.'[9] Can this estimate be trusted? Bancroft reviewed books because the *Monthly* paid him by the word and because his connection with the periodical brought him to some degree of prominence. Yet it is probable that he denounced Bienville's book because Bienville was a foreigner with no known qualifications. Moreover Bancroft may have thought this harsh treatment would please his grub-street editors who themselves knew little about medicine. Bancroft may also have been catering to the taste of readers by offering them what he thought they wanted to hear about this impolite subject.

The anonymous reviewer of the *Critical Review* derided Bienville and *metromania* with equal zeal.[10] He challenged Bienville's obeisance to Ancient medical theories, arguing that 'the Moderns' – especially his won contemporaries in England – had overtaken all others. He denounced the idea of a geographical determination of illness, in this case the notion that nymphomania is influenced by warm climates where women bask in luxury, and he demanded to know how the deviation could proceed from 'image in the mind' to 'carnal act'. But his real objections are less abstruse. He actually declares what his counterpart at the *Monthly Review* deigned not to say: namely that Bienville has no credentials *ex cathedra*, that he is an imposter in medical publication, that he is veritably to be aligned with mounte-banks and other charlatans. Yet the anonymous reviewer is far from accurate himself: he reveals his own inadequate credentials when proclaiming that 'this is not the first treatise on the Nymphomania'. It may not be the first book to *mention* the *furor uterinus* but it is certainly the first tract or book devoted *exclusively* to this condition. The reviewer's inaccuracy may even transcend such flaws; he errs further by assuming that the French version of 1771 and the English translation of 1775 are two different works.

If indecency of subject matter was the genuine cause of disapprobation, Bienville's book of nearly two hundred pages nevertheless contains very little material of a disreputable nature. Medically speaking, it certainly is far from disreputable. Bienville's theories are traditional, based on the medical progress of two or three generations. Bienville's medical argument embodies many of the assumptions of Dr George Cheyne's *The English Malady* (1733), a treatise about 'melancholic hysteria' which does not mention nymphomania though it glances obliquely at it in a couple of passages. Yet there are considerable differences between the two that exceed the philosophic or logical domain: each author has a different sense of the relation between cause and cure – a crucial connection in psychosomatic theory – and each approaches his implied audience differently, Cheyne by politely refusing to approach any matters except the English weather directly and Bienville by laying bare his subject at the outset and then proceeding to account analytically for the formation of its chief symptoms, in this case the all-powerful imagination. Cheyne's, to be sure, is the wittier book; better written, better informed, even something of a potboiler. But more than information is at stake here and there are, additionally, differences between the British and French reading publics: by the 1770s French society – especially Parisian readers – had abandoned much of its public refinement and was eager to learn 'the truth' about sexual customs and private taboos. Deviants such as hermaphrodites and nymphomaniacs were of especial interest.

There are other differences too, some pointing to the progress of five decades of European civilisation from the 1730s to the 1770s, others more narrowly signifying the theme about erotic sensibility I consider so pertinent here. Doubtless Cheyne and his predecessors understood very well that hysterical women crave the male phallus because their imagination is inflamed and diseased. By 1730 one could read about such cases in medical literature and even view them in serious scenes of Greek erotic art as well as comic imitations. One could read about them, moreover, in Boccaccio's tales and in Aretino's sonnets, and gaze at them in Giulio Romano's illustrations of these and many other Italian sonnets. Surely the representation and availability of this body of literature and this mass of art cannot be what distinguishes the medical sensibility of Cheyne and Bienville. Nor is theirs an awareness marked by a different reading class: for if the aristocrats and wealthy classes of the 1730s had constituted a ready market for high-class erotica, by the 1770s that class had filtered downwards to include the middle class and again expanded to incorporate less sophisticated erotica. Both groups, early and late, knew about the past, knew that the Middle Ages had been permeated with disturbed young women said to be possessed by evil demons causing them to crave for male insertion; and they had yearned to such a degree that only the rites of exorcism, when successfully administered, could salvage them and bring them back to a healthy reality. Both knew something about the anatomic disturbances of the *uterus* or matrix, even about its damage at birth; and both physicians were aware that not only accidents caused late in life but violent insertion as well and occurring at any stage could cause irrevocable harm. Both men probably could have recited ancient dogma to the effect that the damaged uterus wanders through the female body wrecking organs and tissues as it marches, even if earlier ages, the Middle Ages, could not. Ilza Veith has studied these theories of the 'wandering uterus' in a useful though limited book entitled *Hysteria: the History of a Disease* (Chicago, 1965) and her results confirm this description of information available to both groups of doctors. But what is only implicit in Cheyne is explicit in Bienville – a sense of the erotic as essentially a province of the imagination – and on this difference my whole argument about erotic sensibility depends.

Moreover, the great proliferation of treatises on venereal disease in the eighteenth century – by Boerhaave, John Douglas, Jean Astruc, John Armstrong, Jourdain de Pellerin, J. Profily, J. H. Smith and many others – influenced popular conceptions of the erotic as well as works on melancholy. Aberrant sexual conditions, although they are usually the primary subject of these treatises, are referred to over and over again. There is a sense in which that age – 1700–90 – was ready to consume the history of *every* disease, and every history included abundant material about other conditions. Significantly, then, Bienville does not classify his chosen condition under mania or hysteria; nor does he mention even once the rampages of the uterus as a cause of nymphomania or delineate those proverbially old wanderings. Bienville's new proximate cause is 'the imagination': therefore he staunchly classifies the condition as psychosomatic. If such classification and implicit reasoning had occurred a half century later (in 1825) it would be less significant: there is nothing radical in 1823, for example, about Dr Andrew Jacob's *An Essay on the Influence of the Imagination and Passions in the Production and Cure of Diseases.*[11] Fifty years earlier it would have been, for in Bienville's 'leap' is seen a significant theoretical difference: the attribution to 'the imagination' of a whole class of sexual disorders. Before this approximate time – the second half of the eighteenth century – the imagination was credited with certain mechanical powers, but it was not considered the agent of sexual aberrations.

For such a giant step Foucault tied ribbons around Bienville's neck in the *Histoire de la folie* (Paris, 1961) and, later, in the first volume of the *Histoire de la sexualité* (Paris, 1976). While 'the leap' appears minimal to us it was not in the late eighteenth century, for Bienville rejects an entirely mechanistic view of the nervous system and shatters the notion that a condition such as nymphomania can arise from an anatomically damaged nervous system. He does so just at the time that a materialist science of Eros – we would call the science sexology, a word unknown in Bienville's time – is becoming established. Bienville does not consider the damaged brain to be a cause of nymphomania or contend that a 'diseased imagination' confounds the physiological–anatomic brain; he rather argues that the brain and imagination influence each other in some reciprocal but unspecified way. Such a mechanistic view *manqué*, defective as it appears to a late-twentieth-century audience, may also seem ambivalent to us: is it mechanistic or not? we ask. The question is simplistic in both epochs: Bienville conceived of the imagination as a less vitalistic force than we do, and the reply he would have made is that the imagination is *both* mechanistic and vitalist at the same time, a position not radically different from our stance today.[12]

This seeming defect of logic notwithstanding, Bienville's line of reason in the *Traité* is not so consistent as I may be implying. On the one hand he is a child of the times and has been enticed into the subject – nymphomania – through a concurrent fascination for the concept of masturbation. He may even have hoped to generate his own theory of masturbation.[13] Bienville wants to correlate the male and female genital zones, and then, further correlate both to the imagination in the brain, doing this as if all three categories were of a parity. This is the point in which Bienville's logic weakens: he believes that all three are more similar than dissimilar, though he nowhere says they are identical, a position we can understand only if we comprehend to what extreme degree 'the imagination' had been mechanised by 1760 or 1770. Therefore, one of the main arguments of the *Traité* is the notion the 'external objects inflame and excite the uterus', causing it 'to masturbate as a result of its vast concentration of nerve tips'.[14] During this time the vast accumulations of nerve endings succumb, the female erotic zone craves – as if it had a mind of its own – enactment of the fantasies it has enjoyed in solitude, and by a process of repetition the habit directly leads to the aberrant condition Bienville designates as nymphomania.

Given this quasi-mechanistic view of the imagination which Bienville's successors, the Romantics, rejected in favour of a far more vitalistic and organic model, Bienville's thinking is rational but not altogether logical. Stimulation leads to action, action to habit, habit to disorder. The chain seems inevitable when viewed this way, except that the nerves are rhetorically personified and endowed with such active mental properties that they cannot help but carry out the demands of the nerves, fibres and animal spirits in the erotic zone. For Bienville the physiological nervous system of man is one vast inter-connecting chain; what one part knows all the others soon will, and the reader who peruses the *Traité* need not read very far to learn that by the term 'nervous' Bienville actually means something akin to 'endowed with extraordinary mental or cognitive capabilities'. So 'nervous', for example, are the anatomic extremities of the male and female erotic zones in Bienville's conception, that he cannot conceive of any physiological or mental state in which either has no contact with the brain, although he would probably have had to concede that if the central spinal cord were destroyed communication would also be impeded. The vast nervous chain would be intercepted.

The clue, then, to Bienville's theory of nymphomania is his whole sense of the role of masturbation in nervous human life: only by cessation of it can the physician hope to cure the patient afflicted with nymphomania. Ten years before Bienville published his *Traité*, Tissot, already mentioned, had brought out a compendium entitled *L'Onanisme; ou dissertation physique sur les maladies produites par la masturbation*,[15] an anthology demonstrating the hundreds of pernicious effects produced by this allegedly vilest of human activities. Dr Tissot attributed almost every form of sexual deviance and lunatic derangement to it. Understandably he was catering primarily, although not exclusively, to a Swiss Protestant readership who wanted to hear this position stamped by the authority of a revered medical expert. But the ascription, while reaching the masses who thoroughly endorsed it, was unoriginal. More than fifty years earlier, in 1707–8, there had appeared on the other side of the Channel an anonymous *Onania; or, the Heinous Sin of Self-Pollution, and all its Frightful Consequences, in both Sexes, Considered*.[16] This book was to be reissued in more than fifteen editions before the century wore out. Each decade in that epoch had naturally produced its own fantasies about masturbation and these vary widely. The doctors mythologised in one vein, writers of fiction in another; for satirists masturbation was the scapegoat, the ultimate sin with which to charge every deceitful knave or fool. But when Bienville discusses masturbation, it is with a difference: he views it as crucial to the development of nymphomania – as a symptom – but never considers it as the genuine cause. To cure nymphomania the imagination must be directly confronted; no substitute for this confrontation exists; no medicines, no cures. Twice Bienville suggests that even religion is of limited value when compared to the magisterial importance of curing the diseased imagination.

Thus we discover in a nutshell in Bienville the essence of the first psychosomatic theory of this sexual disorder. The implications of Bienville's theory, if interpreted, are consequential for the rise of erotic sensibility in Europe; not only as a way of theoretical thinking and as a manner of informal speculation about art but as another example of the ways that science reinvigorates art.[17] What remains is to relate nymphomania to erotic sensibility; and, more properly, to connect a particular development in the history of medicine with the evolution of aesthetic thought in the eighteenth century.

Before the eighteenth century there was no science of the erotic;[18] no neurological or physiological explanation of basic drives and urges except of those observed among the brutes. But in the latter case a distinction was always being drawn between the instincts of brutes and those of mankind:[19] consciousness, especially the self-reflectiveness man possessed and with which God had endowed him for a particular reason, eternally provided man with the ability to harness his basic drives and direct them upwards to the contemplation and worship of God. In general, erotic *thinking* about man was still based on religious notions. Even more erotic art of the early eighteenth century had been religiously inspired, and the physiological revolution of the previous period – discovery of the circulation of the blood, the new brain theory, the new empiricism based on an mechanistic neurology, the innovations permitted by the use of the telescope and microscope – did not drastically alter this state of affairs. As late as the seventeenth-century *fin de siècle*, a sexually hysterical woman is still labelled 'possessed demonically', not called nymphomaniacal or discussed in physiological, neurological or other medical terms.[20] This comes later, in the second half of the eighteenth century; but in the earlier period – up to about 1740 or thereabout – she is thought to possess a sick uterus that wanders through her body in search of a healthy resting place.[21] This is why learned treatises up to this approximate period continue to be written about 'the sick mother', from Renaissance times a synonym for the hysterical female organ.[22] But there exists as yet – in the 1740s – no consensus about a 'diseased imagination' aggravated by masturbatory practices as the main cause of her ravings. This transition from a religious to a scientific explanation has profound consequences in the early eighteenth century for the theory of medical insanity; and although there are traces of the scientific explanation in the psychosomatic theories of Willis and Sydenham at the end of the seventeenth century,[23] those made in the middle of the eighteenth are more substantial and capable of launching a new science of erotic behaviour.[24]

The new thinking about nymphomania in particular and sexual aberration in general could not have been formulated before the eighteenth century for reasons primarily dependent on current theories of the imagination. For European philosophers and scientists then first began to isolate the imagination from metaphysical and theological speculation and scrutinise its operations according to the laws of mechanics.[25] For a while in the late seventeenth century the imagination was treated no differently from other organs in the body: the heart, the liver, the bowels, even the genitals. This approach, based on the belief that all operations of the mind functioned no differently from those of the body, reached a peak in the 1690s. At this time the iatromathematicians (sometimes called iatromechanists) went on a rampage and quantified virtually all motions within the body, as Newton had traced the motions of the planets in the same period in the *Principia* (1687).[26] Swift's satire on 'the mechanical operations of the spirit', first published in 1710, assumes widespread knowledge of the iatromechanists and of the even more recent mechanisation of the imagination.[27] Later on in the early eighteenth century in England, literary forms other than satire present this view of the imagination to the literate layman. Akenside's *Pleasures of Imagination* (1744), a long didactic poem written by a brilliant medical mind who was also a gifted poet of the second rank,[28] is a perfect example. Contemporaries of Descartes and Hobbes a century earlier would have concurred that the imagination was important; they never would have dreamed it could assume such proportions.

Yet I think we ought not to expect theory in one realm – metaphysical speculation about the imagination – to be automatically applied to another – that of sexual disorder. A time lag was necessary for the ideas to take hold, and thinkers who could make the necessary links had to appear. After the 1730s in England, physicians and physiologists adjusted their thinking to the trend attributing medical conditions to dysfunctions of this mechanical and now more powerful than ever imagination.[29] During the previous half century, 1690–1740, disease was ascribed to a number of disturbances ranging from chemical imbalance to malfunction of the non-naturals, as Jeremiah Wainewright had shown in his *Mechanical Account of the Non-Naturals* (1707). In the period before 1690, the tendency to supply secondary or supernatural religious causes to explain many forms of disease was widespread, if not preponderant.[30] But by the time when Bienville was forming the ideas he developed in maturity – the 1750s and 1760s – the imagination had, as it were, already enjoyed its revolution and was now considered by physicians the main cause of sexual disorder. Yet it would be wrong to believe that all physicians had been persuaded, or that medical theory in the mid eighteenth century was an isolated subject incapable of influencing other subjects.[31] To the contrary, medical theory as well as practical medicine then entered into daily life no less than it does today; it was then also influencing aesthetic thought about the emotions and passions in portrait painting as well as shaping ideas about the function of sexual energy in the act of artistic creation (as is evident in a number of key passages in *Tristram Shandy*).[32]

Even the demonstration of this connection between medical theory and aesthetic thought may not satisfy those who ask what the connection is between the medical history of nymphomania as found in Bienville's *Traité* and the formation of a science of the erotic. Despite the knowledge that ultimately any satisfactory answer must take into account a sound philosophy of intellectual history, surely more than the history of ideas is at stake here. More crucial is the acknowledgement that the cults of sensibility everywhere evident in mid eighteenth-century literature actually touched on aspects of daily life.[33] On this development everything stands or falls, even the possibility that sexual knowledge can be condified into a science.

It is necessary then to recognise that four categories of explanation have been manufactured here: a medical condition called nymphomania, the medical history, a state of human consciousness called erotic sensibility, and the actual cults of sensibility.[34] Mid eighteenth-century authors would have understood these categories (though they did not generate them themselves) with the possible exception of the third, for they were so close to erotic sensibility that it may not have been patent. Because the manufacture of all four categories is important for a science of sex, it is crucial to understand that Cullen, a Briton, not Bienville, generated the first scientific theory of nymphomania and, moreover, that a correlative theory for male *satyriasis* would not appear until the nineteenth century.[35] Cullen was the imaginative thinker whose intuition permitted him to recognise the cause of many nervous disorders, not merely of nymphomania;[36] yet if Bienville had not published the *Traité* in 1771, more time would have elapsed before Cullen's hypothesis about the nervous origins of nymphomania could make its way into the public domain, and perhaps even a longer time until it would filter down into popular culture. Furthermore, Cullen was a pious Christian[37] and a revered member of the Edinburgh scientific–university community; but it is one thing for a thinker without status *ex cathedra* to generate a theory about sexual disorders in a polite age and quite another for a leading university professor who is also a member of the Anglican Church. Bienville's religious predilections, unlike Cullen's, are unknown; and while he lives in a country where toleration extends to nearly everyone – Holland – his treatise is addressed primarily to the Continental medical community which was then not so receptive to a theory about the 'imaginative origin' of sexual disorders as Bienville may have liked. Finally, if Bienville was indebted to Cullen, Cullen himself was indebted in his philosophical assumptions to a long train of empirical philosophers extending downwards from Locke and Hume to the Scottish school.[38] Philosophers, rather than physicians, had evolved in the eighteenth century a concept of the autonomous and quasi-mechanical imagination, and this is why psychology grew out of philosophy and why the eighteenth century is the crucial period for the genesis of the province of learning.[39] Cullen, himself one of the most philosophical of eighteenth-century doctors, inherited from his teachers in Scotland, and later on from his wide reading in philosophy,[40] a notion of the mechanical imagination which he later applied to his theory of the origins of nervous diseases. This notion is the background and basic assumption of his concept of nervous diseases.

The relevance of medicine to erotic sensibility therefore becomes patent when the relations of medicine and philosophy in the eighteenth century are scrutinised, and when Cullen and Bienville are located in the continuum of thinking about sexual problems as medical disorders. Both are more advanced than their predecessors, not more progressive in any absolute sense – according to the notion that science, like history, moves toward a perfect end – but more advanced in their knowledge of 'the nervous origins of imagination'.[41] Both men maintain precisely what Blake will deny: that the imagination is anatomically as real and discernible as the liver or spleen, except that is lodged in the brain and therefore incapable of being considered in isolation of it. (In one passage[42] Bienville astutely asks if we do not actually feel our imagination in our head. Some of us may try to locate it there and discover nothing, but there are others, Bienville notices, who do!) The imagination is healthy or diseased: it shares these states with other organs. The diseased state arises from circumstances ranging from physiological defect such as brain damage to poor economic conditions such as the malnutrition of paupers. When such a diseased imagination focuses its energy, Bienville argues, on external erotic images – the literature or art of pornography – the brain and its vassals the nerves carry these impressions to the whole organism. Every part of the body feels it, especially where there are large clusters of nerve endings. The greater the accumulation of nerves, the greater the sensation. Now the uterus, Bienville reasons,[43] is one of these but it is not the only one; so are other highly nervous zones in the female body, as is the tip of the penis in the male. The nymphomaniac, in contrast to less 'nervous women', exercises her uterus, according to these physicians, more than other women. It is also possible that she may have a greater number of nerve clusters to begin with, but by wedding her already nervous imagination to the most highly nervous zones of her body the nymphomaniac sets a psychological motion in process that results in unremitting sexual craving.[44]

Erotic sensibility, however, is nowhever to be found in Bienville's *Traité*, neither as a term nor as a way of thinking, for it is not a type of medical learning or scientific knowledge but a manner of applying scientific precepts to an aesthetic domain. As such, erotic sensibility can be found in the novels of the Marquis de Sade, for example, or in the poetry of Lord Byron, but not in the medical treatises of Cullen or Bienville or any other eighteenth-century physician. Yet no one should conclude from this contrast that a terrific number of steps are required to move from the one to the other: science often paves the way for art, albeit indirectly, and then sometimes allows it (art) to view itself as a science both in theory and practice.[45] At the very least it permits the view that art is based on absolute principles of craft (Plato's *techné*) or knowledge (Da Vinci's anatomical learning), and that as a consequence its laws are scientific. In this sense erotic sensibility is the science of the artist who meditates reflectively on his images, forms and ideas. There are cases, of course, in which the chain is reversed; in which the artist paves the way for the scientist by helping him to imagine his ideas in the first place: one thinks of the representations of 'flying chariots' in the Renaissance that later formed the basis for modern aeroplanes,[46] and still later of Blake's myth about the powers of Energy that antedated the concept in physics by many decades.[47] Most commonly, though, the sequence is from science to art: as in the emerging science of the erotic during the last two centuries, or what has become known in modern parlance as sexology.

There were, to be sure, nymphomaniacs before Bienville scrutinised them; but there was no reflective meditation about them. It is possible that Bienville could have published the first treatise on nymphomania a hundred years before he did – in 1671 – but it is not likely. Philosophers prepared the way in the seventeenth century; doctors complemented their speculative hypotheses in the first part of the eighteenth. Now the forest was cleared for writers like Sade – who learned, incidentally, most of what he knew about erotic activity and thought from philosophers and physicians – to write *Justine, Sodom* and other narratives that employ an aesthetic which for lack of a more elegant phrase I call 'erotic sensibility'. There is a good reason why Sade plundered the writings of then recent philosophers and physicians like Bienville and Cullen for whole pages at one time, often inserting them into his fiction without any apology or hesitation.[48] In Sade's estimate the narrative would not be disturbed by such wholesale intrusions; it could only be enhanced because the audience expected it and wanted to be educated by it.

A single example from *Justine* must suffice for lack of space. It occurs in the section of Part One in which Dubois explains to Justine the act of love. He has already told her why certain people are more passionate than others. Now Dubois himself turns to Justine, and compares her to himself.

> ... Because you are weakly constituted; thus you have but small desires, faint pleasures, scanty whims. But such a mediocre grade is not permissible at all in a person constituted as I am. And if my own good fortune can continue to thrive by the calamity of others, it is only because I find in this misfortune the unique stimulant that strongly pricks my nerves. Also, after the violence of this shock, it more surely induces to pleasure the electric atoms which pass through the hollow tubes in the nerves.[49]

At this point Dubois's discourse grows more technical; the reader wonders with what aims other than didactic ones (surely it is neither ironic nor comic) Sade could have written it. The passage could be lifted from a page in Bienville; indeed it was extracted almost verbatim from d'Holbach's *Système de la Nature* (1770). I take the liberty of quoting it at length because it well illustrates how outright scientific discourse (Cullen, Bienville, d'Holbach) under the right circumstances becomes artistic narrative. Separate this passage from its context, and it could derive from a contemporary text on physiology. It is even replete with ready-made scientific sources: for example, the reference to La Martinière, Louis XV's First Surgeon-in-Charge, who himself wrote several physiological works.[50] The references to nerves, animal spirits, sensations and the five senses are better suited to Cullen's or Bienville's treatises on nervous diseases than to the most erotic tale ever composed. But Sade, a writer who controls his sources, must have believed it had a place. Though lifted from d'Holbach it is found in *Justine*, Sade's most popular novel and a work of erotic sensibility *par excellence*:

Let us now give an analysis of the nerve. The nerve is the part of the human body which resembles a white string, sometimes round, sometimes flat. It usually begins at the brain; it issues in bundles or fascicles, symmetrically arranged in pairs. There is no part of the human body more interesting than the nerve. It is the kind of phenomenon, said La Martinière, that is the more admirable as it appears less likely of action. Life and body – indeed the entire harmony of the body as machine – depend on the nerves. All sensations, knowledge and ideas derive from it; it is, briefly, the centre of the whole human structure. The living soul is located in it, that is to say the principle of life. This same principle dies out among animals, or at least declines in them, and is reduced to mere matter.

The nerves are imagined to be tubes carrying the animal spirits into the organs to which they are distributed. These same nerves report to the brain the impressions of external objects on those organs.[51]

Immediately following this passage is an explanation of the necessity of nervous activity for erotic pleasure, this material also lifted from d'Holbach:

An intense inflammation of the nerves excites to an extraordinary degree the animal spirits that flow into the nerve tubes which, in turn, induces pleasure. All pleasures comes this way. If this inflammation occurs on the genitals, then the pleasure is remarkable. This also explains why we enjoy nervous activity in areas close to the genitals; and why we delight in receiving blows, stabbings, pinches or floggings. These animal spirits produce painful as well as pleasureable shock; almost by virtue of the mental sensation one has received, and as if mind triumphed over matter. Much follows, Justine, from all this: the sphere of one's sensations can be remarkably extended, as I have already suggested. These principles are sound and philosophic; not to believe in them is to yield to antiquated notions.

Now the point to be noticed is that these passages are from Sade's *Justine*, a novel, not from one of Thomas Willis's treatises on the brain, or a French anatomist's discourse on the nervous system, or Bienville's *Traité*. The important point to be gathered, as I suggested at the outset, is that there are differences between Boccaccio's nymphs and Sade's, not patent distinctive between the figures themselves but differences in the author's narrative art: in his inclusion of certain kinds of scientific or pseudo-scientific material. Moreover, in the works of Sade brothels and bawds seem to exist only to be explained in technical language, as in the passage just cited. And something of the same principle is true in paintings in which the central erotic figure – Miss O'Murphy in François Boucher's erotic painting by this name – is diminished by more interesting surroundings. If one compares, for instance, Boucher's central figure, a rather unexciting creature, with Rubens' Andromeda in *Perseus and Andromeda*, the contrast is striking.[52] Boucher's exterior surroundings command more attention by virtue of their colours and enticing texture, than does the conventionally rotund and somewhat flaccid Miss O'Murphy.

Sex in the eighteenth century then is not so simple a subject as it seems, at least not unless the cultural historian merely wishes to document particular occurrences and instances. Sex then, as today, enjoyed certain theoretical prerogatives, though these are far less evidence than the 'case histories' which the social historian can isolate and document.[53] Moreover, technology and religion also influence writers and painters whose aim is to be scientific about Eros and erotic activity, and this is especially evident in cases in which a preponderantly secular or even atheist milieu prevails upon the artist. All of us have heard many times in the twentieth century from authors such as Georges Bataille, Herbert Marcuse, R. E. Masters, W. O. Young, K. Price, Leonard de Vries, that 'the erotic' is not merely an art but a science, and that some of its deepest secrets are intermingled in the economic domain that covers the acquisition and loss of property.[54] But a question is still begged, namely, when the idea that sex and sexuality were first thought to be capable – as were the other sciences of man – of becoming scientific: hence a *science of sex*. By gazing briefly at the very first treatise on nymphomania, I have attempted to chart out a little-known chapter in the rise of 'erotic science', a subject now deemed important enough by contemporary readers to compel bookshops everywhere to carry whole sections of this type of reading. Future students of the science of sex in cultural history who genuinely hope to study the rise of the phenomenon would be imprudent to do so without recognising its eighteenth-century origins.

Notes

1. See William Cullen, *Synopsis nosologiae methodicae* (Edinburgh, 1769), 324. My aim in this essay is not to provide a summary of thinking about nymphomania in the period – others who are better acquainted than I am with the social history

of the age can do that – but rather to study the conditions (philosophical, physiological, anatomic, aesthetic) under which a sexual category such as nymphomania arises. Important information about the eighteenth-century legacy of the 'mania of the uterus' is found in Jean-Marie Goulemot, 'Fureurs Utérines', *Dix-huitième siècle* (Paris, 1980), No. 12 special issue: *Représentations de la vie sexuelle*, 97–111. Those interested in the appearance of nymphomania and other sexual behaviour in the period will gain some sense from Anon, *The Bloody Register; a . . . Collection of Most Remarkable Trials for Murder Treason . . . and Other High Crimes from the Year 1700 to . . . 1764, Inclusive*, 4 vols (1764).

2. A brief survey of the many editions is useful: two editions appeared in French in 1771, the one already mentioned and another containing 164 pages, larger print, different capitalisation and punctuations, but with an identical preface. In 1772 another edition in French was published by M. M. Rey in Amsterdam, this one containing 178 pages. In 1778 Rey printed a 'Nouvelle Edition' containing xxii pages of introduction and 179 pages of text. In 1784 yet another edition appeared in French as 'published in Mentz' (*sic*) and containing 203 pages of text; but the only known surviving copy of this work, in the British Library, was destroyed by bombing in the Second World War, and I have consequently been unable to inspect a copy. In 1789 another edition appeared in French as 'Printed in London', though no printer's name is given on the title-page, this recension also containing xxviii pages of introduction and 198 pages of text. This edition may have been followed by an undated version sometimes called 'the Sixth Edition' printed by 'Vanderauwera' in Brusells, containing 96 pages of text and no introduction, a copy of which is in the Bibliothèque Nationale. Two English translations of Bienville's treatise are known: one published by Bew in London in 1775, another c. 1840, a copy of which is in the Wellcome Library in London. In 1777 van Padderburg published a Dutch translation in Utrecht; in 1782 a German translation appeared in Vienna as pubslished by S. Hartl, followed in the next year by an Italian translation by Graziosi in Venice. No Spanish, Hungarian or Danish translations are known. Selections from Bienville's treatise are quoted verbatim in P. Dusoulier's *Avis aux jeunes gens des deux sexes, ou l'on trouve réunies les observations . . . de M. de Bienville dans son Traité de la nymphomanie* (Angers, 1810). Modern studies of nymphomania such as E. Podolsky and C. Wade, *Nymphomania* (New York, 1961), I. Wallace, *The Nympho and other Maniacs* (New York, 1971) and A. Ellis, *Nymphomania* (New York, 1964) do not mention Bienville because of the authors' lack of awareness of eighteenth-century theories of sexuality. A reprint of Bienville's *Traité*, with a preface by Jean-Marie Goulemot, was published by Editions Le Sycomore, Paris, in 1980.

3. L. G. Michaud, *Biographie universelle* (Paris, 1855), entry under Bienville, and August Hirsch (ed.), *Biographisches Lexicon der hervorragenden Artze* (Berlin, 1929–35) 1, 529. Nor is Bienville listed in A. L. J. Bayle and A. Thillaye, *Biographie médicale par ordre chronologique d'aprés Daniel Leclerc, Eloy . . . 2* vols (Paris, 1855), or in Morel de Rubempré, *Biographie des médecins français vivans . . .* (Paris, 1826), dictionaries of biography in which one would expect to find the author of such a popular treatise included.

4. See the 'Observations de l'Imagination par rapport à la Nymphomanie', in the first edition (1771) of the *Traité*, 125–54. All English translations in this essay are my own.

5. I documented this development in 'Science', in Pat Rogers (ed.), *The Context of English Literature: the Eighteenth Century* (1978), pp. 153–207; 'Nerves, Spirits and Fibres: Towards Defining the Origins of Sensibility; with a Postscript, 1976', *The Blue Guitar*, II (1976), 125–53 (*Enlightenment Crossings*, chap. 5); and in 'Psychology', in G.S. Rousseau and Roy Porter (eds), *The Ferment of Knowledge* (Cambridge, 1980), 143–210 (*Enlightenment Crossings*, chap. 4). See also Jean Starobinksi, *Hostoire du traitement de la mélancolie des origines à 1900* (Basel, 1960), and J. V. Baker, *The Sacred River: Coleridge's Theory of the Imagination* (Baton Rouge, 1957). Dozens of works written in the eighteenth century make the same point; it is impossible to provide even a partial list in the space allotted here, but for works in two different genres see Z. Mayne, *Two Dissertations concerning Sense, and the Imagination, with an Essay on Consciousness* (London, 1728), and L. P. Poulten, *Imagination: a Poem* (1780). These physiological traditions culminate in J. P. Marat's *Philosophical Essay on Man*, 2 vols. (1773), especially in such passages as this one (11, 54): 'the mechanical power of the imagination is not confined to any particular organ, it is extended over the whole body'.
6. See Bienville, *Traité*, 156–7. My references, though given in my own literal translation, are to the first French edition.
7. Ibid., p. 174
8. See the three paragraphs below.
9. *Monthly Review*, LIII (Sept. 1775), 275.
10. *Critical Review*, XXIX (Mar. 1775), 252–3.
11. Published in Dublin by C. P. Archer. Even in the 1720s several medical works attribute physiological defect to the mechanical power of the imagination, but I have found no works that consider the imagination to be the pre-eminent cause of sexual disorder; see, for example, James Blondel, *The Power of the Mother's Imagination Examin'd* (London 1729), and Daniel Turner, *The Force of the Mother's Imagination upon her Foetus in Utero* (London, 1730).
12. Although the subject is more complicated than it seems. For discussion of blends of mechanism and vitalism in eighteenth-century medico-physiological thought, see Elizabeth Haigh, 'The Roots of the Vitalism of Xavier Bichat', *Bulletin of the History of Medicine*, XLIX (1975), 72–86. Hans Driesch's *History and Philosophy of Vitalism*, trans. C. K. Ogden (1914), is now outdated but contains further useful material.
13. This was a common pastime of the period among medical thinkers. See A. Comfort, *The Anxiety Makers* (New York, 1969), *passim*; E. H. Hare, 'Masturbatory Insanity: the History of an Idea', *Journal of the History of Mental Science*, CVIII (1962), 1–25; R. H. MacDonald, 'The Frightful Consequences of Onanism', *Journal of the History of Ideas*, XXVIII (1967), 423–31. R. Hunter and I. Macalpine, *Three Hundred Years of Psychiatry 1535–1860* (1963), 349: 'by the nineteenth century the sequence masturbation–venereal excess–venereal disease–nervous disease–insanity was firmly established not only in the lay but also in the medical mind despite an occasional rational approach like that of John Hunter (1786) to these emotionally charged matters'.
14. Bienville, *Traité*, 78. Further commentary on the quantites of nerve tips in the sexual zones of the human body is found in William Rowley, *A Practical Treatise on . . . the Breasts* (London, 1772). In *Deformity: an Essay*, (London, 1754), William Hay, himself a dwarf, asks if the nervous system, and also the

erotic zones, of deformed persons differ from those of normal men. Here, in brief, is an aesthetics of deformity.

15. Lausanne, 1760. Material about Tissot is found in Theodore Tarczylo, *'L'Onanisme de Tissot', Dix-huitième siècle* (Paris, 1980, No. 12, special issue: *Représentations de la vie sexuelle*, 79–96, and in J. M. Goulemot (n. 1 above).

16. This book was in a fifteenth edition by 1730 and a nineteenth by 1759. The first edition may have appeared as early as 1707/8.

17. See Wylie Sypher, *Literature and Technology: the Alien Vision* (New York, 1968); I. A. Richards, *Science and Poetry* (1926); G. S. Rousseau, 'Literature and Science: the State of the Field', *Isis*, LXIX (1978), 583–91 (*Enlightenment Borders*, chap. 8).

18. A 'science of the erotic' is necessarily subsequent to a theory of the 'natural philosophy of mind' – what we would call psychology; this subject arose precisely in the middle of the eighteenth century when the erotic dimension of human behaviour was being incorporated into it. Works that deal with the matter include: David G. Loth, *The Erotic in Literature: a Historical Survey of Pornography* (New York, 1961); David Foxon, *Libertine Literature in England 1660–1743* (New York, 1965); John A. Atkins, *Sex in Literature* (London, 1970); Phyliss Kronhausen, *Erotic Fantasies: a Study of the Sexual Imagination* (New York, 1970). Jean Hagstrum's important recent book, *Sex and Sensibility: Ideal and Erotic Love from Milton to Mozart* (Chicago, 1980), says that its 'main interest is not in sources or context but in meaning', p. 253; as a consequence the role of science in relation to the rise of sensibility is omitted (see also n. 44 below).

19. As Swift imaginatively did in *Gulliver's Travels* and as did contemporary physicians; see, for example, Thomas Morgan, *The Mechanical Practice of Physick* (London, 1735), 287–8; Robert Douglas, *An Essay Concerning the Generation of Heat in Animals* (London, 1747), 65–6; recent studies include E. Tuveson, 'The Origin of the "Moral Sense"', *Huntington Library Quarterly*, XI (1948), 241–59; L. C. Rosenfield, *From Beast-Machine to Man-Machine: Animal Soul in French Letters from Descartes to La Mettrie* (New York, 1941); and J. A. Passmore, 'The Malleability of Man in Eighteenth-Century Thought', in Earl Wasserman (ed.), *Aspects of Eighteenth-Century Thought* (Baltimore, 1965).

20. I. Veith, *Hysteria: the History of a Disease* (Chicago, 1965), but for an example from the 1690s, see David Irish, *Levamen Infirmi:... Concerning melancholy, frensie, and madness...* (1700), 94.

21. Veith, *Hysteria*, 126–47. By the time Erasmus Darwin writes about 'Erotomania' in *Zoonomia; or, the Laws of Organic Life*, 2 vols (London, 1794–6) II, 353–5, the concept of the wandering uterus has disappeared.

22. Veith, *Hysteria*, 121–4.

23. See L. J. Rather, *Mind and Body in Eighteenth-Century Medicine* (1965); L. King, 'Soul, Mind, and Body', in *The Philosophy of Medicine* (Cambridge, 1978) chap. 6, pp. 125–51; G. S. Rousseau, 'Psychology', in *The Ferment of Knowledge* (Cambridge, 1980), p. 169 (*Enlightenment Crossings*, chap. 4).

24. This is gleaned by comparing Bienville's treatment with Jacques Ferrand's earlier work on the same subject: *De la maladie d'amour, ou maladie érotique* (Paris, 1623), trans. in 1640 by E. Chilmead as *Erotomania, or a Treatise discoursing of the Essence, Causes... and Cure of Love, or Erotic Melancholy* (Oxford). Edward Synge's *Sober Thoughts for the Cure of Melancholy* (1742) need not be

considered in this context: whereas it deals with 'love sickness' in medical terms, its point of view is theological, not scientific.

25. The trend is seen at mid-century in Frank Nicholls, MD, *De Anima Medica* (London, 1750), and is the thesis of my study of 'Science and the Discovery of Imagination in Englightened England', *Eighteenth-Century Studies*, III (1969), 108–35; see also: P. Gav. 'Newtons of the Mind', *The Enlightenment: an Interpretation*, 2 vols (New York, 1966–69), II, 174–85, and R. Schofield, *Mechanism and Materialism* (Princeton, 1970), 134–6. For an example of pre-mechanical thought about the imagination written by a physician, see L. J. Rather, 'Thomas Fienus' (1567–1631) Dialectical Investigation of the Imagination as Cause and Cure of Bodily Disease', *Bulletin of the History of Medicine*, XLI (1967), 349–67.

26. For an example of the application to medicine, see E. Ezat, *Apollo Mathematicus: Or the Art of Curing Diseases by the Mathematicks* (1965); for the body-politic see J. T. Desaguliers, *The Newtonian System of the World, the Best Model of Government* (1728); for music J. de Vaucanson, *An Account of the Mechanism of ... playing on the German-flute* (1742).

27. See especially the passage beginning 'To proceed therefore upon the Phaenomenon of Spiritual Mechanism ...' in A. C. Guthkelch (ed.), *A Tale of a Tub ... by Jonathan Swift* (Oxford, 1965), 271.

28. The mechanical images are described in A. O. Aldridge, 'The Eclecticism of Mark Akenside's "The Pleasures of Imagination"', *Journal of the History of Ideas*, v (1944), 292–314. Robert Douglas, the scientist, considered Akenside so brilliant in poetry and medicine that he dedicated his *Essay Concerning the Generation of Heat in Animals* (1747) to him.

29. But not yet to sexual disorders. See n. II above.

30. C. Hill, *The World Turned Upside Down: Radical Ideas During the English Revolution* (1972), 287–300, and K. Thomas, *Religion and the Decline of Magic* (1971), 787–90.

31. J. P. Marat observes at many points in his autobiographical *Philosophical Essay on Man*, 2 vols (1773), that physiology led him to medicine, and medicine to the anthropological study of man.

32. Critics of Sterne are beginning to take account of this nervous aspect: see F. Brady, '*Tristram Shandy*, Sexuality, Morality, and Sensibility', *Eighteenth-Century Studies*, IV (1970–1), 41–56, and James Rodgers, '*Tristram Shandy* and Ideas of Physiology' (PhD thesis, University of East Anglia, 1979).

33. Especially in *la vie d'amour* and its medical explanations; James Adair, MD, explains in his remarkable *Essays on Fashionable Diseases* (Bath, 1786), 13–14, how the influence operated: 'Upwards of thirty years ago, a treatise on nervous diseases was published by ... Dr. Whytt ... Before publication of this book, people of fashion had not the least idea that they had *nerves*; but a fashionable apothecary of my acquaintance, having cast his eye over the book ... derived from thence a hint, by which he readily cut the gordian knot—"*Madam, you are nervous;*" the solution was quite satisfactory, the term became fashionable, and spleen, vapours, and hyp, were forgotten.'

34. I explain the rise of the four categories in 'Literature and Medicine: the State of the Field', *Isis*, LXXII (1981), 406–24 (*Enlightenment Borders*, chap. 1).

35. Johnson defines the word in the *Dictionary* (1755) and quotes a passage from John Floyer's *The Preternatural State of Animal Humours* (1696), but satyriasis,

the sexual condition, is not discussed in medical literature until the mid-Victorian age; see J. C. Bucknill and D. Tuke, *A Manual of Psychological Medicine: containing... Treatment of Insanity* (1858).

36. J. Thomson, *An Account of the Life, Lectures, and Writings of William Cullen, MD*, 2 vols. (Edinburgh, 1859), 1269–70, 344.
37. A. L. Donovan, *Philosophical Chemistry in the Scottish Enlightenment* (Edinburgh, 1975), 277–8, has produced some evidence to doubt this assertion but perhaps not enough to prove that Cullen, like Hume, was actually a sceptic.
38. Thomson, 1, 279–80.
39. Rousseau, 'Psychology,' in *The Ferment of Knowledge*, 143–209 (*Enlightenment Crossings*, chap. 4). While eighteenth-century thinkers would have called this subject 'secular natural philosophy', there is no doubt of their awareness that they are generating a new province of learning about 'psyche–logos'. To credit the nineteenth century for the rise of psychology is to locate the development far too late.
40. Moreover, Cullen had succeeded Whytt, 'the philosopher of medicine', as Professor of Medicine at Edinburgh, and was thoroughly familiar with his theories. See Thomson, 1, 100–16, for documentation.
41. Anti-empirical Blake denied, of course, that imagination had anything whatsoever to do with memory or the nerves; see Blake, *Milton*, 41:3–5, and D. Ault, *Visionary Physics: Blake's Response to Newton* (Chicago, 1974). But Blake notwithstanding, Bienville leads to a whole series of nineteenth-century hypotheses about 'nervous artistic creation' in which the artist's nervous maladies and sexual misadjustment enable him to compose; see Bernard Straus, *Maladies of Marcel Proust* (New York, 1980), and J. E. Rivers, *Proust and the Art of Love* (New York, 1981).
42. Bienville, *Traité*, 157–8.
43. Bienville, *Traité*, 52.
44. J. Hagstrum's *Sex and Sensibility* (n. 18 above) is excellent for its fusion of ideas about sex and sensibility, but the author overlooks the physiological connections, such as those implied in the psychology of the nervous whore or deviant.
45. Herbert J. Muller, *Science and Criticism* (New Haven, 1964).
46. See M. H. Nicolson, *Voyages to the Moon* (New York, 1948), 150–200.
47. M. D. Paley, 'The Sublime of Energy', in *Energy and the Imagination: a Study of the Development of Blake's Thought* (Oxford, 1970), 1–29.
48. Important reasons are supplied by Jean Deprun in 'Sade et la philosophie biologique de son temps', in *Le Marquis de Sade* (Paris, 1966), 156–93.
49. This and the following passages cited below appear in *Justine, Oeuvres Complètes de Sade* (Paris, 1966–7), VII, 107–9; the intentionally literal translations are mine.
50. Germain Pichant de La Martinière (1696–1783), whose many surgical and medical works appeared in the form of 'Mémoires' sent to the King.
51. Sade, *Justine*, 108.
52. The erotic aspects of this painting are discussed by Sir Kenneth Clark in *The Nude: a Study in Ideal Form* (New York, 1956), 208.
53. L. Stone, *The Family, Sex and Marriage in England 1500–1800* (1977), and R. Trumbach, *The Rise of the Egalitarian Family: Aristocratic Kinship and Democratic Relations in Eighteenth-Century England* (New York, 1978).

54. G. Bataille, *Death and Sensuality: a Study of Eroticism and the Taboo* (New York, 1962), 164–96. Although the assertion is impossible to document in a brief space, Edmund Burke, the alleged prophet of conservatism, was also alive to the connection of the nervously erotic and the economic; but his most recent biographer, I. Kramnick, may go too far in *The Rage of Edmund Burke* (New York, 1977) when he argues on p. 188 for a manically repressed Burke was also 'paranoid and homosexual'.

7
Discourses of the Nerve (1989)

By the mid-1980s I had been working on the nerves for two decades, concentrating primarily on the increasing discourse surrounding them in the eighteenth and nineteenth centuries. What would a broader approach entail with a larger time span for comparison and more focused on the brain? The brain had come into its own in the 1980s in ways unseen in the more confident 1960s, as the bibliography of this book makes evident. The neuroanatomical optimism of the 1960s was overthrown in the localization of the 1980s, suggesting that identification of particular anatomical domains would be a longer – and more arduous – process than had been imagined earlier. Contemporary neurosicence had been exploring cortical localization for a long time; new was the invigorated plasticity and emphasis on the synapse that suddenly made the Enlightenment discourses I had been immersed in spring back to life.

But what, I kept wondering, was plasticity? What fundamental difference to consciousness and cognition did it make if the action was centered in the synapse? A software/hardware divide kept suggesting itself analogically at that time (the 1980s) at a moment when computers promised so much and the neuroscientific debates centered on brain localization led to theories of Artificial Intelligence. Which wiring was 'plastic': hard or soft? Were both? The neuroscience I was reading (M. S. Gazzaniga, Antonio R. Damasio, J. E. LeDoux, Robert M. Young, Patricia S. Churchland, and especially Gerald Edelman's Neural Darwinism *published in 1987) intimated that a revolution in thinking about the pliable brain and its consciousness (not simply memory and altered states of awareness) was occurring.*

The new view was that the synapses in the brain were not hard-wired but malleable, continually being shaped by experience and capable of effecting neuroanatomical change as the result of altered experience. The old categorical view that 'we are our anatomy' was at best incomplete. It was rather that – with certain biogenetic provisos and very crudely put – 'our neuroanatomy is,

*at least in part, the product of our experience.' Yet the Enlightenment con-
stellation of nerves, spirits and fibers had anticipated this new, and more fluid,
synaptic software. Several revolutions had occurred, to be sure, in the rise of
neuroscience from c. 1700 to the 1980s. However, the current view promoting
synaptic flexibility within cortical localization, and involving experience
as crucial to the eternally changing state of the synapses, seemed remarkably
proximate to the eighteenth-century brain revolution from Willis forward.
Experience captured as sensation, and the animal spirits conceptualized as
the body's fundamental marking material, had been a basic tenet for the
nerve doctors in the world of Willis and Locke, Cheyne and Adair.*

*But how did Willis and his disciples make the leap to privilege brain
neuroanatomy in ways resurfacing in modern neuroscience? This essay
was written while contemplating that question in the context of the historical
span from the Ancients forward, and while taking a 'long view' suggesting
that Enlightenment sensibility – especially its imaginative literature and Scottish
philosophy – intuitively understood that it should preoccupy itself with (what
we call) the neurophysiology of consciousness. It was not, of course, parallel to
our neuroscience. That would go too far. Nor did it anticipate it: all science
moves more or less in flow. It rather intuited – in a loose sort of way but far
beyond any prior historical period – the plasticity of the body's nervous system.*

*The Enlightenment emphasis on experience as the crucial ingredient in
higher associated thought is self-evident to all readers who have combed its
vast annals 1660–1820. Experience is inscribed everywhere in its philosophy
throughout Europe. But the more I assessed nervous discourse of the eighteenth
century, and sought to provide adequate contexts for understanding it,
the clearer it became that whatever their neuroanatomists had propounded, the
British popularizers (Cheyne, Kinneir, Adair, Trotter, et al.) appropriated
the theory primarily for the purposes of social hierarchy, even exclusion. Nerves
and spirits were thus a type of caste system by the late eighteenth century. Put
otherwise, if Le Cat entered the racial debates on the field of nervous physiology
and Bienville the sexual wars there, Samuel Richardson and other writers of
'sensibility,' continuing to Jane Austen, the Romantics and the Victorians,
embedded the paradigm in the fabric of their plots and the texture of their
narratives. As time marched on, both doctors and writers conjointly shared in a
view that the most captivating people were 'suffering nerves' because they were
inherently well-bred and refined. You 'were' your nervous states: the more
nervous, the more marked your breeding and the higher your social class and
resulting status. Here then was neuroscience transformed to social sensibility,
as so many polite popularizers aimed to explain to their inquisitive readers
wondering where they stood on the social ladder. Also, the more you were
afflicted by nerves – as in so many ailing artists – the greater your talent with*

a price to be paid either in chronic ailment or abbreviated life. It was only a short step from this neural sensibility to Coleridge's poetry as madness of a higher type, and then to Nietzsche's Apollonian/Dionysian divide and Thomas Mann's visionary tuberculars.

This nervous-class basis also had monumental implications for gender. In the aftermath of Willis and Cheyne all sorts of upper-class males were suddenly configured as 'nervous': sensitive, sympathetic, emotionally fired up, empathic, melancholic, hysteric in their tears and tantrums, willing to die for the validity of their passions, sexually differentiated along the lines of heightened sensations of pain, and even – as one semi-pornographic writer of the early nineteenth century stated – by virtue of the possession of a greater number of nerve tips in their male genitalia the better to experience sexual pleasure. This neural sensibility of males was something new. The Appendix to this essay listing dissertations on differentiated nervous states – especially among male hypochondriacs and hysterics – demonstrates to what extent the nerves were now implicated in the mammalian condition itself: not merely as civilized Europeans in cosmopolitan capitals but also among wild animals in the jungle with complex nervous systems. Sensibility – especially an evolving Darwinian sensibility – was not merely a literary phase of the late eighteenth century, labeled and unpacked by critics like Northrop Frye in their aim to understand the roots of Romanticism. Sensibility was also a hallmark of Western civilization: if necessarily a classification of literature studied as a complex label by eighteenth-century scholars then also a truth about the perpetual crisis of the human experience revealed through one's own consciousness. To the degree that sensibility involved reflection on own's emotional state, it was tantamount to consciousness itself. The brain and its synapses had evolved in such a way, as neuroscientist Jean-Pierre Changeux has currently demonstrated,[1] to ensure this optimal neural heritage for all living creatures in the Great Chain of Being, to echo Pope's phrase in An Essay on Man, *who learn from experience.*

The expansive endnotes in this essay permitted me the amplitude to explore byways of these early modern neural frames: for example, to speculate on the strange reticence of historians of science and medicine to speculate about the effects of Thomas Willis' imaginative leap; to comment on the precipitous eighteenth-century decline, and then the strange alteration, of melancholy as found – for example – in Sterne's Tristram Shandy; *to notice to what degree nervous discourse had become by c. 1760 a rhetorical vehicle to contain the age-old mind-body debates; to rethink the magisterial early modern transition*

---

[1] *As I write this headnote his new book,* The Physiology of Truth: Neuroscience and Human Knowledge, *has just been announced as forthcoming in 2004; see the bibliography.*

from passions to emotions not merely in chronological and geographical terms but in the light of this neural legacy; to suggest that the tumultuous eighteenth-century milieu of ideas and practices had been the formative epoch in gendering the nerves; finally, to watch the 'nervous style' in English literature crest in the developing eighteenth-century novel and the sturdy muscular prose of critics such as Joseph Addison and Samuel Johnson – each of these topics worthy of a book in itself.

Discourses of the Nerve (1989)

I call this two-part chapter 'Discourses of the Nerve', having developed it as part of a larger project on the discourses of mind and body on which I have been engaged for the last few years (Rousseau, 1976); but in different settings and circumstances than this one it could as well have been called 'The *Product* of Literature and Science', for – as you will see in Part 2 – I am concerned as much with the *product* of our activity, with the finished object, the eventual discourse, the narrative produced at the end of the process we are beginning to call 'Literature and Science', as with the discourses themselves.

1 The discourses of mind and body

Between the late Renaissance and the start of the nineteenth century, the troubled theoretical relations of mind and body were encoded in dozens of different discourses. These were layered, or tiered, in the way Foucault and other contemporary discourse theorists have demonstrated (Foucault, 1973). Four of these discourses dominated the models of the early European naturalists, who were not yet scientists in our modern sense. I call these the discourses of melancholy, hysteria, the nerve, and sensibility, despite the vexed and thorny problems of definition and taxonomy, and the symptoms that were then diagnosed under the perplexing umbrella of hysteria (see Appendix A). Ultimately, all were discursive attempts to medicalize the imagination, then believed to be a crucial activity for the analysis of cognition and consciousness as well as mental pathology (Rousseau, 1976). Despite their differences, their overlaps and reciprocities, their rhetorics and their grammars, the four discourses had this medicalization in common; but no sooner were the discourses analyzed than the endeavor's implicit and inexorable meta-physic of mind and body became apparent. Before the early seventeenth century, mind and body remain demonically encoded; afterwards, they gradually reside in the domain of the naturalistic. Given the relatively swift diachronic transformation this is a crucial epoch for relations of mind and body. But by crucial I do not mean privilege, in the sense of privileging this transitional era over others: the nineteenth and twentieth centuries are, after all, equally crucial for other discourses. Nor do I mean privilege in the sense of suppressing their contrary discourses (i.e. *counter*-nerve, *counter*-melancholy, all those supernatural discourses that discourage secular taxonomy and empirical observation) or privilege as achievement, in the sense that Cartesian dualism or Lockean associationism can be said to have altered the fundamental paradigm of mind and body the Enlightenment inherited from the Renaissance. Yet during this early period the four discourses of mind and body broke away from the older, supernatural forms and embedded themselves in new discursive forms and genetic structures, some of which will soon be called 'scientific' literary forms by the Royal Society; and they endured until the end of the Enlightenment when another break – Romanticism (or call it Lamarckism or proto-Darwinism) – will again change the way the ampersand in mind *and* body is construed.

The reciprocity of mind and body is also intriguing as an analogue of the problematic in the ampersand of Literature and Science, even though there is no contemporary subject, no recognized field theory, of mind and body. One has to create a subject for it oneself: a narrative, a rhetoric, a story line, even to grasp what role the footnotes on mind and body can contribute to a field for it, just as Aldous Huxley, a much underestimated pioneer for Literature and Science, had to invent a narrative for a book replete with the discourse of mind and body, which he eventually subtitled 'a study in the psychology of power politics and mystical religion in the France of Cardinal Richelieu', and whose main title is *The Devils of Loudun* (1952).

But what *is* privileged here are the competing discourses of mind and body, their overlaps and reciprocities, their mutual rememberings and forgettings, on the rationale that we might learn something by analogy from the ampersand in their conjunction, and – furthermore – because cultural history has never had a subjective and social, as distinct from a logical and internalist, history of their reciprocity (see Porter). So this approach through overlapping discourses entails Literature and Science as an antidote, at least, to the impoverishing incompleteness of existing historical explanation. When we consider the social antagonisms that mind and body have served, is it any wonder that we should want to inquire after their ideological relation? The profiles of mind and body have been widely disparate in the popular imagination, body having suffered miserably at the expense of mind.[1] But body had been a trope for centuries; what changed in the Enlightenment was an altogether new level of metaphoric energizing: just the opposite of what the old school of historians of the Royal Society (R. F. Jones, Morris Croll, even Marjorie Nicolson) would have us believe about science and the 'plain style', about the constitution of a new genetic literary form called 'the scientific paper', or 'the scientific essay': plain and straightforward, without embroidering language in any way, laying out its parts like a mathematical proof – Q.E.D. – and then 'embodying' this thing, this amorphous knowledge called 'science'. I do not privilege either mind or body here as a completeness axiom for philosophers or historians of science but as a alternative approach to an inquiry they alone have mapped diachronically,[2] and I do so in the attempt to constitute a category, a subject, a field for it. Today it is a cliché to maintain that no one has construed this dualism seriously for over a century; it was not so before 1800, when both were terrifically mythologized. The ways they continue to be mythologized afterwards, is an area of debate this paper hopes to open up.

Anyone with an ear close to contemporary neurophysiology or sociobiology could surmise that there is no dualism: only the monism of matter and motion – vulgarly, that there is brain and essentially nothing but brain, as the philosopher Thomas Nagel has admirably argued (Nagel). But is brain mind or body? More importantly, is intellection mind or body? If my doctor tells me I must have a heart or kidney transplant, I do not think I will lose my identity as George Rousseau. But if my neurologist informs me that I must have a brain transplant – assuming we have reached a more advanced stage of medicine – I wonder what of me will remain, this despite the knowledge that the mental functions permitting intellection are not *all* located in the cerebellum. Here the logic gets thorny, if for no other reason than that the issue of selfhood and identity is intimately tied to these matters, as even the philosophers and scientists of early modern Europe recognized. No sooner do we claim brain is body, than the resonances of *brain as mind* clamor for attention on grounds that we cannot be identities that are less than our own consciousness. Many academic groups (not merely philosophers) privilege mind over body, even if they do not subscribe to its dualism. Among cultural and social historians it is clear that body has had a bad press since the Renaissance (see Porter; Gallagher and Lacqueur; Bottomley; Turner), so that even if philosophers demonstrate that the dualism of mind and body no longer exists today in any rigorous sense, on an important *non-philosophical* level the split is not a myth but an institutional reality.

Until we can account for why this divide was produced and continues to be maintained in many quarters (just like the divide of Literature and Science); until we trace out what interests the proponents of mind versus those of body have had; we are unable to abolish its existence by fiat. As a topos mind and body extended everywhere into Renaissance discourse: in sermons, in poems, in prose romances; in didactic treatises, in moral memoirs, in theological tracts; and – of course – in the anatomical and physiological discourses that figure in Appendix A.[3] The literary component of mind and body was also extensive. As the Renaissance waned, the four discourses gradually assumed the non-ecclesiastical (or secular) charge of the reciprocities of mind and body, just as colleges in our time have proved to be the landlords of arts *and* sciences (that problematic yet influential ampersand again) when viewed conjunctively.

1.1   The discourse of melancholy

Diachronically, the discourse of melancholy was the first to develop. It was an old discourse by the time Burton built his giant edifice on its ruins. Under less able pens or less polymathic minds than those of Burton in holy orders – minds like those of Timothy Bright, Andre Du Laurens and Felix Platerus (all physicians), or the divines who warned about the signs of religious melancholy – melancholy's rhetorical techniques and genetic encodings differed, but its semiotic remained the same. Melancholy took the chemical humors, especially bile and bilious fluids in the body, as its main sign, enlisting madness and derangement as its signifier, stressing that the sign was material substance and signifier predictable human behaviour.[4] To the degree that its authors were primarily physicians interested in therapy and cure, it is not surprising that sign and signifier should have been privileged in this way, nor odd that melancholy constructed a symptomology based on the derangements of the four humors by the determinants of climate, geography, terrain, weather, topography, and national boundaries. The diagnosis of melancholy was in this sense an interpretation of signs predicated on symptoms. Yet what is extraordinary about the *discourse* of melancholy – as distinct from individual symptoms or individual case histories – is the way melancholy was encoded in discursive genetic forms and then privileged among competing discourses as the embodiment of the cultural myth of mind and body. 'Embodiment' is, of course, a problematic concept, presenting many types of metaphysical hurdle to the theorist of discourse; it is not a problematic we can resolve here. But set the diachronic dials to 1550 or 1600, or even 1650: no other discourse of mind and body has been so universally mythologized.[5]

Whether as Hamlet the gloomy Dane, or in Taylor's 'Lamentation' of 1618 (see Taylor), melancholy captures the European imagination everywhere. As all words and things then revealed a morality and correspondence, so too did the melancholy of the doctors. It was the given belief that the malady resides as much in the mind as in the body. This dual residence is uttered in the aura of terror and wonder, as if this awareness had been the great scientific discovery. Doctors insisted that melancholy was no one pathological state but a more general condition of imagination. There were thought to be so many types of melancholy that it can be said of melancholy that it becomes the signifier of something *other* than biliousness; something akin to, but as yet not identical with, social status. This juncture is where the internal contradictions of the discourse become patent, for few human conditions have generated such metastories and scientific critiques.

Historically, both a literary and a scientific discourse never developed; the bifurcation is a figment of our post-Kantian imagination.[6] Dr. Bright learned as much classical rhetoric at Cambridge as any philologist; he invokes the trope of *partitio* as often as Burton, Bacon, and the non-physicians.[7] Burton not only heightened this use of classical rhetoric, but encoded his own encyclopaedic tendencies into the vehicle of the 'anatomy'. His *Anatomy* becomes the repository of the whole world (allegedly because bile and madness are endemic), encoded in a literary vehicle (the anatomy) whose essence lies in disjunction and fragmentation. The more Burton searches for the causes of the signifier madness, the less he can find them. All he discovers as he ferrets out truth from words and things, is chaos and confusion. In the process he himself seems to become mad and take on the attributes of the very signifier (madness) whose cyphers he claims to be decoding.[8]

But if the discourse of melancholy was *neither* essentially literary nor scientific, neither natural nor supernatural, because these intersections had not yet been fixed, its agenda was egalitarian *vis-à-vis* gender and sex, especially because the Renaissance perceives human imagination as androgynous. There is no sense of a hierarchy of genders other than that which the culture inscribes. Unlike other discourses then, melancholy conceals no hierarchies *vis-à-vis* gender, any more than it veils the author's ideology. This is why, later on, Swift could model his *Tale of a Tub* and Sterne *Tristram Shandy* on a neutral Burton whose only ideological agenda is the genial madness these authors cultivate through their mad hacks and disoriented narrators. In Bright and Burton, in Du Laurens and Platerus, males as well as females are mad, and female bile remains the same as male. Unlike the status of hysteria, the agenda here entails universal structures, a human rather than a genderized humoral chemistry, with Hottentots and Indians, for example, as prone to this humoral chemistry and subsequent diagnostic as the Queen of Portugal and Prince of Wales.

The ideology of the doctors also merits attention: paradoxically, they claim to be inside and outside the discourse at the same time, enmeshed in the signs they read by virtue of their professional skill, but nevertheless apart from the madness diagnosed. (I have not yet found a single female physician who wrote about melancholy and therefore I continue to invoke the male pronoun.) He legitimizes his writing by implied beneficence: that somehow the whole culture benefits from his 'scientific' expertise because almost everyone is depressed. This legitimation reveals the sequence of his power: to see in order to interpret; to interpret in order to diagnose; to diagnose in order to cure. No one (high or low, male or female) is marginalized. His power lies at the center because the whole world is confused, chaotic, mad; and the physician has dis-covered, he claims, the objective source of the chemistry of madness. In bile (whether black, white, yellow, or green), not in the mystery of the Godhead or the elusive human imagination, in bile lies the secret of the human Book of Nature. Governments rise and fall, human are born to die, but bile, it appears, remains a constant: it is an extraordin-ary mythology, claiming to be as 'scientific' and 'dispassionately objective' as anything the Ancients and the Renaissance ever pro-duced, but in fact it was rhetorical, polemical, ideologically loaded, anything but dispassionate. The medicalization of imagination was a groping for the unknown through the tangible of the bile. No wonder that Freud and the psychoanalysts of the early twentieth century revolted against this agenda and generated another myth of depression to replace it.

1.2 The discourse of hysteria

Renaissance and early European hysteria differed, although its tropes were stylistically no less neutral than those of melancholy, no less teleological. But hysteria developed mechanisms of concealment for its teleology – this despite its value-laden, ideological, and rhetorical agendas and despite its own self-conscious claims to objectivity. In some ways hysteria was the nighttime of melancholy. Its ideologies thrived on distinctions of gender, its male proponents having privileged male anatomy over the allegedly weak female genitalia from the earliest known pronouncements at the time of Hippocrates and Soranus of Ephesus. We should probably not overlook within this context that not a single male received the 'stigma' (the sacred stigmata) between the Middle Ages and the twentieth century, between Francis of Assisi and Padre Pio, though hundreds of women did.

Like melancholy, the discourse of hysteria aimed to show how endemic among females the signifier of hysterical behavior was, but here similarity ended, its agenda and program having a different purpose, as did its discursivities and diachronic appearances. What the Greeks meant by the 'wandering womb' or the 'suffocation of the mother' (*hysterike pnix*) bears little resemblance to the medical hysteria of the Renaissance despite the conventional histories of Ilza Veith and, before her, in France, the larger one by Georges Abricosoff, histories that will now no longer do.[9] Until the Enlightenment hysteria is said to be the female condition of the 'wandering uterus', significant for medical doctors when disordered or deranged. This mania is the 'furor uterinus', the swelling of the so-called rising womb, the 'furor' that Enlightenment physicians will exalt as the cause of the newly defined condition nymphomania (the mania again delimited to women: 'nymphs'); not actually wandering like a displaced organ, or floating like a self-contained island throughout the female body and within its rivers of blood, but rising – it was then said by the doctors – so high within its own position and thrust so extensively into the territories and domains of the other organs, into their fissures and crevices, their unoccupied spaces and open fields (this open field is the metaphor that continues to be used), that one might as well consider it a 'wandering uterus'. When inflamed and swollen it will be described as 'suffocating', stifled, choked; especially on the spatial grounds that there is no unoccupied vacancy left for its swellings to permeate.

Foucault thought hysteria was the bridge to modern psychiatry for the way it marginalized the female body, but he penetrated right through its claims to objectivity. 'This inferior [female] body which Sydenham tried to penetrate "with the eyes of the mind" was not the objective body available to the dull gaze of a neutralized observation' (Foucault, 1970, pp. 125–6). In the discourse of hysteria, this female body was a moral and ideological imagining. Foucault's study of medical texts led him to this insight. Yet, unlike the more egalitarian melancholy that privileged bile within the humoral derangement while devaluing the rest of the anatomy, hysteria was *par excellence* the suppression of the *female* body in her most vulnerable genital organ. Hysteria, the antidote to melancholy, was the discourse that dominated gender and sex in early Modern Europe, its concealed program the *inferior* (as well as interior) female body no matter to what degree it made altered states of consciousness its explicit agenda. Its more urgent subjects were the allegedly soft spaces and useless flaccid interiors of women (so much for scientific objectivity in the face of female bodies which, if they are anything, are more durable than male!). Thomas Sydenham, the English physician and medical partner of John Locke, will strive to alter this status in the 1680s for reasons that have not yet been explored (see Dewhurst, pp. 36–46). Sydenham will develop a new 'hypochondriasis' as the male equivalent of female hysteria and defend the bifurcation of this condition into male and female versions on scientific and objective grounds, whereas in point of fact the gender revolution of the Restoration lies behind his privileging of the male body in this new way. In brief: to be hypochondriacal in the 1680s and 1690s entails a new means of setting males apart from other males.

After ca. 1700 hysteria will again be transformed as its sign and signifier depart from its representations in the fifteenth-century *Malleus Maleficarum*; its discursive monopoly withdrawn, so to speak, as it marginalizes *upper-class* women in the name of spleen, vapours, 'hyp' (the vernacular term in the eighteenth century for medical hypochondria), nerves. But until then – until Willis and Sydenham – the discourse of hysteria serves to explain to women, however diverse its rhetorical strategies, why they are failures as a result of their anatomy; more specifically, as a result of their irrational cravings, erotic appetites, soft daily routines, idleness, unrelenting boredom, and especially their capacity for sorrow and suffering. And it reassures men, conversely, that they are immune from such frailty, having been anatomically privileged. This is why seventeenth-century printed annals are so remarkably permeated with discourse debating whether men could contract the never-fatal condition. Yet Sydenham's transformation of hysteria begs for attention, especially the way he re-anatomizes it in the hypochondrium or intercostal nerve behind the lungs. By so doing, Sydenham inadvertently exposes the follies of the enterprise of hysteria among his predecessors, and displays his own phallocratic biases to the effect that if men *must be hysterical*, we doctors can demonstrate anatomically why it must be a lesser sort of hysteria. It is no wonder then why Foucault believed that after the lunatics who were sent away on ships of fools, hysterical women (which is practically to utter a tautology) had been the most marginalized group of the population of early Modern Europe (see Foucault, 1970, ch. on hysteria, *passim*).

Yet if the agenda of hysteria and its research program in Europe's medical schools sought to remythologize and mechanize the uterus in the name of suppressing women by explaining their record of failure to them, its genetic encodings also differed. Unlike melancholy, hysteria thrived on the binary opposites of male and female versus the gender-free chaos that pervades the melancholic's view of the world. Different too were hysteria's oppositions and the deterministic mythologies it sought to revive, especially the view of women as sinners, chained to their bodies, enslaved to their (suffocating) wombs, while men were 'scientifically' demonstrated not to be so, all this occurring during an alleged scientific revolution in anatomy. Hysteria will explain this legacy of female sin through the fiction of an organically 'suffocating mother'. When this explanation fails, in the eighteenth century, the discourses of sensibility, equally fictive if less organically vivid and less pictorially dramatic than hysteria's will provide another approach equally grounded in anatomy. Small surprise then that our twentieth-century Marxist historians, like the seventeenth-century Puritans, should have been so profoundly attracted to the centuries in which hysteria flourished so exuberantly: a continuous record – so the male doctors claimed – of female sin.

The fictions of the discourse of hysteria were rhetorically charged. If melancholy had been 'anatomized' (Burton) in a genetic vehicle that sprawled, reducing itself to chaos and confusion, the discourses of hysteria had no such generic constancy. They could adapt to the Latin couplets of a sixteenth-century versifier, as well as the prose analytics of such practicing physicians as Thomas Willis, Charles le Pois, the Thomas Sydenham already mentioned, or the dialogic inventions of Bernard Mandeville (see Appendix A, 1711).

It is folly to erect artificial disciplinary boundaries for hysteria when none existed at the time of its Renaissance genesis and seventeenth-century transformations; all the more anachronistic, given the common ground of fiction and ideology, to surmise that these discourses of hysteria belong exclusively to the province of medicine. Was the Bernard Mandeville who wrote *The Fable of the Bees* as well as the treatise on hysteria primarily a doctor or a writer? The question is as bad as the false dichotomy, doubly so because we have been brow-beaten to believe that the doctor should be the more privileged because he tells greater truths than creative or discursive writers. And who was more objective, Mandeville-the-writer or Mandeville-the-doctor? Let us be more mercilessly explicit about what is at stake in these admittedly rhetorical questions: in which academic department should Mandeville's discourses be studied today, given that fiction, ideology, and rhetoric constitute their essential discursivity? If medicine had not been privileged to the degree it has in the last three centuries, the mumbo-jumbo of hysteria as a label to describe almost every state of altered consciousness (altered from what?) would not have had to await the twentieth century for illumination of its conceptual weakness. We would understand better than we do today why psychiatrists now claim to see few 'hysterics' of *either* gender.[10] Female anatomy remains a constant, but magically, it seems, female hysterics have all but disappeared. Of course cultural circumstances and social arrangements alter, and no one wants to impoverish history by castrating its differences, or doubt that Willis and Sydenham (and later on, Breuer and Freud) did indeed treat women with hysterical symptoms, but the fictions of hysteria, especially its biological determinism and techniques of rhetorical persuasion designed primarily for positivistic male readers, also need deconstructing, demythologizing, opening up.

1.3 Discourses of the nerve

We should not expect the discourses of the nerve and sensibility to be more dispassionately or objectively disposed than their predecessors or more sequentially patterned in the sense of source and influence on later discourses. What changes are the socio-cultural exigencies that place these discourses under stress, especially as they borrow from each other and forget what is inconvenient in the appropriation.

The discourse of the nerve arises when the animal spirit (already conceptualized in the sixteenth century but not yet influential) becomes the obsession of the mechanists in the middle of the seventeenth century in the aftermath of Cartesian physiology. As mechanistic and vitalistic ideologies then vied for privilege as the fundamental explanation of micro- and macrocosmic structures, and as they elicited a competing ideology of materialism during the eighteenth century, the nerve emerged as the signifier of every psychological theory of human behavior barring mystical and animistic ones; universally invoked by doctors, empiricists, moralists, ethicists, physiologists, and even diet mystics like 'lettuce-and-seed Dr. Cheyne' whose *English Malady* depends entirely on the nerve for its existence. By the nineteenth century and the era of Sir Charles Bell and, much later on, Ramon y Cajal, neurophysiology is already inscribed as the fundamentum of all theories of learning and knowledge. Whether in studies of the reflex action, cerebral localization, the autonomic nervous system, or the integration of these, the nerve remains the base unit. Victor Weisskopf, the physicist, has epitomised what the nerve has meant for human destiny when viewed in geological time, and Dr. Fred Plum, the neurologist, has put his bias more succinctly if also more positivistically: 'there is brain and only brain; ultimately, everything reduces to brain and can be explained by brain' (see Plum).

The reduction can be interpreted in many ways and endorsed or maligned depending upon one's attitude to positivistic theories, but it has been pointed out to Plum and his colleagues that they should read philosophers like Thomas Nagel (already cited), who would assist them to understand their own models of privileging. But the disputes between our twentieth-century philosophers and brain theorists notwithstanding, the nerve was first privileged in the Enlightenment to a degree it had certainly not been in the discourses of melancholy and hysteria, where there is rarely a sense of human beings as essentially 'nervous creatures' (in the hundreds of pages of Burton's *Anatomy* the nerves are almost never invoked).

The agenda of the discourse of the nerves is always a mediation between mind and body, less so between mind and brain; perhaps this is why its subtexts are always so hierarchical. Those who claim to pursue nerves empirically are aware of their mechanistic, materialistic, or vitalistic philosophies. They are always radical dualists or radical monists but rarely middle-liners. Their agenda often attempts to relate brain to body and thereby to construct a natural philosophy of the soul; their research programs show them as less than dispassionate, as they respond to political crisis and social exigency. Now I know that the word ideology was not coined until 1801 by Destutt de Tracy; the nerve researchers of the Enlightenment were ideological nevertheless. When we ask, what did those 'scientists' think they were doing, we eventually return to the same binary pairs of mind and body. Not all nerve research of the Enlightenment can be reduced, of course, to linguistic structures or concepts. The eighteenth-century experiments made for reflex action, for example, have little to do with language in their origination, were even more mathematical than they were linguistic, which reminds us of our enterprise here. Because we who advocate an interdisciplinary Literature and Science worry who our real audience is, we try to entice all potential audiences in the name of common assumptions; try to persuade in just the ways the linguists assure us we typify ordinary speech acts, especially the common assumption that scientific models are just another set of models to describe nature's laws: ultimately neither more nor less accurate than competing models, rarely free of value or ideology when set into discursive narratives, and certainly no 'truer' than any other fictions. And yet the two aspirins that relieved my headache on the airplane yesterday are not rhetorical, ideological, value-laden or polemical aspirins until I start *talking about them*. When we return to the basic unit of the nerve – the animal spirit – we glimpse to what degree nervous discourse was itself language-bound, for the sheer number of these discursive projects is remarkable (see Appendix A for examples).

Before the nineteenth century, the 'spirit' is the sign of all the discourses of the nerve, even in those I call counter-nerve (mysticism, hermeticism, the Paracelsian and Behmenistic tradition, Stahl's animism, Swedenborg, Blake *et al.*). As William Empson long ago noticed, spirit is a word lending itself to heavy metaphorization and rhetoricity, but my point is that 'spirit' also invited ideological privileging whether in written or spoken forms. In one sense the worst accident that ever befell physiology was its intimate association with animal spirits. Yet when we recognize that *all* Enlightenment physiology was 'nervous' and necessarily grounded in the spirit, we can see why a critique of nerves developed by those who claimed to be excluded from its social privileges. To the storytellers – the Sternes and Smolletts, the Diderots and Austens, who partook in this critique – the excesses prompted by the metaphoricity of nervous discourse were sometimes risible. But the doctors and physiologists, the Hallers and Whytts, the Cheynes and Cullens, could also be myopic to excess, or if not patently myopic (which much of the evidence affirms) then so enmeshed in the discourse as a cultural and professional 'product' that they cannot get rhetorically outside it. Enlightenment doctors, convinced of their utility to medical science, reduced human behavior to spirits, fibers, and nerves in their writings, creating a lush jungle of metaphor, which when brought under the lens of linguistic analysis proves to be a value-laden, subjective, and passionate labyrinth. Literary critics who till this era have enjoyed pinpointing the risibilities of these excesses, noticing how often animal spirits are terrifically personified, metaphorically embroidered to run on tracks and roads like carriages and trains, stop and go, are impeded by accidents, even heat up and catch cold (see Rousseau, 1976; Myers).

In Sterne's *Tristram Shandy* the animal spirits are the key to the riddle of the hero's dilemma: accounting for Tristram's messed-up personality (as the opening paragraph indicates), as well as his tragically (and by contrast at times comically) defective sense of time. Sterne derides their linguistic excesses from the start, and in one sense he deserves praise for curbing those so-called 'objective' Enlightenment doctors and physiologists. But in quite another sense, in his Rabelasian gene as it were, the Sterne who seems to be a sleepwalker in the world of Bakhtin's Rabelais is curiously both attracted to and repelled by the animal spirits, his ambiguous state of mind a perfect specimen of the fluidity of discourse when it labors under the anxiety of appropriation found in Sterne. More specifically, when someone like Sterne is attracted by its regenerative energy to fire up his own failing imagination (Sterne was seriously ailing as he composed *Tristram Shandy*), which was all too prone to embrace the radically vitalistic notion that life exists in its most irreducible form, in the life force of the animal spirit. And Sterne, quite unlike the very different Defoe and Fielding, makes the great imaginative leap that becomes the seed of *Tristram Shandy* as he himself lingers in the throes of consumption, in the depletion of his own vital life force. Recoiling from the animal-spirit treatises he had felt impelled to reject on purely linguistic grounds, they now allure and attract him – paradoxically – by their own regenerative power over him. Here then is the 'appropriation' and 'translation' of discourse through the strange tension of attraction and recoil. Not the anxiety of influence but the anxiety of *appropiration*. Not the influence of science *on* literature (those tortured arrows going in both directions) but the act of creation, in Koestler's sense, while in the imaginative heat of appropriating from another discourse, by importing and then exporting, by blurring, as would Godwin and Priestley later on, the dualisms of mind and body through the essential, irreducible life force that could regenerate him. The Tristram who self-consciously wonders if his book is 'wrote against any thing' can supply one answer only: 'If 'tis wrote against anything, – 'tis wrote, an' please your worships, against the spleen' (IV. xxii). Little wonder that a consumptive valetudinarian should set his Book of Life 'against' spleen, the seat of hysteria and hypochondria. The nerves, as both Sterne and Tristram knew, conveyed spleen's juices to his endangered melancholic imagination, as well as permitted the laughter – the true Shandyism – that kept him alive.

For us in this forum whose critical task is no different from all other forms of serious criticism, it is not merely the anxiety of appropriation that beckons but the larger cultural ways these doctors privileged the nerves. It was preeminently social and cultural privilege, deriving from extraordinary rearrangements then occurring in the Ancien Régime and, more specifically, between the genders. I claim no collusion between the scientists and the upper classes then, but the more I study this record, the more I grow persuaded that the agenda and research program amounted to a process resembling collusion. In the narratives and metacommentaries of the doctors (unlike the Sternes), there is usually one, unequivocal goal: the assertion of class distinction and gender boundary. Whichever route their discursive practices took, and whatever the degree to which these doctors and scientists remain inside or outside their discourses, the teleology of the agenda remains constant: to demarcate social classes and gender boundaries. This is why *female nerves* are treated as a separate category, and why the *female body* then breaks off so drastically from the male. In this sense the discourses of the nerve reveal an ideology similar to that of hysteria but with this difference: while hysteria sought to mythologize the *womb* as a means of carving out a notion of female gender, that of sensibility mythologizes, even demonologizes, the *nerves*. Both are anatomic, organic approaches yet with this distinction: if both genders have nerves, males have no wombs, females no penises. Perhaps a phallocractic difference underlies the gap between the discourses of hysteria and the nerve, but if so, it was not because the scientists were generating a neutral philosophical mechanism or vitalism, but rather as an urgent response to rearrangements of gender.

By now it is a cliché that in 1750 or 1800 many more people were writing than had been in 1550 or 1600. Those who wrote discourses of the nerve amount to at least tenfold the number as those in melancholy or hysteria, so we cannot expect the same isomorphic uniformities or similarities of view. The discourses of the nerve are more necessarily pluralized; within their ranks difference and marginality appear everywhere. Seizing 1800 as a convenient dial, a crisis of representation begins to appear among the scientists who now privilege their own discursivities above the agenda of the nerves itself, some preferring narrative in anatomy and physiology, while others endorse analytic commentary, critique, or polemic. Genetic encodings have also penetrated into every type of literary form: dialogical (as in Mandeville), discursive (as in Cheyne), poetic (as in Flemyng, the author of a long Latin epic poem published in 1747 called *Neuropathia*), satiric (as in Sterne), didactic (as in Jane Austen), fragmented (as in Coleridge), and so forth. As discursive commentaries the discourses of the nerve assert social stratifications, while their concealed hierarchy remains the supremacy of mind or body. As metacommentaries, they range from the serious (as in political or economic treatises that metaphorically appropriate their nervous agenda, i.e. 'body politics') to the playful and the satiric, as in Sterne's 'well-dress'd gentleman', who charmingly turns out to be a mere homunculus appropriated from the preformationists and epigeneticists. Always at stake in the agenda, no matter what its rhetoric, is a definition of life in terms of the dualism of mind and body, or soul and body, and a bias about what it means to be 'human' in distinction to the brute kingdom. In this sense there is no failure of nerve (*pace* Peter Gay who invokes this dramatic phrase in his study of Enlightenment culture – no pun intended). Enlightenment scientists *as well as* Enlightenment writers (it will not do to think of them separately or to read backwards, anachronistically, our own post-Kantian disciplinary boundaries) have appropriated the largest question of all, the question about life, but language has entrapped them. Small wonder then that Boyle, the great chemist but also often the most astute of commentators on the contemporary cultural scene, should write, unprovoked while in a meditative mood, that 'those great transactions which make such a noise in the World, and establish Monarchies or ruin Empires, reach not so many Persons with their Influence, as do the Theories of Physiology' (Boyle, pt. ii. p. 3).

1.4 The discourse of sensibility

Sensibility is fourth, a shorthand label for irritability, excitability, the passions in all their jumble and diversity, those vital exciting powers that Hunter and Cullen and Brown will exalt and the Romantics and Germans later on, a concept, however loose and imprecise, nevertheless familiar now to literary historians and historians of science. We also know it as the prelude to Romanticism, a label, a shorthand, for the European movement diachronically subsequent to Classicism. Northrop Frye called the literature of England between approximately 1740 and 1800 'an Age of Sensibility'. But the literary historians have no story to tell about the origins of sensibility, except to say that it first appears in the Restoration, in French novels like Madame de Lafayette's *Princess of Cleves* and in sentimental plays like Steele's *The Conscious Lovers*.[11] Yet imaginative literature seems to configure sensibility before science appropriates it, perhaps as imaginative literature idealizes heterosexual love. And literature will continue to transform sensibility into the nineteenth century, when novelists like Jane Austen, critical of its pretensions but appropriating its tropes altogether differently from the Sterne who by then seems to have inhabited another world, another cast of mind, turn it upside down and make it the centerpiece of *Sense and Sensibility*.

Yet it would be false to think, as some literary historians do, that sensibility was only a *literary* discourse. On the contrary, if it began there, or first appeared there in written discourse, it was soon imported into scientific writing. Albrecht von Haller, the Swiss Protestant physiologist, made sensibility the centerpiece of his physiology, claiming its complete dependence on nerves when he wrote that there can be neither sensibility nor irritability without them.[12] Haller maintained that only those parts of the body supplied with nerves possess sensibility – thereby privileging the nerves extraordinarily – while irritability is a property of the muscular fibers. It is not merely the rigid opposition Haller establishes that commands attention but the privileging of the one (sensibility and the nerves) over the other (irritability and the fibers). For his dual scheme and his sustained claim that this original observation was based on years of rigorous experimentation Haller was attacked; even so, his hypothesis was more encompassing than this, was embedded in a discursive prose work every bit as polemical and ideological as Boyle's pronouncement about physiology. 'Nerves', Haller wrote, 'are the basis for brain and sensory impressions, for all human passion and reason, for emotion and feeling, for higher associated ideas and principles, for the thoughts of monarchs and the legislations of parliaments' (Haller, 1786, p. iv). This appears remarkably close to Victor Weisskopf and Fred Plum when they maintain, almost deterministically, that we owe virtually everything we are to our nervous apparatus.

Haller may have been satisfied to decree physiological laws, but his colleagues wanted 'sensibility' to be the basis of an entire approach to life, an ethic in itself. Haller considers the primitive sensibility of sensory impressions merely as the microcosm of sensibility's greater role in the macrocosm of human affairs, as it was displayed in human sympathy, empathy, benevolence, virtue – all the cults of sensibility in the moral realm. This was a remarkable doctrine even for an optimistic era, and it was developed by Europe's finest scientists after Newton, Boerhaave, and Leibniz. Of Haller's sensibility Condorcet wrote in the *Dictionnaire Encyclopedie*: 'The work in which Haller published his discoveries [about sensibility] was the epoch of a revolution in anatomy' (Condorcet). But by 'anatomy' Condorcet signifies something other than what we do today; by anatomy he suggests a gaze, a way of seeing the entire world around him: anatomy as picture, image, vision, and – of course in view of its vestiges from the Renaissance – anatomy as dissection. Here is Charles Bonnet, the Genevese biologist claiming to be 'in contact with another great man, who was soon to make in physiology the same kind of revolution that Montesquieu did in politics: I speak of Haller' (Savioz, p. 155). Again, physiology signifies something much grander than the delimited domain we mean today, its orbit and sway both temporally and spatially, in Bachelard's sense, more extensive. The catalogue can be extended. Louis Figuier, the nineteenth-century medical doctor in Paris and prolific popularizer of the entire realm of natural science, continued to write about Haller's sensibility as if it had been the scientific revolution of the Enlightenment, *the* new way of grasping interior time and space (Figuier).

Haller's hypothesis was quickly translated from Latin and German and exported from Middle Europe. Robert Whytt, the Newtonian professor of medicine at Edinburgh University whose work on reflex actions remains a classic of neurophysiology, imported Hallerian sensibility to Britain, challenged it, popularized it, debated it; so did Cullen, John Brown of Brunonian medicine and, in France, virtually all the physiologists.[13] The optimistic claim was that Haller's theory of sensibility would do for the human body what Newton's calculus had done for the spheres. Today Haller's physiology is not news to the historians of science, but his charged language and social ideology is, as well as the cults of sensibility that developed around it or concomitantly with it. Like Newton, Haller knew he must expound his laws in ordinary language as well as in mathematics; he was aware how much of his success depended on the way he charged a single word: *sensibility*. The more lay culture popularized sensibility, the more scientists wanted to debate it, or transformed its original Hallerian conception to such a degree that a new version emerged, as was the case for Wordsworth at the very end of the eighteenth century. At stake here were not the metaphorical excesses and rhetorical flights of animal spirits, but the wide appeal of a doctrine claiming to be objective and unideological, yet one whose ethical implications were evident and unacceptable in several camps; moreover, a doctrine whose language (and I trust that this language is the point of interest to us) – whose language persuades and cajoles by appealing to those readers who *already* think they possess delicate nervous systems that will produce just the type of sensibility Haller describes.

How then did Hallerian sensibility progress from a physiological doctrine to the shorthand label for an entire literary movement? The route is more complicated than there is time for here and includes the tension of social arrangements then, as well as anachronistic tendencies of literary historians when they study a previous era. Even so, the scientific doctrine of sensibility was soon called upon to legitimate class distinction and gender difference. On the surface sensibility's research agenda was an early scientific positivism examining nerve tips, dermis, skin texture and color, reflex action and reflex arc in relation to norms of refined behavior, while attempting to demonstrate that acceptable social behavior is *physiologically* predetermined. Sensibility was in this approximate sense a type of eighteenth-century sociobiology. This is why fictional figures such as Sade's Justine are portrayed as having the most exquisite nerve tips in the anatomic nether region where they count for a woman of her propensities, and why – by contrast – her poor Hottentot cousin Venus, in Africa, has the thickest, roughest dermis on earth, presumably with very few nerve tips in it. Yet when sensibility's ethic is exposed, as it will be around the turn of the nineteenth century, the configuration of anatomy and character alters. Blake's women (to the degree they possess this gender) – Ahania, Enitharmon,

Vala – are not fibrous or nervous at all; Blake reserves nervous sensibility for his men, as if to reverse, even in gender distinctions among mythological figures, the prevalent scientific traditions of his time (see Hilton, ch. V, pp. 79–101).

The iconography of sensibility, like that of hypochondriasis, displays male sensibility as often as female. In this sense sensibility is somewhat less polarized *vis-à-vis* male and female gender. Its signs are also more varied: not merely the strength or flaccidity of the animal spirit or fiber within the nerve, but now those *in addition to* the texture of the dermis, tears of the eye, hot flush of the cheek, and natural expressions of blushing, weeping, crying, swooning, in brief all the anatomic manifestations of emotion and feeling, sympathy and empathy. A weeping or blushing male becomes the signifier of extraordinary delicacy and exquisite sensibility, in the Enlightenment thought to be quasi-hermaphroditical, in the nineteenth century evidence of dandyism or decadence.

Yet there is this difference. In the discourses we conventionally call imaginative and philosophical, the feelings and emotions are privileged over the reason and intellect as an index of sensibility, this while Haller and the physiologists are developing a so-called 'scientific' theory to substantiate the maxim that the more sensitive the nerves, the more sensitive the person. But in what sense then was sensibility an objective or dispassionate, a neutral or value-free discourse? The narrowly conceived (i.e. internalist) physiological advances that came out of sensibility were admittedly considerable and irrefutable so far as I know; but as soon as they become encoded in discourse they demonstrate another teleology. All writing, like all language, is of course metaphorical and rhetorical, and what must especially interest us here are not the general means of persuasion in a so-called 'scientific' text or discourse of sensibility, but the particular strategies used. In the immediate case, Haller's tropes are anything but neutral reports of the experiments he conducted. His discursive practices reveal what was new in the discourse of sensibility when compared to the older one of the nerves. Looking back from our gaze, we might say a confrontation was building between the forces of innate physiology versus those of environment. But if the discourses of the nerve continued to purport – as our own brain theorists do today – that we *are* our neurophysiology, that we *are* synapses, that we *are* our brain – sensibility develops the same possibility while affirming class distinction and gender difference. We are dealing here with doctrines of sensibility that had *already* been in the public domain *before* the Hallers and Whytts conducted their experiments, but all along my point has been that the physiological discourse of sensibility, as distinct from other writings on sensibility, was not immune to these social factors. It assumed them, it incorporated them, it embedded them into the tropes of its discursivity. Anatomy and physiology were – to be sure – among the 'softer' sciences, as even Ramon y Cajal would concede when interrogated by his positivistic-minded colleagues in physics and mathematics. These subjects could be assimilated more easily than physics or mathematics; and sensibility was just the kind of doctrine (like Freudianism or Jungianism) that lent itself to popularization prone to distorting it from what it originally was among the physiologists. Yet even under Haller and Whytt, some of the best of its Enlightenment theorists, and among dozens of other mechanists and vitalists, materialists and anti-materialists, and all those who are not easily labeled or cubby-holed, its social biases are evident.

2 Conclusion: the product of literature and science

It is time to take stock and see to what conclusions our argument has taken us. One dimension is clear even *without* conclusions: the topic sufficiently vacillates between theory and practice, between critique and metacritique, to indicate the presence of an inescapable tension. Again, this has to do with privilege of the one over the other – in this case narrative or story over critique – and the sense that theory, or metacritique, has not earned its keep, so to speak, unless validated by a prior practice.

Furthermore, two conclusions can be drawn without much reflection as they directly follow from my analysis of these four discourses. The first is that as I have tried to decipher these discourses I have continued to ask myself two main questions about what the method can yield, as well as inquire what these scientists (when they *are* scientists) thought they were doing. The more I isolated these two questions, the more I had to concede how much of the *activity* of science is not linguistic, or at least attempts to break away from the constraints of language. That science today is entirely unable to break away is, I presume, one of the main rationales for the development of Literature and Science as an autonomous discipline. But in the end it would be as detrimental to our purpose (however that purpose will eventually be defined among those who work in Literature and Science), as it would be unfair to the European scientists who generated these discourses, to pretend that *all* their activity was language-bound, or that they could confront reality in linguistic categories only, as if language were the only way their brain could process a reality external to themselves.[14] Detrimental, furthermore, to pretend that there were no scientific instruments, measuring devices, mathematical procedures, laboratories, diagnoses, therapies.

The second conclusion is that in the process of self-conscious reflection on my method of privileging discourse over instrumentation and therapy, I recognize that I present this terrain in an iconoclastic and possibly even disturbing manner to professional physiologists and neurologists, for whom language often signifies impediment rather than clarification, or at least an intermediate grid whose clarificatory functions are minuscule in comparison to those of laboratory investigation and especially of predictability and falsifiability. Disturbing, moreover, to some professional historians of science and medicine who find this threshold too all-inclusive for what they consider to be 'science'. Instead of approaching this material through the conventional disciplinary filiations (i.e. by viewing these subjects biographically, within a diachronic flow of the history of physiology, or as recognized *specific problem cases* to be solved for specific moments, such as the history of melancholy in the sixteenth century, or the history of hysteria in the seventeenth); instead of the broad contextual approach of literary scholars who until recently rarely looked at the domain of science as a valid influence; instead of all these, I offer a more or less synchronic analysis that gazes from afar at a vast bulk of discursive writing without privileging *literary* discourse over *scientific* discourse, or for that matter any discourse over any other, and by construing *literary* criticism as essentially similar to every other type of incisive criticism. What I do privilege is *this particular body of discourse* within a wide array of competitors in the period of early modern Europe largely on grounds of neglect, and I must be held accountable to demonstrate the consequences of the neglect. That is, I must justify that the neglect impedes historical understanding by a type of incompleteness axiom, as well as explain my refusal to privilege metacommentary and metacritique of these discourses in their own era over ordinary commentary and ordinary critique. I must at least be willing to show why the narrative of a physiologist in 1700, for example, is neither more nor less indicative of, neither more nor less value-laden than, the species I call the discourse of hysteria, or the discourse of the nerve, than a metacritique or metacommentary on one of these narratives.[15]

The conclusions that are less immediately obvious also derive from my approach but implicate as well, it seems to me, our activities in Literature and Science. They are, so to speak, the metacritique of the discourses of the nerve as it was worked up for this *particular* occasion, and they address Literature and Science in the abstract by taking stock of this historical moment. In both these senses, my approach focuses on the *peculiarity* of Literature and Science, construed either as the collective activity of a community of scholars or – alternatively – as a developing discourse that by now merits commentaries and metacritiques. By peculiarity, I mean the sense that there is something inherently different about 'performing' or 'doing' Literature and Science in an age of specialization when compared to 'doing' traditional literary criticism or even traditional philosophy. But this essential peculiarity also includes the perhaps equally alien notion – or so it must seem to many who have been trained to believe otherwise – that when scientific ideas are reduced to their verbal (i.e. literary, rhetorical, discourse) representations and encodements, scientific discourse must be treated no differently from any other type of discourse; Galileo's pronouncements deliberately cast in a fluid and vernacular Italian rather than in the accepted cadences and rhetorical Latin of the time serve as one example among many. And – what must again seem rather alien and peculiar to many contemporary scholars – the peculiarity that when science is viewed as a *discourse* rather than as a set of logic-rational inferences classified into the taxonomies we traditionally call organized academic science (biology, chemistry, physics, medicine, etc.) or accepted laboratory procedures, its 'meaning' is neither more nor less transportable or transferable than the meaning of literary and philosophical discourse. This matter of the transportability or transferability of scientific thought, in comparison with literary or philosophical thought, is not a subject about which many scholars have cared to comment. The reasons would seem to be too obvious to belabor.

The consequences of the 'moment' also require standing back, so to speak, in another way, for the two derivative 'taking stocks' entail the content and assumptions of a Literature and Science talk. But *have* I just delivered a Literature and Science talk, and how can I know it if there are as yet neither regulations nor accepted conventions for the genre? I might know it by the product of the talk, and all along I have been invoking 'products' and 'yardsticks of measurement' as if I knew in advance what these products and yardsticks had been productive of. But I cannot determine whether I have fulfilled my obligation here if I do not know what 'product' Literature and Science *should* be making, in the sense that the old New Criticism was promoting a specific set of principles determined by an a priori agenda. And if I do not know at any point what that product should be, then it follows that I *cannot* know what a Literature and Science talk is.

Should this product be a new theory itself, developed because scientific approaches have been brought to bear on humanistic domains and because the scientists themselves have overlooked this domain? Or, may the product of Literature and Science be something less grandiose, such as insights into a body of discourse (as the one I have been discussing), or, even less grand as insight into a specific author or single text? If the latter (i.e. the insights) then the scientists who seek to produce this product for Literature and Science will have to initiate themselves into literary theory, for I see no way of constructing this product at any level without a modicum of theoretical grounding, in my case having profited from semiotics, speech-act theory, and discourse theory.[16] And the literary critics – at least those who work in periods before 1750 or 1800 – will have to become historians of science, or at least historians of science *manqué*, if they hope to master this body of discourse and the interpretations that have been endowed upon it over the decades.

But if the former – if the product must become a *new* theory in and of itself, which has not been developed because there have not been numbers of scholars who worked professionally in the interstices of Literature and Science, and therefore have been unable to develop a chaos theory or productivity theory, a completeness theory or paradigm theory, all these being hypothetical examples – then we are going to have to shift our notions of the diachronic aspect of the enterprise. And we will need to eliminate (to be brutally frank) three quarters of the papers usually presented in forums called Literature and Science, *mine included*, as I have certainly not developed any new neurological theory, or any other type of theory here, except to suggest that discourse theory serves the purposes of, and opens up fields for, Literature and Science when placed under certain specific conditions.[17]

To conclude, if the product of Literature and Science must be a *new* theory in and of itself, then a product such as 'discourse theory' at large or the 'speech-act theory' of the linguists has greater claim to being a genuine product of Literature and Science because it was generated by humanists (that is, those trained to interpret the languages of texts) who applied a scientific approach to a hitherto humanistic enterprise, than the product of all those who have not yet discovered a new general theory as in the examples I have just provided. More tersely put, if the former, then the only way to test whether someone engages in Literature and Science is first to ask them what they are working on and then what their *new* theory is. Finally, if the former (i.e. the new theory), then we will have to address a whole range of hard questions about the professional institutionalization and implementation of our subject, which most of us have not even begun to formulate, let alone place before deans.

Robert Scholes has written in the opening sentence of his book on semiotics that 'the humanities may be defined as those disciplines primarily devoted to the study of texts' (Scholes, p. 1). What will our professional scientists think when they discover that they are 'scientists' at home but 'humanists' within the Society for Literature and Science? – an appropriation, an importation, an ambivalence, a tension, which, if anything I have been suggesting is valid, is all to the good in relation to the creative act, but which, when we have all returned to normal academic business, presents hurdles and obstacles to appointment, promotion, and tenure. And what do we say to the theorists among us who, flaunting the new discourse appropriations that disenfranchise the conventional disciplines, point out that our discussions do not concern Literature and *Science* but Literature and *Literature*, or *Rhetoric and Rhetoric*, or Tropes and *Tropes*; that is, not a conjunction at all where the force of emphasis lies on the weight of the ampersand in Literature *and* Science, but a leveling out, a reducing of all things into words, as if this newly formed society (SLS) should have been called 'Society for Literature and Language', *sans* the science, as we extract the linguistic part of science and banish all the rest, as if all the human brain can do is process reality in linguistic categories.[18]

And what will our deans and provosts say when we justify funding for our next sabbatical leave on grounds that we humanists want to fill in the interstices between Literature and Science, forgetting the weight and ambiguity of the ampersand as well as the directions of the arrows for the moment, and overlooking the unassailable fact that the concept of science has been very greatly transformed over the centuries, as philosophers of science *manqué* or historians of science *manqué*, and that we are the *only ones competent* to accomplish this because we humanists know how to interpret texts? Will they reply, but I thought you were in the German department, or the French department? And what will be their retort to the scientists amidst us who, perhaps more naively than us but more practically, continue to ask why all these words are needed rather than well-stocked laboratories?

Finally, considering the social realities of the world we inhabit today, its corruptions and injustices, its political schisms and political polarities, what do we think we are doing in a newly founded Society for Literature and *Science* – here let us indeed emphasize the linguistic, rhetorical, and even the demagogical dimension of science – when women continue to be oppressed in the name of inferior anatomy and physiology even if we no longer call them inferior interior spaces, and when races, religions, and sexual alternatives are condemned by a new American fundamentalism, a new American moral majority, a proliferating neo-Right, which some say control billions of dollars and manipulate millions of minds, even the young American minds who sit in our college classrooms but who are not learning because a prior manipulation has prescribed that the free investigation of science is forbidden in a Christian world, and that literature is valuable only to the degree that it has been both censored and censured?

Will we reply that Literature and Science (again forgetting the very problematic ampersand and the epistemological dilemmas posed by the direction of the arrows of influence) is actually neither a new brand of humanism, as George Sarton claimed half a century ago,[19] nor a plugging of the interstices of literature and science, however construed, but more . fundamentally and more practically, a 'science criticism', just as there is music criticism art criticism, literary criticism, sports criticism, criticism almost of a journalistic type, where the author need not explain much more than the newspaper journalist?

I would like to think that Literature and Science could remain pluralistically receptive to all these approaches for the time being, an open field without fences or no-trespass signs, at least until it can explore the institutional and the academic implications of its recent blossoming. But I realize that a desideratum for this type of pluralism may be untenable in academic settings where turf counts for much, where the student's precious time is often too limited to study these ancillary matters, where the conventional university departments have the weight of decades behind them, and where for the one hundred and thirty scholars who convened in Worcester in October 1987 and who believe in building these bridges and creating these new taxonomies, there remain thousands – literally thousands – who vigorously disagree with us.

Appendix A

Note: The list below provides a very eclectic sampling of the types of discourse in each category, and is not intended to suggest that the list is representative of the category, let alone complete in any way.

The Discourse of Melancholy:

1586	T. Bright, *A Treatise of Melancholie*
1597	A. du Laurens, *Discourse de la conservation de la veue: des maladies melancholiques*
1600	Nicholas Breton's *Melancholike Humours*
1600	F. Platerus, various treatises on the melancholic fever
1623	E. Ferrand, *Maladie . . . melancholie erotique*
1691	T. Eggers, *A Discourse Concerning Trouble of Mind and the Disease of Melancholy*
1716	R. Baxter, *The Signs and Causes of Melancholy*
1723	W. Stukeley, *Of the Spleen*
1727	W. Harte, *Religious Melancholy*
1742	E. Synge, *Sober Thoughts for the Cure of Melancholy*
1765	A. C. Lorry, *De Melancholia et Morbis Melancholicis*

The Discourse of Hysteria:

1623–39	Charles le Pois, various treatises on the signs in medicine and the causes of hysterical maladies
1670	T. Willis, *Affectionum quae dicuntur Hystericae et Hypochondriacae*
1682	T. Sydenham, *Epistolary Dissertation on Hysteria* (Eng. trans. 1696)
1704	M. Alberti, *De morbis imaginariis hypochondriacorum* and *Dissertatio medica de hypochondriaco-hysterico malo*
1711	B. Mandeville, *A Treatise of the Hypochondriack and Hysteric Passions*
1725	R. Blackmore, *Treatise of the Spleen and Vapours, or Hypo and Hyp*
1729	N. Robinson, *A New System of . . . Hypochondriack Melancholy*
1732	T. Dover, *Hypochondriacal and Hysterical Diseases*
1777	P. Pomme, *On Hysteric and Hypochondriac Diseases*
1777	J. Berkenhout, *A Treatise on Melancholy and Hypochondriacal Diseases*
1783	E. P. C. de Beauchene, *De l'Influence des affections de l'âme dans les maladies nerveuses des femmes*
1788	W. Rowley, *A Treatise on Female, Nervous, Hysterical, Hypochondriacal, Bilious Disease . . . with thoughts on madness and suicide*

Discourses of the Nerve:

1733	G. Cheyne, *The English Malady*
1738	D. Baynes (Kinneir), *A New Essay on the Nerves and the Doctrine of the Animal Spirits*
1740	M. Flemying, *Neuropathia; sive, De Morbis hypochondriacis et hystericis*
1765	R. Whytt, *Observations on the Nature . . . of those Disorders commonly called Nervous, Hypochondriac, or Melancholic*
1768	W. A. Smith, *A Dissertation upon the Nerves*
1744	J. F. Isenflamm, *Ueber die Nerven*
1778	D. Smith, *A Treatise on Melancholy and Nervous Disorders*
1784	A. Tissot, *Traité des Nerfs*
1787	J. M. Adair, *Essays on Fashionable [Nervous] Diseases*

The Discourse of Sensibility:

1750	M. Sherlock, *Letters on Several Subjects* (Letter 3, 'Of Sensibility')
1751	R. Whytt, *An Essay on the Vital and other Involuntary Motions of Animals*
1751	D. Diderot, *Elements de Physiologie*
1753	A. von Haller, *A Dissertation on the Sensible and Irritable Parts of Animals* (Latin 1753; English translation 1755)
1759	A. Smith, *Theory of Moral Sentiments*
1761	H. Smith, *Essays Physiological [the physiology of sensibility]*
1768	L. Sterne, *A Sentimental Journey*
1769	A. Tissot, *An Essay on the Diseases Incident to Literary and Sedentary Persons*
1770	P. H. D' Holbach, *Système de la Nature*
1772	M. Helvetius, 'La Sensibilité Physique,' in *De l'Homme*
1773	Anon., *Sensibility. A poem, written in 1773*
1775	J. P. Marat, *De l'Homme, ou des Principes et des Loix de l'Influence de l'Âme sur le Corps, et du Corps sur l'âme*
1787	Mrs. Harriet Thomson, *Excessive Sensibility* (a novel)
1798	G. Canning, *Pieces on sensibility* and *The New Morality*
1811	J. Austen, *Sense and Sensibility*

Notes

1. Another essay is necessary to document the point. Suffice it to say that some redress has occurred in the last decade, although the just published *Oxford Companion to the Mind* (Gregory) has no plans to add a companion volume on the body.

2. The philosophers have written at great length about the mind–body problem, but in most instances have not done so diachronically. A specimen of the kind of question they ask is found in Matson. For a different approach see Rather.

3. See the headnote to the Appendix. The taxonomical crux is genuine and cannot be dismissed: I divide this vast hulk of writing into four discourses for the purposes of diachronic convenience (i.e. arranging the four chronologically) and – more significantly – in order to demonstrate how the intrusion of 'nerves' changed each of the discourses after the mechanization of the nerves (Cartesianism) in the middle of the seventeenth century. There were theories of the nerves, of course, *before* then (Galen, Hippocrates, Vesalius, Fernel et al.) but the nerve was not considered to be the source of all life as it would be in the aftermath of Descartes. A more valid objection is why there are not *three* discourses without that of the nerves, each existing in a pre- and post-nervous (i.e. or pre- and post-ca. 1650) phase. The main reason is that the discourse of the nerve assumes a life of its own by 1650 or thereabout, and I have wanted to show its distinctness from the highly derivative but nonetheless different discourse of sensibility. But I hope to amplify this tetrapartite taxonomy some-

day more fully in *Fire in the Soul: The European Medicalization of Human Imagination*, a book in preparation.

4. My approach fuses this semiotic attitude with speech-act theory; the utility of both, considered individually and then in conjunction, is evident when one compares it with a more conventional historical approach.

5. If the question is asked: what is the competitor to the discourse of melancholy? the reply is the discourse of hysteria (not hypochondriasis, which was amalgamated with hysteria at the end of the seventeenth century). The literary tendencies (discursive practices, genetic encodings, rhetorics, ideologies, value-structures, gender differences) of each discourse are different despite certain overlaps dictated by their both being generated within the same culture; for this reason few authors who write the one discourse also write the other, as can be demonstrated bibliographically. For further evidence of the disparity of these discourses, see Jackson. Like Jackson, I have found few treatises that combine melancholy and hysteria: for hysteria read on.

6. I.e., as a consequence of Kant's *Critique of Judgment*, after which time it became nearly impossible for anyone philosophically aware who had read and understood Kant to construct an adequate and determinate *theory* of any discourse, let alone of several discourses, on grounds that no discursive project ever fully knows itself.

7. For this trope at the time, see Vickers; for the curriculum there, see Winstanley.

8. See Hodges, whose approach diminishes the role of the imagination.

9. See Abricosoff and Veith. There were, of course, many others, including Cesbron (1909), Bianchini (1931), Howe (1944), Brain (1963), Wajeman (1982), and Trillat (1986). Helen King has provided a revisionist argument to show that the Greeks meant something entirely different by 'suffocation of the mother' (*hysterike pnix*), and that hysteria as we know it is a modern, Renaissance invention; see her unpublished PhD thesis, 'Medicine and the Ancient Greeks', Cambridge University, 1986.

10. Some explanation of the reasons are found in McGrath, especially pp. 152–61.

11. On the vexed matter of its origins and of the rationale for discussing origins at all in a problem of this type, see my discussion of the primary as well as secondary literature on the subject (1985).

12. Haller, 1755, p. 690. This is the anonymous English translation based on Simon Tissot's French translation of 1755 and now printed by O. Temkin as a supplement to the *Bulletin of the History of Medicine*, IV (1936), pp. 651–97. In the introduction Tissot claims: 'The great discovery of the present age is SENSIBILITY and IRRITABILITY, described in the following treatise' (1755 edn, p. iii). For Haller (like Goethe) as a splendid example of the man of literature *and* science before Wordsworth and Coleridge conceptualized this type in the now famous preface to *Lyrical Ballads* (1800; 2nd edn 1802), see Guthke and Rudolph.

13. For Haller and Whytt, see French; for Cullen, Thomson remains invaluable; for Brown and Haller, see Neubauer; for Haller and France, see Lesch; also important here is Dewhurst and Reeves.

14. This processing of all reality into linguistic categories is a crucial matter in system building, and would make an excellent topic for a future conference

of SLS, especially if poststructuralist critics could talk to brain theorists and if deconstructionists could discuss the matter with those interested in artificial intelligence.

15. The historical moment constantly changes. In 1975–78 it seemed essential to assess the influence, however ambiguous it may then have seemed, of French critical theory on traditional approaches to Literature and Science; see Rousseau (1978). I surveyed the field as a literary historian, very much alive to the discussions then raging among historians and philosophers of science about the immense problem of demarcation in the two realms; see, for example, Allen G. Debus's inaugural lecture as the Morris Fishbein Professor of the History of Science and Medicine at the University of Chicago. By 1980 various groups were discussing literature and medicine in ways that seemed more vital than the older and more traditional Literature and Science; for discussion see Rousseau (1981).

16. I acknowledge my indebtedness while being aware of the theoretical flux of these approaches and the thorny problems they continue to pose, but I would have to write another paper to explain my relation to them and how their assumptions inform my practice here. Although my grasp does not lie on the cutting edge of work in these fields, I am nevertheless unembarassed to relate that I have profited from Pratt.

17. My sense of discourse differs from Timothy Reiss in that I view the appropriation of discourse from other domains than its local one as ongoing after ca. 1700, the period where Reiss stops; nor do I believe that what Reiss calls 'a discourse of modernism' is in any way identifiable, let alone completed, by 1700, requiring as it does the transformations of discourse ca. 1800.

18. This approach may appear to be that of the trickster but the name of anything is crucial to its entire subsequent development; for the essential, even quintessential, act of naming, see Hochberg.

19. See Sarton, who begins: 'My views on the history of Science were first published in Brussels in 1912 and those on the New Humanism in 1918' (preface, p. ix). It has taken the better part of a century to absorb them.

8
Towards a Semiotics of the Nerve (1991)

This essay accompanied its predecessor and was written at the same approximate time, i.e., around 1989–90. Here I aimed to interpret the nerve as the sign, or diagnostic, of something other than itself, almost as if the stones could speak about an unseen world in which mind informs body. By 'diagnostic' I meant ultimately the intuition: not diagnosis as the revelation of any biomedical signs but, rather, what nerves and spirits held out, in history, or for an evolving human condition. Hence I configured its protean human shapes, especially the Enlightenment creature of sensibility in a state of heightened, lowered, flattened or deranged nerves: as melancholic, hypochondriac, hysteric, depressive. The legacy remains: the human condition is fundamentally melancholic.

But I could see that the age-old mind–body debate peered out. It was impossible to avoid it, just as it continues to reappear in contemporary neuroscience disguised under other names. The four discourses I isolate here (see below) proved to be no more than the cultural framing of four historical epochs: melancholy in the Renaissance, hysteria and hypochondria in the early modern period, nervousness in the Enlightenment, and so forth. Other frames could be found. Yet I was sustained by their presentist appearances, especially the endurance of these mind–body debates into our own time. Another framing apparatus struck me: set the chronological dials to circa 1700 and the four discourses were more proximate to each other than they became in the nineteenth century. Previously their boundaries were blurred, almost non-existent. Now, as time advances, they become demarcated. Yet their shared mind–body dilemma continues into our own time under different images and signs. A new radical monism today, confident that neurophysiological explanations are the whole of the truth, assures us that both sides of the divide can be explained by brain mechanisms. Of course, they cannot.

Disciplinary concerns also continued to vex me, especially the way that academic subjects were paradoxically growing further apart in contemporary research cultures despite a rhetoric of interdisciplinarity and lip-service to wed (loosely) literature and science. Everyone claimed to be an 'interdisciplinary scholar' but no infrastructures would support them for economic reasons. Still, the real matter here was the earlier problematic of two decades (1968–88) woven on a tapestry of four interrelated discourses. In my four approaches to 'the nerve' each avenue suggested something else. Each appeared to contain a sign system for some segment of the human condition vis-à-vis cognition and feeling, reason and passion, knowledge and fantasy; yet each was diminished without pondering its binary Other. These truths seemed far more engaging than any labels attached to historical epochs or literary movements. The chronological convergence of a Scottish Enlightenment and the European cults of sensibility were, of course, noteworthy. They too were labels. But the diachronic appearance of the two at the same approximate time (Enlightenment and sensibility) must not be suppressed or overlooked. Yet even Freud or Proust, a century later, would have understood themselves in these previous terms of nervous sensibility, each having invoked this language of the older post-Willisian discourse. Was Proust not every bit a 'Man of Feeling'?

Consciousness, however, was one realm: a large part of the mind and soul of man. What, I wondered, was the mechanism of consciousness if every historical epoch set about to discover itself under the sign of something else: whether as melancholy, hysteria, hypochondria, or sensibility in this huge chunk of European discourse or – now – in our cortical localizations and exquisitely nuanced synapses? I began to think that the quest might ultimately not be scientific (i.e., neurophysiological) but religious: the condition of the soul as it marches through cultural revolutions (Industrial, Marxist, Bolshevik, Chinese), sectarian clashes and – more recently – accelerating technological change. If the latter approach has any validity, then our purely neurophysiological search for the sources of consciousness, enabled by this new technology, merely becomes another religion among dozens of prior ones: merely the latest in a long series – all of which returned me to the transformative eighteenth century.

The gap between the apparent fact of consciousness and a higher nervous sensibility captured in so-called 'nervous maladies' is my subject here. So much of the poetry of the Enlightenment grappled with this subject in its own clumsy ways and embedded this tension in its verse forms. Erasmus Darwin is a prime example, unifying the natural kingdom and poetic universe into one. He would have found these discussions and their predicaments redemptive. That is why I have cited him. He would not have chided me for noticing that a frog's brain differs radically from a human being's; therefore, a frog's consciousness and imagination is of another order. But what are we to say about the human condition, lorded over by a brain containing trillions of nerve cells, when it is evident that throughout history it has continued to be nuanced under the sign of an enduring – as Dürer might have said – melancholy? The whole of classical music, not merely Beethoven and Tchaikovsky in their famous 'melancholic serenades', embraces these voices of melancholy. Are nerves, and the brain they serve, unrelated to melancholy? What, anyway, is the melancholy of a frog? And what are the human consequences, apart from prelapsarian sin, of an evolution of melancholy anyway? Like consciousness itself, the Age of Saturn has been more or less continuous. A strong case can be made that we are still living under its sign.

Culture Viewed in Geological Time (1991)

It may appear odd to begin a chapter in *European* cultural history with a poetically sweeping statement culled from the work of a contemporary physicist whose research has earned him a place in the pantheon of science. All the odder when our discussions today centre so pre-eminently on local versus global knowledge, relativism rather than absolute reason, and regional ideologies – especially now in the sciences – rather than on immutable laws of nature derived from cold reason and logic; but I trust my drift will soon become clear in relation to this seemingly remote opening. In *Knowledge and Wonder: The Natural World as Man Knows It*, Victor Weisskopf explains how organisms developed in history and what the accretion of a particular function meant for the differentiation between the bacterial world and man:

> The nervous system is perhaps the most important innovation in the progression from the bacterium to the higher species. Nerves are long strands of special cells that, like telephone wires, transmit messages from one place to another . . . The brain itself is a complicated tangle of an enormous number of nerve cells, as many as ten billion, which are interconnected and arranged in a way we do not yet understand. But this tremendous unit of nerve cells is able to react to the stimuli coming from the outside. It can think and feel.[1]

Having established that man's nervous capability is what differentiates him from all other living forms and creatures on the earth, Weisskopf describes what this nervous apparatus has meant for human destiny viewed over geological time:

> The greatest step forward in this trend for better coping with environment was the development of the nervous system. This is a special combination of interlocking cells capable of transmitting stimuli

from one part of the unit to the other. Thus, coordination became possible between the functions of different parts. The most important innovations made possible by the development of the nervous systems were the sense organs. They are special cell accumulations that are sensitive to messages from the external environment such as light, sound, pressure, smell, etc. The messages received are transmitted through connecting nerve cells to other parts of the unit so that the unit is able to coordinate locomotion and other reactions to the outside conditions.[2]

Then Weisskopf demonstrates the utility of this co-ordination capability in prehistoric times:

As a result, the units could react on changes in the environment in many ways that were most useful for the protection of the individual and for the acquisition of food. The structure could move toward light; it could recognize food by its smell or its shape; it could avoid danger by moving away or by protecting itself when a large object approached. Our unit acquired what we call a 'behavior'.[3]

But if nervous co-ordination permitted man to communicate with the outside world and escape the prison of his solipsistic self, as well as acquire the first versions of a behaviour, the further organisation of this unit allowed for a brain of a particular type with retrieval capabilities previously unknown in geochemical history. The transition was so consequential that it is worth quoting Weisskopf's lucid description of it at length, especially for a non-scientific readership possibly unfamiliar with this segment of the primitive history of our race:

The development of a nervous system was so useful and effective that any mutation or sexual combination leading to a larger or more intricate nervous system gave rise to increasingly successful units. Thus a continuous evolution toward an increase in nerve cells began, and led to the formation of a brain. This organ is an accumulation of a large number [i.e. over ten billion] of interconnected nerve cells capable of storing the effects of the stimuli that the unit has received. The storage was the beginning of what we call memory. An action that previously has had good results with respect to food intake or avoidance of pain is kept in memory and repeated readily if similar circumstances recur. Obviously the ability to 'remember' such situations was an enormous asset for our units and helped their struggle for survival under difficult conditions. It supplied the ability to learn from experience.[4]

Here then is early man – the first experiential organism who learns by remembering:

> At the beginning, such memory and learning mechanisms were not very complicated. With modern electronic equipment one can easily construct a device with a 'nervous system' that remembers past situations and determines its actions on that basis. A machine controlled by a modern computer may serve as an example.[5]

Yet Weisskopf's example now captures the essence of the unit's organisation, as it soars to its main point:

> A system of interlocking nerve cells is in many ways equivalent to a system of interconnected electronic vacuum tubes or transistors. A device with a few thousand transistors can perform most impressive acts of remembering situations and avoiding them later on. But, in fact, the brain of even an insect is a more complicated device. It contains ten to a hundred thousand nerve cells. The human brain has as many as ten billion; it is infinitely more complex than any man-made computer.[6]

Weisskopf's summary of the development of life hangs on this event of brain formation, as we ourselves stand reeling over the implication that a mere insect has a more highly developed brain than Weisskopf's system of interconnected electronic vacuum tubes, and that man's brain is infinitely more complex than any man-made computer. One of these implications is the boundary – the almost ineffable demarcation – before and after the nervous event. To think, analogously and in the context of another boundary (this one linguistic), as the poet W. H. Auden wrote in 'Venus Will Now Say a Few Words':

> Think – Romans had a language in their day.
> And ordered roads with it, but it had to die.

So let us imagine there was a time when all reactions to the external stimuli were *completely* and *entirely* determined by chemical structure without the advantages of memory and learning. Chronologically, more contemporaneously to think that when Vladimir Horowitz moves us to tears he achieves all this solely – or almost solely – by the use of nerves and brain; that he is not feeling or interpreting Scarlatti or Schumann in some unique way, not achieving his successes at an intellectual level occasioned by greater understanding, but better co-ordinating, better summoning the synapses of his own nervous apparatus. This disparity is the one so consequential, it seems to me, for the cultural history of the Enlightenment, remote as nerves and synapses may appear at first glance to readers expecting a cultural history based on sociopolitical institutions and the ideas that surrounded them.

The nerves in Early Modern European culture

This analysis of life over millions of years will instil wonder in any observer, but tells us virtually nothing about recent history: the last few thousand years, the last few centuries. Weisskopf demonstrates to what degree this nervous organisation rested entirely on combinations. His very language explains that: connections, interlocking communications, transmission, co-ordination, function, efficiency. It is clear from his metaphors that the more nervously complex the organism became, the better 'combined' and therefore more efficient, its processes of organisation. And it is evident from his language that everything modern man celebrates is in some way the direct consequence of this nervous capability. Nervous capability, to echo Keats, was therefore meta-regional and meta-national: it transcended parts of the earth and countries, affecting the whole species rather than a part. Anything indigenous to the nations – the English, French, Dutch or Germans – would be insignificant in comparison to it; and the fact that England has been an island, France more peninsular, and Germany more landlocked, all these pale in comparison to this unassailable point about the development of nervous organisation. The bewildering matter is not merely that *everything* cognitive and functional is neurophysiological, but that literally everything we think and do is ultimately nervous. And then, to think that the entire analysis of inchoate subjects, like high and low culture in the Enlightenment, is necessarily neurophysiological. This is a recognition that does not dawn easily on those (like most of us writing in this volume) who are used to thinking of national and epochal categories. And yet, if we are to garner the point about nervous development, or what might more dramatically be called nervous determinism, we must begin with the far gaze – the distant view – rather than, more immediately, the middle of the Enlightenment. The scientists among us always want to pause, of course, on the plateaus of nervous development and ponder life at the next altitude, so to speak; the historians on its peaks of cultural history. But what is the cultural dimension of nervous development? From Mesopotamia to Munich, Babylonia to Berlin, from Latium to London, nerves and brains have barely been interpreted in their sociocultural contexts. Here Weisskopf and his fellow physicists can offer only limited inspiration. The Greeks had a sophisticated theory of nerves and brain based on invisible spirits no one had seen but which all were certain existed.[7] But, and it is a crucial adversative, they never connected brain with soul, nor soul with brain, the brain then not lodging in the head at all. This view endured more or less unchanged throughout the ancient and medieval world. Every philosopher commented on the subject. Few topics could compete with it, it being the primary proof of God's existence in an anthropocentric universe in which God had to prove to man why he had created him different from other living creatures. Thus motion and matter, soul and body, existed in particular relationships which mechanical philosophy changed, as it created a dualistic order of mind and body, brain and soul, and then, through its mechanistic anatomists, removed the brain to the head. In 1766, the *Annual Register* devoted columns to the nervous theories of an obscure French physiologist, Monsieur Bertin, offering him as much journalistic space as news of war and peace. The *Register* wanted its readers to know that Galen had understood nerves as well as any modern:

> This great man, says M. Bertin, saw very well, upwards of 1600 years ago, that a fluid ought to produce all the wonderful effects which we observe in the exercise of our motions and sensations; and he derived its source from the brain, from whence it diffused itself thro' the rest of the body. If he [Galen] could not see what modern anatomy has discovered, he could still less see those spirits, that subtil fluid.[8]

Notwithstanding his limitation, Galen was credited with the discovery. The philosophical analogy was thus continuous: from the acclaimed Galen and the forgotten Bertin to our twentieth-century Ramon y Cajals and other neurophysiologists, all certain of the vital 'subtle fluid', whether it be mystical substance or electrophysical wave particles. Yet with this difference: among the historical periods, the first to cling to nervous physiology with a vengeance was the eighteenth century, and in some qualified senses it is historically valid to claim that much of the eighteenth-century Enlightenment was one magisterial footnote on nervous physiology, a remarkable attempt to secularise cognition and perception through the brain and its vassal nerves. Other eras demonstrated interest of course, but nothing remotely approaching the eighteenth century's, which, among other activities, naturalised, theologised, demonised, mechanised, climatised, medicalised, internalised, metaphorised and analogised the nerves.

The century's energy in this activity was extraordinary. It, so to speak, naturalised them as it made the brain and nerves the basis of a vast number of research programmes in secular natural philosophy.[9] Philosophised them into a dualism of mind and body from which Western Civilisation has yet to recover, always attributing this mind–body split to Descartes, it being, more accurately, the proof of a late Renaissance nerve craze that set everyone looking for the key to these vital spirits.[10] Theologised them in both the conservative and radical theology of the day, as nerves were said (by the conservatives) to be a beneficent God's physiological gift to a wicked people who needed them for reformation, as well as (in radical versions such as the Boyle Lectures) revelation of God's goodness in endowing his creatures with the unit of organisation they most needed.[11] Demonised them as empiricists and spiritualists continued to endow them with magical and alchemical powers no one had ever seen.[12] Mechanised and vitalised them in countless anatomical and physiological debates then raging all over Europe.[13] Taxonomised them as well, into stronger and weaker nerves, greater and lesser, major and minor, pigmented and non-pigmented, white and black, red and yellow, as Boerhaave and Haller, Linnaeus and Cullen and many others purported in their nosological schemes.[14] Darwinised them too, as Erasmus Darwin suggested when claiming that the nervous system had been evolving all along – in creatures that swarm as well as in higher intelligent animals – and would eventually evolve into something much grander than it was at the end of the eighteenth century.[15] Pathologised them into normal and abnormal states – the state of affairs, so to speak, of the nerves that coloured all human health and determined longevity.[16] And biologised the nerves in the embryological discussions about reproduction, preformation and epigenesis.[17] In all these activities it was conceded, with remarkably little opposition, and as Hume and other Scots philosophers scrupulously argued, that, whatever memory was, the brain and its vassals the nerves could never be far away. If learning and attention require memory more than anything else, we instantly see what an extraordinary homage to the nerves this is. So crucial were these nerves to the complex machine called man (a beast machine of a much more complex type), that the eighteenth century – if one may be metonymic about its diverse efforts in this area – almost worshipped in its temple. The *temple of nerves* is not an inappropriate name for this intellectual house.

This vast theoretical labyrinth had its counterpart in the socialisation of the nerves, as it were, in the ordinary daily life of the time. The nerves were medicalised, academised, globalised, climatised, electrified, genderised and sexualised. Concommitantly, the nerves are engraved in the social history of the day: at spas and resorts, among doctors and patients, among the flourishing cults of sensibility that served first and foremost, to separate the social classes in an era when aristocratic title in itself was insufficient, and when antagonisms between the social classes took more hostile forms than mutual derogation. The nerves were academised as virtually every European medical school, regardless of its reigning beliefs, assigned its students to write dissertations on this subject: at Leiden, Harderwyk, Reims, Paris, Montpellier, Marburg, Halle, Wurzberg, Giessen, Göttingen, Leipzig, Erfurt, Jena, Helmstedt, Dresden, Basel, Padua, Rome.[18] I have found no medical schools where dissertations on the nerves were not written. The nerves were literally electrified as Wesley, Franklin, and Mesmer, each in his own way, set to regenerate one social class or another, high and low, by this nervous use of electricity. The nerves were globalised, nationalised, regionalised and internationalised, as different countries, climates and regions were held accountable for particular states of the nerves and their diseases. For the first time, one could talk of French nerves, Dutch nerves or Italian nerves, while paradoxically being cautioned that, whatever local, indigenous conditions prevailed, the nerves themselves were universal, common to all people on the face of the earth. And the nerves were genderised and sexualised as crucial differences of male and female (such as hypochondriasis and hysteria), caucasian and black, oriental and mohammedan, heterosexual and hermaphroditical, were attributed to this or that nervous strength or defect, or (as Spallanzani and LeCat maintained) to the nerve tips in the dermis of the skin, which extended to the cerebrum and cerebellum of the brain. For the first time in medical history, conditions as different as jaundice and low intelligence were attributed to nervous causes. This stage represented a vast change from the explanations of previous generations and shows us how cultural history can retrieve discourses and cross boundary lines left undiscovered by more traditional disciplines.

The nerves were also internalised and mentalised into the most imaginative processes of which language and art were then capable. In psychology and art criticism throughout Europe, the nerves occupied centre stage, whether in discussions, such as Edmund Burke's of aesthetic sublimity (*Enquiry into the Sublime*, 1757) or more technical debates about the painter's hand as he holds the brush and strokes the palette, and the pianist's fingers. The nerves were also visually represented in a broad repertoire ranging from anatomy textbooks to satirical cartoons, some of which are reproduced in this chapter. The nerves had been versified as a major *topos* in all countries, perhaps nowhere more extensively than in England, allegedly the home of the mad, the insane, in, for example Dr Malcolm Flemyng's long epic poem in six cantos called *Neuropathia*, published in 1747;[19] hundreds of lines of Latin hexameter verse celebrating – as Victor Weisskopf has more recently celebrated in a very different context, as discussed at the opening of this chapter – the utility of the nervous pathways. Most intriguingly for students of language and literature, the nerves were also metaphorised and analogised, as a whole new vocabulary developed consisting of words we hardly recognise today. This was based on flaccidity and tension, acridity and tone, fermentation and putridity, and, in words that are no longer in any modern dictionaries, 'hippohiatrical' functions and 'levigations' of the nerves; on biliousness, chyle, spasmodic colics and unnatural ferments, affections and cachexies, censoriums and climacterics, contextures and exquisite particles, ferments and fibres of the brain, fluxes and effluvia, tubes and performances. It produced a maze of neologisms such as 'black humour' (whose first use is not in a playwright such as Beckett but in a *medical* text of the 1720s),[20] deobstruents, dimoculations, empyr-heumatics, and a repertoire of slang terms and abbreviations whose meanings have long been lost: hypp, hyppos, hyppocons, markambles, moonpalls, strong fiacs, hockogrogles.[21] And – most intriguing to me – it generated an entire vocabulary of neurospasts, as John Evelyn called these puppets on strings, a concept that fascinated the period, especially in the form of electrified neurospasts.[22] Even literary criticism and social commentary were culture-bound by this degree of specificity, as critics Thomas Warton and Samuel Johnson, among many others, sought to define the gradually evolving 'nervous style', which they construed as a particularly masculine type of muscular and sinewy English prose best described as 'nervous'. The developing style was viewed as superior to all others; it partook of the optimism and progress which was the tenor of the day. Within this linguistic activity, generated in an era when language theory was rapidly changing, the *nerve* itself became the sign of a semiotics of analogy.

To invoke the nerves then, whether in medical, philsophical or any other type of discourse, as, for example, Mandeville or Cheyne or Hartley had, was usually to indulge in radical analogising, as both *A Treatise of the Hypochondriack and Hysterick Passions* and *The English Malady* demonstrate, and no linguistic trope was then so important for phil-osophical and scientific analysis as analogy. Hedonistically speaking, the nerves also presided over the seats of pleasure and pain. To claim, as Dubois instructs the Marquis de Sade's Juliet, that all pain is in the nerve, was then to utter a paradigm too well-known for further amplification. As for insanity and madness, even Dr William Battie, who is sometimes called the father of modern psychiatry, long before Pinel and Esquirol held that lunacy lay entirely in the nerves, without recourse to gender. Diseased spirits, diseased nerves, diseased imagination: this was the sequence of the mind's malady – the surest sign in recognizing lunatics. As for the erotic lunacy that springs from love-sickness (states of mind Breuer and Freud were later simply to call hysteria), even this variety was neutralised, as the English poet Anthony Selden wittily wrote in 1749:

> While Grief and Shame her Face o'erspread,
> Upon her Knee she lean'd his Head.
> Then points the Dart, and with her Hands
> The crystal-rooted Film expands.
> But, O! the rack was so immense,
> So twing'd the *Nerve*, and shock'd the Sense,
> He begg'd her, yelling with Despair,
> The fruitless Torture to forbear,
> Confounded with the horrid Pain,
> He storm'd, and rav'd, and rag'd in vain.

Rank and class: the fashionability of nerves

It was predictable, if not inevitable, that nerves would become fashionable, given this intensity of mythologisation, but not democratic – in the sense that nerves had been created equally in all persons – within this neuralising of culture high and low. That the nerves became fashionable even when viewed from the perspective of political and ideological tropes in written texts and spoken interchanges testifies to their vigour and degree of infiltration. For centuries, and certainly since the discussions of Guicciardini and Machiavelli, the body politic had been a commonplace to describe the mysteries of government, yoking together politics with the body – or corpus – of civic polity.[23] In the eighteenth century these tropes alter to 'nervous government', as James Lowde, an early follower of Malebranche and Locke, noted in his tract on man viewed in his anatomic and political dimensions.[24] But, as the political state was nervous, so too was the individual, and the analogies between the nervous one and the nervous many, the small and the large, the private and public, increased proportionally to this new nervous mythology. Nor were social distinctions omitted; if anything, nervous mythology segregated the social rank and file anew and provided all the classes with an important new model of aristocratic life. Nerves also permitted behaviour to be predictable, and enabled an acceptable code of social behaviour at a time when it seemed there might soon be none. The new social and geographical mobility of the eighteenth century obscured worth and station, which now had to be asserted. The easiest ways were external: especially fashion (through clothes) and illness (by disease). These exterior signs, anew, became the tropes of rank and class. If the lower classes emulate the upper in all centuries and live vicariously through them, they did so in the eighteenth, as seamstresses and charwomen, pickpockets and prostitutes like Moll Flanders and Fanny Hill, fantasised that they too could be as 'nervous' as their gentlewomen mistresses. 'My nerves are all unstrung,' a corpulent and robust servant exclaims, emulating her mistress in a minor novel of 1748.[25] But it would be wrong to think this mythology of nervous culture required an *antecedent* scientific theory. Just the reverse is true: Haller's theories of nervous sensibility do not validate the principles of nervous government or nervous aristocratic life, but the reverse. His hypotheses profoundly manifest his *own* nervous culture – in Switzerland, Germany, Europe, the world: no less than Freud's in *fin-de-siècle* Vienna, for science is culture-bound, partaking of the varying degrees of positivism and ideology that permeate every age.[26] So too for imaginative literature so far as this osmosis is concerned, as *Tristram Shandy* is permeated with the same animal spirits and fibres that will become analogised in Scottish treatises on the law and other branches of government. For writers of discourses as diverse as those by Haller and Sterne one assumption is held in common: that the nervous myth is already widely disseminated. Yet, as the nerves produced no mortal disease like consumption or cancer, they were not morbidly feared and no reason existed to demystify or demythologise them to the same degree. The more mysterious the vital fluid (the unseen spirit within the hollow tube) – so the inner voice of the upper classes might have spoken could they have been conscious of these complex processes – the better for all those now reasserting their class station in a society when doubt about this hierarchy was pervasive. It was a curious collusion of sorts between those who generated (formulated and articulated) physiological theory and those (political figures, business people, professionals) who led the masses.

Gentility and delicacy were thus reinvigorated and rearmed with new tropes. For snobs, parvenus and social climbers, the way to rise was simple: be hyppish, be nervous, be bilious, be rich. The sequence was plain. If consumption was then the disease of poverty and deprivation, nerves was the condition of class and standing. How odd that this cultural shift should occur just when the sign of upper-class authority in England and France was perceived image rather than power, and when the genders were growing theoretically further apart than they had been in many centuries, as the historians of sex and the family have shown us. If it was fashionable to blush and weep and faint, it was even more glamorous to present with nervous symptoms. It was smart to associate with those who were also nervous; better still to be in a constant correspondence, as so many fine ladies then were, with others equally nervous. To be nervous was to be romantic; to be romantic was divinely wished among those middle-class newcomers who until recently never thought they could be. Even suicide was considered a viable alternative, and suicide then was widely held to be an act, as Hume and others claimed, leading directly from diseased nerves – the *sine qua non*, the precondition – of deranged imagination and certifiable lunacy.[27] In all this tumble of social differentiation was a sense that the upper classes were writing themselves out, as it were, of chronic nerves; but the very opposite is true. The reader's penetration *into* the book *Clarissa* coincides – literally as well as conceptually – with penetration into her own body; the more Clarissa writes, the more nervous she becomes.[28] This is why Adam Smith could believe that sympathy cemented civil societies as well as trade imbalances among nations, and what is sympathy if not a predisposition of the nerves associated into higher states of feeling and ideation?[29]

The social ramifications of this newly nervous culture were not to be minimised, as whole professions, if not entire genders (women), realigned and reasserted themselves under the weight, in part, of nervous myth-ology. If the rich felt themselves obliged to prove their difference, what better way to assert difference than by looking different and frequenting different places? Bath and Tunbridge, Harrogate and Scarborough, Llandrindod and Llanrwyst, not to speak of the Belgian spas (especially Spa) and the Aix-les-Bains, and the various 'Maria-bads' (Maria baths) of Central Europe that then came into their own, as did the professions of doctors, physicians, apothecaries, nurses, midwives, quacks, mountebanks, empirics who now catered to the whims and whimsies, the caprices of the neurasthenic rich. 'He was the greatest of the nerve doctors, Dr John Makittrick Adair was to concede about his conspicuous rival of the previous generation, Dr George Cheyne[30] – a bewildering claim in view of the universal jealousy of doctors and when Cheyne's exorbitant fees are rehearsed. For all these reasons it makes sense to invoke a newly constituted category called 'nervous discourse', without worrying excessively about charges of engaging in unitary label-making, and especially if we define this term ('nervous discourse') as a network of contradictory narratives and explain what is meant by the semiotics of the nerve.[31]

We are searching, then, for an understanding of this pervasive sign – the nerve – rather than its disciplinary diachronic history. We want to fathom how it is that a particular sign arises within a particular culture and under what specific conditions. But we will not be satisfied by its logocentric legacy only: those traces left in words alone. We want a more holistic picture that conjoins words and things; that separates discourses rather than disciplines (history, language, science, medicine); and that will explain the nerve in the light of the whole picture of civilisation then, not merely one of its manifold components. Given the structures of authority that obtained in the eighteenth century and the fluidity of the social classes, it was predictable that the upper class should have contrived these mechanisms of separation with virtually no opposition from the medical profession, which considered itself an intrinsic part of the newly defined rich. Medicine and ideology – the only ideology most doctors knew was the ideology of the pound; they were the last collective group to oppose the new nervousness on philosophically rational or inductively empirical grounds. Of course they battled among themselves, just as other professionals do today, especially when ideologies come into conflict; some had their scruples, their principles, their philosophy. But viewed from without, from this distance of time after the lapse of two centuries, they appear to have been somewhat in league with the rich, as mi' lady this had her private doctor that, who – so common in the diaries of the time – 'was called to her side for nerves every day this week'.[32] This dynamic *sorting-out*, this renewed separation of the rich, precipitated a reinvigorated emulation by the lower classes; these social processes lie at the heart of the cults of nervous sensibility and the mythology they engendered.

If the demarcations between the social classes were then fragile and requiring of these disparate kinds of reaffirmation, so too (although of a different parity) were the body's anatomic nerves, as doctor after doctor pleaded for a 'tonic strengthening' of them, a phrase that means little to cultural historians today unless it is decoded and glossed.[33] The principles and methods of tonic strengthening baffle us even when anatomic and physiological equivalencies are taken into account. To strengthen the nerves would seem, linguistically, to be a metonymy for toughening the character or (as North Americans might say in the vernacular) getting one's act together. In our quasi-macho western-world twentieth-century culture, the nerves remain genderised and imbued with the sense of femininity. In the eighteenth century, nerves were palpably real, although often metaphorised for excessive effects. To be sure, there were some who condemned this fragility and its attendant nervous physiology, who claimed that nerves were not so all-important, but their voices were somehow muted and their numbers waned as the most fashionable persons hastened to the great nerve doctors of the day.[34] There was no actual conspiracy or collusion in this nervous tidal wave: it swept through the culture as so many other waves have. To distance it and see it for what it was: this is the hard task for the cultural historian. To think that it would *not* take its toll on the imaginative literature of the day remains the foible of our internalist literary historians who read at the level of individual texts only.[35] No wonder, then, that Henry Mackenzie, for example, who understood men and women of feeling as well as anyone, and who fictionalised his beliefs about the new creatures of sentiment he saw as heroic, wrote to a friend plainly: 'this is an Age of Sensibility'. And it is little surprise that Jane Austen believed she was responding to the aesthetic of an era when endorsing it under her own imprint as a generation of 'Sense and Sensibility'.

Town development and urban planning were affected as much as anything else. One can imagine the Baths and Tunbridges, the Harrogates and Scarboroughs under these ruthless circumstances of gain and profit. Here, lurking everywhere, were not merely doctors and diseases, rich patients and their undiagnosed conditions, but pump rooms and assembly halls, booksellers and musicians, architects and landscape gardeners, charities, drums and routs: all commanding vast amounts of money and wealth. Not to have weak nerves and a delicate constitution was to be out of place here. Lady Luxborough (an addict of Bath who was also Bolingbroke's sister) assured her constant correspondent, the poet William Shenstone: 'My disorder…turned to a fever of a slow kind, chiefly nervous, attended with pains in my bowels, which, added to want of rest, have weakened me so much, that I have not yet crossed my room.'[36] Shenstone replies: 'As your Complaints are *entirely nervous*, [you] must have an undoubted tendency to further it!' The exchanges of Mrs Ralph Allen and Mrs Beau Nash – not to mention their lesser female epigoni – were similar, written in what seems a kind of code language to us now.[37] Selina, the wealthy Countess of Huntingdon and another exclusive Cheyne patient, was not much better off, sunk, as she almost always was, in nervous depressions of religious origin, even after the completion of her methodist chapel.[38] Little was different in fiction: in *Evelina*, Lady Louisa

Larpent despairs when her lover leaves her; Lovel comforts her: 'Your Ladyship's constitution is infinitely delicate,' only to evoke Louisa's retort, ' "Indeed it is," cried she in a low voice, "I am nerve all over!" '[39] The sign is even evident in the most ephemeral gibberish of the time. The silly farrago that called itself a *Register of Folly . . . written by an Invalid*, while Smollett and Christopher Anstey were still at Bath in the 1760s, trivialised these ailments by personifying them in absurd couplets:

> Tho' I own I am sorry such trifles still flurry
> My Ogilby-nerves, to produce such a hurry.[40]

Other 'valetudinarian guides' provided their own versions of nervous mythology then cranking out a million-pound industry. One finds criticism of the developing system in certain quarters – the objection that nervous mythology is merely a hoax to induce consumption of pills and potions, doctors and diets, resorts and retreats in a culture that was already consuming extraordinary amounts of commercial produce – but no substantial objections that changed anything. An anecdote, whose veracity has never been corroborated, makes the point: near the end of his life, Samuel Johnson described the sense of diseased or weak nerves as 'medical cant',[41] but the objection was a drop in the bucket, as those learned who inveighed against the trade in nervous drugs. Actually the repertoire of nervous therapies has yet to be compiled: not merely quack nostrums but the dozens of panaceas ranging from what we would call health food diets (Cheyne's regimen of 'lettuce, see, milk, and wine') and balneological treatments (whether of cold water or hot, mineralised or sulphuric) to James Graham's celestial beds and aphrodisiac cures for the nervously sterile and barren (one wonders why Sterne never took his Tristram to Bath as Austen does Lydia), but also Mesmer's electric shocks and animal magnetisms which flourished in the salons of the rich, and which were calculated to stimulate even the most withered nerves of any person.[42] These nervous nostrums themselves would fill pharmacopoeias, extending from analeptic and assa foetida pills; from Anderson's drops and William Tickell's aetherial spirit and Ward's aether pills, to Raleigh's confection, Bishop Berkeley's tar water, Dover's drops, Pierre Pomme's recipes, John Hill's wild valerian roots, Mrs Stephens's juleps, Backer's cure of dropsy, to the more generic class of pearl and closet cordials, plain Nantz, quicksilver, calyx of zinc, acid of lemon rind, rhubarb and magnesia, not to mention hyppish medicines like extract of saturn (whatever that was pharmaceutically), prescribed in John Goulard's 1777 *Treatise on the Effects of . . . Lead* and Malcolm

Flemyng's (the physician-writer already mentioned as the author of *Neuropathia*) tartars of mercury a few decades earlier, and without considering the effect of all these preparations on the royal patent office and their implications for profit and gain.[43]

The social milieu of nerves: class and gender

It was in this atmosphere and social ambiance that the literature of 'sensibility' (as its writers *themselves* called it) flowered, and it seems to me essential to be alert to *their* nomenclature. For the decoding of the sign is as much an analysis of the word as anything else. Semiological interpretation requires close linguistic interpretation viewed within the contexts of the culture: anthropological, sociological, historical, etc. What then was this literature of sensibility? Whether viewed in its Smollettian version in *Humphry Clinker* (1771), a version tied geographically to a spa locale as this geographically vivid novel opens, or its many other prose and poetic forms during the second half of the eighteenth century, the literature of sensibility itself broke off from its earlier versions of a hundred years before (in England as written by Milton and Dryden, in France by Marivaux and Crebillon), and became highly medicalised. Smollett's irritable hero Matt Bramble attributes his misanthropy to 'the nerves of an invalid, surprised by premature old age, and shattered with long suffering', only to have his nephew Jery Melford pooh-pooh Matt's nervous explanation and substitute another nervous one: 'tender as a man without skin, who cannot bear the slightest touch without flinching.'[44] Bared nerves indeed, but if the nerves, as our contemporary physicist has shown us while taking a very different gaze, endowed humankind with the most sophisticated communication system imaginable, they nevertheless mandated these reasserted social stratifications when the upper classes came under pressure to prove who they were. It was predictable that novelists like Smollett (a practising physician, medically trained and thoroughly conversant with the scientific debates about nerves) would directly respond to these transformations.[45]

Their economic dimension has received less attention than one thinks, for the cost of sensibility, in whatever version, was significant. The hordes of quacks who hung on and about the spas were many, yet even the tribe of certified doctors and apothecaries in Bath (not to mention the fashionable Continental watering holes) was large and continually proliferating. These were no ordinary doctors but the shamans of the rich. When the Duchess of Northumberland returned to Sion House, having visited her cousin, the Duchess of Newcastle, at Bath, she exclaimed surprise at 'the wigs and golden canes everywhere about the parades and crescents'.[46] For centuries there had been rich patients for doctors to treat; what differed now was the way the doctors went about diagnosing. The point is not – as some medical historians have shown – that medical practice suddenly altered under the weight of new evidence or theory, but that a cultural myth engulfed medical theory *itself*, privileging the nerves and exalting them as never before. The physicians' status was as important as the patients'. Bath was, of course, a unique type of resort, but even there in 1740 – for example – all but three of the physicians had Oxbridge degrees, the exceptions being two from Leiden, and one, actually Dr Cheyne himself, from Edinburgh.[47] The Bath Corporation and the hospital it regulated placed some controls on this medical undergrowth, but fees and therapies, and what we would call the quality of the care given, remained almost exclusively in the physicians' hands. The doctors quarrelled among themselves, and entered into paper wars calculated to raise their socioprofessional status and increase their number of patients, allegedly because 'nervous diseases' were on the rise. Even the Prime Minister, Robert Walpole, had come to Bath to be treated 'for nerves' by the famous doctor and President of the Royal Society, Richard Mead.[48] But the development of nerves was not exclusively British, emphasised as it was in this island culture. S. A. Tissot, the Swiss Protestant physician whose claim to fame was, in part, his anti-masturbatory campaign, had been as explicit about the role of nerves in human life and human culture as any Bath doctor. His *Essai sur les maladies des gens du monde* and his *Traité des nerfs et de leurs maladies* spoke loudly and plainly about the nervous culprits.[49] Tissot's approach to nerves was anthropological: to classify medical conditions by nations and regions. But its symptomology was culturally identical with his British colleagues', and, while Tissot was being translated into English in the 1760s (works such as his *Essay on Diseases Incident to Literary and Sedentary Persons*), Robert Whytt, the Scottish physician and professor of medicine in Edinburgh, was lecturing to medical students in Britain's now most forward-looking university about the crucial significance of nerves, as well as training a generation of physicians who would invade Bath after Cheyne. By the time the American colonies revolted in 1776, Tissot's works in English had been widely disseminated and Whytt had become a household name, especially known as the 'nerve doctor'. Tissot's account appeared to be the authoritative medical explanation: how people grew ill and under what conditions. Its subtext was otherwise: be nervous, be fashionable, be sought-after; the paradigm was a prescription, an equation of social truth that could as easily be reversed. *Au fond* Tissot's rhetoric had its intended effect of hurling any aspiring social climber, or one who had already arrived there, into 'nerves'.

Early anthropologist though he was, Tissot had not explained the sociology of the phenomenon: how nerves considered as an ethic, or even ideology, caught on. This was left to James Makittrick Adair, a Scottish doctor, educated under Whytt at the University of Edinburgh, who became Cheyne's successor in Bath at the end of the century and the wealthy eccentric Philip Thicknesse's adversary.[50] Adair churned out book after book on nervous conditions and their treatment, especially under his own regimens for recovery; but in actuality these were self-advertisements calculated to make him appear the discoverer – and healer – of a new province of medicine. Whether true or not (and there is some validity to his claim that he had genuine expertise in the field) he stood to profit. He justified his literary productions on the semi-original grounds that medicine itself was a *social* institution:

> Should any of my fashionable readers express their surprise at meeting with a dissertation on fashion in a medical essay, my reply is ready; that as medicine, as well as some other arts, is become subject to the empire of fashion, there can be no impropriety in considering by what means this has been effected.[51]

Impropriety there was not, but plenty of distortion. Claiming modestly that he hoped to chronicle the rise of fashionable disease in the eighteenth century, Adair's descriptions are as revelatory for our semiotic purposes in decoding the sign of the nerve as anything Cheyne wrote. Always acknowledging his teacher's famous essay on nervous diseases of 1764, Adair, like Cheyne, served up to his readers the explanations they wanted to hear:

> Upwards of thirty years ago, a treatise on nervous diseases was pub-lished by my quondam learned and ingenious preceptor DR. WHYTT, professor of physick, at Edinburgh. Before the publication of this book, people of fashion had not the least idea that they had nerves; but a fashionable apothecary of my acquaintance, having cast his eye over the book, and having been often puzzled by the enquiries of his patients concerning the nature and causes of their complaints, derived from thence a hint, by which he readily cut the gordian knot – *'Madam, you are nervous'*; the solution was quite satisfactory, the term

[nervous] became quite fashionable, and spleen, vapours, and hyp, were forgotten.[52]

It is as if a single word, magically, had such transformative power. Although forgotten among the denizens of Lady Luxborough's circle low spirits and nervous ailments extended further back than to the time of their mothers and aunts. The cultural historian has a longer gaze and sees how similar are Sydenham's hysteria, Mandeville's hypochondriasis, Robinson's 'spleen', Cheyne's 'English malady', Haller's 'sensibility', Whytt's 'nervous diseases', and now – in the 1780s – Thomas Coe's 'bilious conditions' – the maladies of the bile – on which Adair claims to build his own theory.[53] But neither medical symptomology nor a glossary of linguistic transformations (from the language of the hyp to the byzantine neologisms of 'bilious concretions' and 'bilious solids') makes the crucial point as much as *fashionable disease*. Cheyne had justified the spread of nerves on grounds of the progress of expanding civilisations: 'We have more nervous diseases,' he wrote in the *English Malady*, 'since the present Age has made Efforts to go beyond former Times, in all the Arts of Ingenuity, Invention, Study, Learning, and all the Contemplative and Sedentary Professions.'[54] By the 1780s, though, Adair abandoned progress for fashion; for him in the 1770s, disease was romantic, glamorous, idealised: proof of difference. Yet it must not remain static, because tides of taste change, as does geography in an era of unprecedented social mobility. What Adair recounts about its taxonomic transformations from melancholy and spleen and hysteria to biliousness and now nerves, Dr Thomas Dover had described in 1732 – a half-century earlier – of its *geographical* spread:

At first, the *Spleen* was said to be the entire Property of the Court Ladies; here and there indeed a fine Gentleman was pleas'd to catch it, purely in Complaisance to them. Soon after, Dr. *Ratcliffe* [sic] out of his well-known Picque to the Court Physicians, persuaded an Ironmonger's Wife of the City into it, and prescribed to her the Crying Remedy of carrying Brick-dust; the City Physicians took the Hint; and the Country Doctors remov'd it into the Hundreds of *Essex*, whence a learned Academick brought it with him to *Cambridge*: Soon after it was heard of in the Fenns of *Lincolnshire*, and it crossed the *Humber* in 1720. The Contagion [of spleen] has at last extended itself into *Northumberland*.[55]

On one level Dover's delightful bagatelle amounts to little more than an eighteenth-century version of modern communication theory, i.e. how information spreads, even through the shires and backroads of eighteenth-century England; on another, its infectious wit is too silly to be discredited: the same fictive play of mind apparent in Pope's *Dunciad*, as the goddess's roll-calls echo round the town. Yet this social genealogy of spleen appears in a purportedly serious medical work, Dr Dover's *Treatise on Hypochondriacal and Hysterical Diseases* (1732), aiming to explain why disease is so fundamentally genderised; the specific reasons why men develop hypochondria and women hysteria; a book every bit as didactic as Cheyne's *English Malady* or Adair's *Fashionable Diseases*. But, like theirs, Dover's also assumes the force of the unwritten assumption: the notion that, although both men and women have nerves, females' nerves are the more delicate, the more sensitive. We see what this unwritten paradigm was if we look later to the nineteenth century, even to Charcot and Breuer and Freud. In the nineteenth, women will be said to be more prone to hysteria than men, not so much out of any innate anatomic difference that translates into unavoidable physiological process, but rather because they labour to conceal their feelings. This suppression, usually of sexual desire – it is said – hurls them into unavoidable hysteria, in the eighteenth century, the pathologically normal body does the work of the feelings. Cheyne's and Adair's patients are not nervous because they have *suppressed their emotions*; quite to the contrary, all their hysteria arises from bodily conditions, especially loosening and tightening (flaccidity and tautness of the solids and fluids) in the nervous system.[56] So the point is not so much that male nerves are less sensitive and fragile than those of the female, as that states of mind such as hypochondria and hysteria are entirely predetermined by the body. It is small wonder, then, that the most elusive states of mind – imagination, creativity, genius, even memory – would be medicalised under the weight of such emphasis on the body.

Furthermore, when almost everyone in the late eighteenth century was, it seems, aping the fashions of the wealthy and great under the need to assert class distinction, why should *disease* be omitted from the variety of methods? The phenomenon of emulating aristocracy *to this extreme degree* is itself unparalleled before the eighteenth century; why then should the splenetic fits of Queen Anne or the Duchess of Marlborough go unnoticed, when, as we have seen in virtually every middle-class novel of the period from Defoe to Fanny Burney, most lower-class females expend extraordinary amounts of energy observing the mannerisms of 'gentlewomen' and then emulating them (or pretentiously copying them, as the case may be)?

What needs decoding and dismantling, then, is not the tone of Dr Dover's astonishing and exorbitant passage about nerves invading

England's fens and marshes, as Dover had recounted, but the embedded, unwritten assumptions that inform it. This is the genuine work of a semiology that hopes to retrieve the significance of a particular sign within lost cultures. Yet the systems of all these fashionable 'nerve doctors' – from Willis to Cheyne, from Boerhaave to Mandeville, from Garth to Mead, Boerhaave from Haller to Tissot, from Hartley to Cullen – can never be dismantled until we can isolate and identify why these thinkers could uniformly claim, as Adair does here, that 'no part of the physiology has engaged the attention more, or reaped greater consequences in our time, than the nerves.'[57] No one today wants to revert to those odious Hazlittian 'spirits of the age' that characterised epochs often by the most simplistic of labels, or regress to Basil Wiley's loathsome world views and world pictures. Certainly no one should claim that this was an Age of Nerves, as if upper-case 'Ages' were discreet things whose boundaries could be charted, as previous cultural historians have argued for Ages of Reason and Ages of Passion. But for the upper classes, much social differentiation *did* lie in the nerves, just as nervous philosophy and sympathetic reasoning resonated with meaning as a separating out mechanism for the Adam Smiths, the Humes and the Scottish moralists: the same resonance that causes imagination theory to become so highly medicalised in the eighteenth century,[58] and, alternatively, that prompts and anonymous writer in 1744 – a dunce perhaps? – to think that there could be a readership for a prose work he hazards to call *The Anatomy of a Nervous Woman's Tongue: A Medicine, A Poison, A Serpent, A Fire and Thunder.*[59] We will soon see who this developing readership was.

The fantasies inspired by this nervous mythology also require decoding and dismantling. They are as intriguing as the projects of Fellows in the Royal Society trying to ascertain what the vital nervous fluid was and attempting to dissect and reproduce the nerves. Yet if the constellation of illnesses passing under the rubric of consumption was then the *fatal* English malady – the disease from which people actually died – nervous sensibility was the *life force*: the vital *je ne sais quoi* of the upper class; the spring of vitality and creativity that set it apart from the *hoi polloi* but which could also bring it misery in the form of depression, accidie, indolence, as in Thomson's castle, or lassitude in the shape of the noonday demon, the waste of sloth.[60]

The remarkable fillip of the semiological version of analysis is that it brings passages to life that would otherwise remain dead. Entire hulks of writing that were formerly consigned to the closets of obscurity or the shades of hermeticism suddenly gather meaning. One such place is the correspondence of Richardson the novelist and his female confidante.

Mrs Donnellan consoled her friend, the gendered novelist, in words that could be the epigraph of this chapter:

Misfortune is, those who are fit to write delicately, must think so; those who can form a distress must be able to feel it; and as the mind and body are so united as to influence one another, the delicacy is communicated, and one too often finds softness and tenderness of mind in a body equally remarkable for those qualities. Tom Jones could get drunk, and do all sorts of bad things, in the height of his joy for his uncle's recovery. I dare say Fielding is a robust, strong man.[61]

Robust, that is, unlike soft Richardson! This is no mere 'attempt to console Richardson for his perpetual ill-health', as Ian Watt long ago suggested,[62] or, if such consolation, certainly not merely a mindless female comfort, but the clearest indication – the very semiotic I have been attempting to identify and isolate for scrutiny – of a veritable revolution in social thinking. Mrs Donnellan's unstated premises are the mythologies of the age which can be schematised and epitomised as follows:

1. the soul is limited to the brain;
2. the brain performs all its work through the nerves;
3. the more exquisite and delicate one's nerves – morphologically speaking – the greater the ensuing degree of sensibility and imagination;
4. upper-class people are born with more exquisite nervous anatomies; the tone and texture of their nervous systems are more delicate than those of the lower classes;
5. greater nervous sensibility makes for greater writing, greater art, greater genius.

It is a rather odd sequence, alien from the pluralistic habits of mind so firmly ingrained in late twentieth-century culture. Difference, indeed, is the chief signpost of cultural semiotics: not to read these differences often entails vast loss. Even when the poet Pope, whose self-delusive but nevertheless ironic, dying words were 'I was never hyppish in my whole life,'[63] lamented 'this long disease, my life', and complained in his letters about 'this crazy constitution', his complaint indicated difference. Chronic nerves were living emblems of gentility and delicacy. There was no reason to avoid them, no virtue in doctors concealing them from their patients, as in the cases of cancer and consumption. Philander Misaurus, and alias for a Grub Street hack writing for 'females of parts' in 1720, advised them: 'When sharp, fermenting Juices (not easily miscible) shall meet, and by their furious Contest, cause cruel Twitchings of your nervous Fibres; comfort your Heart, and be extreamly pleas'd.'[64] Though technical and couched in hard words, the advice was heeded, particularly among those anxious about their status. John Midriff, another fashionable nerve doctor like Dover and Adair, wrote a long book in 1720 for those 'who have been miserably afflicted with these Melancholy Disorders since the Fall of the South-Sea, and other public stocks.'[65] Forty-eight years later, long after the emotional turbulence caused by personal economic loss in the bubble had subsided, Daniel Smith, a deist well-read in Newtonian philosophy, was found expounding the same ideas in a medico-theological idiom: *A Dissertation upon the Nervous System to show its influence upon the Soul.*[66] Nervous mythology was not a credo for one generation, but a type of discovery whose positivistic energy was so great that it has carried down through the late twentieth century, as our neurophysiological activities make evident.

To return to our Enlightenment, the physicians of the fashionable and smart sets were swift to isolate 'nerves' as the cause of all disturbance in their patients, almost as if they had identified the crucial gene of ailment and distress. Indeed, the discovery of nerves, unlike the diagnosis of cancer or consumption, was cause for celebration, as in Boswell's case, who touted his melancholy far and wide.[67] Nerves were neither the signs nor the symptoms of ephemeral illness but an inherited condition that, if undetected, could lie dormant only for so long. Inherited like wealth or milk-white skin, nerves and their fibres were unique among the organic structures: ingrained, they could neither be bought nor stolen, copied nor caught. This is why Elizabeth Carter, the brilliant letter-writer, ministered to her ailing correspondent Catherine Talbot with the evangelism of a prophetess. 'The low spiritedness...of which you complain, assures me you cannot be well, nor ever will be, while you have the strange imagination that a weak system of nerves is a moral defect, and to be cured by reason and argument.'[68] As we saw in the communication of Mrs Donnellan and Richardson, the sequence here is also crystalline: weak nerves are not the signs of moral defect, as Elizabeth Carter stresses, but the proof of discrimination and delicacy. The patient should therefore never reason herself out of them, as Talbot seems eager to do. Miss Carter even psychologises the nerves here with a corollary about attention: 'I must enjoin you for two months to amuse yourself, and wile away the time, and be as trifling and insignificant...and never during that time to apply to anything that requires close attention.'[69] *Far niente*, an Italianate do nothing: this was the best cure for a condition whose mortal consequence was inconsequential. But it focused attention and concentration: it was the villain, the perfidious culprit; the same steady close attention that invaded the nerves of literary and sedentary persons, as Tissot had shown, and caused them to become neurasthenically depressed. For those prone to the dictates of the imagination, such as scholars, writers and religious types, a life of focused attention was as dangerous as the worst fistula or cancer. So the mythology went; the reality was otherwise: Richardson did not die of focused attention any more than Carter or Talbot (or Horace Walpole for that matter) expired from the fatigue of letter writing.

Ponder it though one may, the nerves would never consume the body, as in other medical conditions. Exempt from contagion, no pollution could be caught from them. 'I am genuinely relieved,' exclaimed Lady Mary Wortley Montagu's sister, Lady Mar, 'to learn that the *worst* is delicate nerves; this I can manage; the excrescences of a diseased liver, stomach, or bowels would be so much worse.'[70] Nervous sensibility was thus a dissemination rather than a disease; an outreach whose extension was the *sine qua non* of fashion. Nerves could be painful, but the pain was said to be mentally lodged and never posed the threat of death to the body. Even the poets, especially Akenside and Armstrong, were amazed at the proximity of pain and pleasure in the nerves. Hence Samuel Garth, the poet of *The Dispensary* and a leading physician at the turn of the eighteenth century, remarked: 'How the same Nerves are fashion'd to sustain the greatest Pleasure, and the greatest Pain!'[71] Within this context of pain the nerves were an anti-gout, often accompanying the real gout, but much more discriminating in its victims. And the proof – in the collective fantasy – that nerves were the supreme life force of the fashionable was that nerves would never lead to death, only to the *fear* of death: *timor mortis*.

An apocryphal story about Samuel Johnson's last days survives from the early nineteenth century, white-washed over the pages of the *European Magazine*.[72] It recounts that, just a few months before his death, Johnson consulted Nan Kivel, the fashionable London physician. Johnson provided his full case history, only omitting *timor mortis*, the fear of death. Kivel seized upon the omission; to this Johnson replied: 'Alas, it is so, it is so...' To which Dr. Kivel rejoined: 'I only wanted that symptom [the fear of death] to make yours a complete case of hypochondriasis, which will only require a little exertion on my part, and rather more on yours, to *entirely cure*.' Johnson soon died, but not because of spastic intercostal involvement, then thought to lie in the seat of hypochondria. The note-worthy aspect is Kivel's confidence in an 'entire cure' despite the diagnosis of male hypochondriasis, and the fact that Regency readers wanted to hear this story mythologised in this way. Other onlookers who could pierce through the flapdoodle of these (often crude) discourses of the nerve often screamed catcalls: sometimes in an altogether different mode, as in Austen's assault in *Sense and Sensibility*, where the hypocrises of the ethic of sensibility are exposed through the characters of the circle hanging on the Dashwood women; sometimes trivially, as for the 1732 caricaturist who claimed, facetiously, in an engraving 'Of the Hypp: The pleasures of melancholy and madness', that 'there are Pleasures in Madness, which the Splenetick, of all sufferers of the nerve, are least acquainted with'.[73] To those who were neither parvenus nor social climbers, nor anxious about their niche in polite society, the vast cultural octopus of nerves seemed a remote social phenomenon, let alone a valid or invalid scientific hypothesis: a cluster of concepts or ideas which never could or would impinge on them. They never grasped what the debate about nerves entailed. For the others, especially the social climbers, and those who were already located in places of high station, nerves had become a way of life, touching on everything important: sex, love, sanity, insanity, and, most crucially, one's social standing.

As the eighteenth century wore on, the myth intensified. Accreted to it eventually was a sense that the reckless and the lecherous, more politely the sensuous and the erotic, behaved as they did out of an innate propensity lodged in the nerves rather than because the nerves had grown diseased in any way. Historically, the fact is that they were eating themselves into the grave, indulging in lechery, and generally wrecking their health with late nights. The discourses of the nerve that constituted the fabric of the mythologies I have been discussing coped with that reality. They especially did so by emphasising that nerves signified no disease of passion, as nineteenth-century tuberculosis would, but were the sign of *passion itself*. Nerves and sex were thus intimately connected, as they would be in the oeuvre of the Marquis de Sade:[74] direct concrete proof that one had been capable all along of understanding passion's kingdom. Those who were not nervous could never respond adequately to sexual desire, let alone at the level of 'toujours la chose génitale'. This concretion of the body is what unites all the discourses of the nerve: Mandeville, Robinson, Tissot, Bienville – the entire company. The doctors' explanations reveal why: nervous persons could contract diseases with devastating somatic effects; in the myth, though, *all* their troubles were nervous, whether they were diagnosed as hypish, splenetic, or as hysterical women in the older nomenclature, or nervous in the newer, or whether diagnosed as hypochondriacal men. Again and again, gender, class and race determined the shape of the sign: nerves *in relation to these aspects.*

More than anything, then, I have been attempting to demonstrate that, as the middle classes at mid-century continued to demand liberty and the rabble the franchise in suburban Middlesex and other boroughs, the upper classes increasingly aimed to set themselves apart. Nerves provided them with a myth about their origins: an aristocratic model of life they could follow.

The formation of the nervous personality in cultural history: sex and gender

By these diverse cultural means, entailing different social practices formed in diverse social institutions, the nervous personality of the nineteenth and twentieth centuries jelled into a type we continue to recognise today: the idealised consumptive type; the romantic poet or artist wasting away and eventually decaying; the neurotic genius who comes almost directly from William Cullen's lecture notes; in women, the creative Crazy Janes (as in Blake's poem) and compulsive anorexics (as in the notorious fasting woman in Tutbury, Staffordshire) who will haunt the nineteenth-century imagination.

The difficult aspect of the coagulation and the stereotype, so to speak, is not its various strands or internal contradictions, but the process of organic formation itself. It was a slow, dynamic growth extending over many decades, eventually lending credence to the belief that those who were nervous partook of a type of vitality nowhere else to be seen, even if the vitality – the Shavian Life Force – itself could only be anatomically pinpointed. This principle of vitality had been not merely medically but culturally grounded; eventually it was imperialistic, extending its vast sway from Dr William Battie (who held that madness itself actually resided in the nerves) and the already mentioned Dr Cullen (whose 'neuroses' of 1768 were the first set of classifications of nervous types),[75] to early nineteenth-century accounts claiming that genius mandated sensitive nerves. Seen whole, it was a cultural wave extending from the nebulous borderlines of sanity and insanity in the Restoration (as Dryden had said in 1682, 'Great Wits are sure to madness near ally'd)[76] to the medicalising of the creative act under the strain of genius, and eventually to the nervous agony of creative writers like Richardson, Cowper and Chatterton, who in their very different ways were inmates of its all-too-familiar prison. 'When I was young,' Theophile Gautier wrote, 'I could not have accepted as a lyrical poet anyone weighing more than ninety-nine pounds.'[77] More than a hundred years earlier, Mrs Donnellan, as we have seen, ventured to apply a version of this angle of vision to hypochondriacal novelists like Samuel Richardson. In brief, what had been reserved for the parades and crescents of Bath and England's newly developing seaside resorts became the way of the world in her towns and cities by the early nineteenth century. But it would be wrong to think of the nervous type itself as a *nineteenth* century development. It developed much earlier. The neurasthenic woman suffering from a myriad of female maladies – from anorexia and hysteria and dementia among other types – certainly flowered in the nineteenth century; but she was not, so to speak, born then. If Crazy Jane and her cohorts infuse the iconography of Byron's and Walter Scott's society, they nevertheless appear throughout the annals of late eighteenth-century discursive (literary as well as medical) literature.

As the nineteenth century evolved, nervous mythology altered, was not eradicated but transformed, as it gradually became apparent that a newly constituted English aristocracy need not distance itself in these same ways any longer. Disease remained the reward, however ironically, for sin – as it was to be for Emma Bovary – but not for mere neurosis or self-indulgence, as it has become in our twentieth century. Yet nervous disease became more localised and specialised than it had been in the eighteenth century. Upper-class male hypochondria practically vanished as a topic of investigation – this under the new nineteenth-century obsession with sexuality as a regulating force and in the belief that females, not males, were incarnations of the senses. The new idea was that males were subject to stress in the workplace that made them prone to a type of melancholy unknown to women, as in Charles Lamb's bizarre 'tailor's melancholy.'[78] Coupled to this male ailment were a whole series of female nervous maladies predicated on a developing science of female hysteria that would culminate in the discourses of Gilles de la Turette and Breuer and Freud. As tuberculosis and cholera competed with nervous ailments for attention and government subsidy, there was less spotlight on exclusively nervous conditions and hence on the older competing discourses of the nerve. Lunacy itself came to be seen as the higher sensitivity, especially after the reforms of Pinel, Esquirol and Charcot; and it too was partly unhinged from its former, monolithic, neurologic base. The intimate bond that had existed between the upper classes and their inherited nervous apparatus was severed but not gone, replaced by a new moralising and psychologising of illness that made disease the symbol of human character, as it remains in our time with AIDS. In the case of fatal disease, which *nervous* diseases had never been, it was a *hamartia* over which no one could triumph, high-born or low. Not even in the neurasthenic versions associated with French decadence was our nervous condition said to be morbid.[79] Proust, like Wilde and other decadents in their Franco-English milieu, linked his neurasthenia to creativity; not to be nervous and ill, he came to believe, was not to write. His long seance in bed was not so different from Wilde's retreats to the seaside, where Wilde wrote in what he perceived to be exquisite solitude, almost always ill, or at least reputedly suffering. And Schumann's reputation, we must not forget, rose after he threw himself into the Rhine. None of these versions was so very far removed from Richardson's neurasthenic creativity, although we have constructed a mental notion that somehow the eighteenth century was immune from these developments; but Richardson's creative condition also flourished under the distress of a nervous ailment which his physician, Dr Cheyne, could never precisely define to his inquisitive patient.

The remainder of my story in the nineteenth and twentieth centuries has yet to be pulled together. The sheer bulk of extant material entails an *embarras de richesse* for the cultural historian; I can only suggest the shadows of its discourses here. The literally hundreds of scientific and parascientific books written about the nerves in relation to man and woman generically viewed, ranging from Thomas Trotter's crucial *Nervous Temperaments* (1807) and John Cooke's *Treatise on Nervous Diseases* (1822) in England, to Broussais' books in France in the 1820s, especially his *Traité de la nervosité et de la folie*, a type of early nineteenth-century Foucaldian approach that stresses the discourses of nervous physiology and links the clinical nerve to its social manifestation.[80] The nineteenth-century development of a theory of character, already evident in Jane Austen, also flowers in De Quincey's *English Opium Eater*, the Brontës, Melville's *Ambiguities* and Carlyle, and will show moral fibre to be lodged in the nerves. Thus in *Mansfield Park*, when extraordinary revelation is about to be made, Austen shrewdly narrates: 'It was not in Miss Crawford's power to talk Fanny into any real forgetfulness of what had passed – when the evening was over, she [Fanny] went to bed full of it, her nerves still agitated by the shock of such an attack by her cousin Tom, so public and so persevered in.'[81] To the end of *Mansfield Park*, nerves hold a key to character, as they do in others of Jane Austen's novels, especially when Sir Thomas is about to pass judgment on Fanny; 'he knew her to be very timid, and exceedingly nervous'. These are no metaphoric representations of anatomy but the thing itself – the literal nerves which will hurl Austen's heroines into some of their most poignant and pathetic moods.

The culture of Europe ca. 1800 is infused with this unspoken paradigm: from the literary representations of nerves in Blake's 'Auricular Nerves of Human Life' in *The Four Zoas*, to D. H. Lawrence's dichotomy between 'a nervous attachment rather than a sexual love' in *St. Mawr*. The aesthetics of sympathy and empathy, from Burke in his *Sublime*, through various transformations in Novalis, Keats ('negative capability'), Jean Paul, and eventually the *Einfühlung* (i.e. empathy) that will form the basis of *fin-de-siècle* psychological aesthetics. The nineteenth-century cults of blushing and tears, not merely the old sorrows of Werther, but now the new sorrows of Charlotte too (as in Charlotte Smith),[82] and the pervasive malaise and ennui such moral tears inevitably induce, a malaise and decay extending at least to the French existentialists of our century. The nineteenth-century theories of the optic nerve in relation to the painter's gaze, evident not only in Turner and the English School but in other painters as well; and the nineteenth-century aesthetics of the imagination, follow so predictably from the eighteenth's and which has claimed that Mozart's music was greater than all other because 'it assaulted the nerves more'. Then there is Coleridge's paean to the nerves as the true sources of growth and organic form: 'They [the nerves] and they alone can acquire the philosophic imagination, the sacred power of self-intuition, who within themselves can interpret and understand the symbol, that the wings of the air-stylop are forming within the skin of the caterpillar...'[83]

All these are discourses requiring retrieval. Among the types of materials that need to be culled and interpreted are nineteenth-century medical 'theories' of the affluent classes, as in Thomas Beddoes' *Hygeia; or... The Personal State of our Affluent Classes* (1802–3), which continues to promote the significance of the nerves, as well as Moritz Heinrich Romberg's famous *Nervous Diseases of Man* (1853, English translation) composed at the peak of Victorian civilisation, George Miller Beard's description of neurasthenia in his now famous 1880 *Practical Treatise on Nervous Exhaustion* as the new disease of American civilization, Abraham Myerson's *The Nervous Housewife* (1920), to the more eccentric Daniel Schreber's – Freud's patient – *Memoirs of my Nervous Illness*. This development reaches a peak during the time when Virginia Woolf claims in her diaries of the 1920s that 'writing calls upon every nerve in my body to hold itself taut,'[84] and Sylvia Plath reveals to Anthony Alvarez just before her death that her suicide will be no swoon, no attempt 'to cease upon the midnight with no pain', but something to be felt in the nerve ends and fought against 'at the level of the nerve tips'.[85] A second wave, so to speak, developed when Charles Darwin and other zoologists of the nineteenth century elevated these same nerves in their phylogenetic studies of the emotions of animals, thereby returning us, as it were, to our physicist, Victor Weisskopf, who views the nervous system with a geochemical gaze and who sees its aesthetic impulse and influence spread over geological time.

There is, of course, an opposition to each of these cultural movements – a counter-culture, as it were, which has not been explored here for reasons of space, extending from Mary Wollstonecraft, who denounces nervous women in *The Rights of Woman* as useless creatures who accomplish nothing, to all sorts of moralists who see in these nervous mythologies and nervous discourses running wild in the nineteenth century no hope for the progress of culture. But the movement is there none the less: pervasive, writ large, indelible. Eventually it will have to be studied, if for no other reason than that neurophysiology, now in such a positivistic and imperialistic phase, continues to claim to have made such strides in our own time.[86]

Conclusion

Let me conclude on a note about significance: the significance of this story, the significance of these different discourses of the nerve, and the crucial matter of privileging these discourses over others. The contradictions of the nerve story are obvious: no inner logics, no agreement among themselves, only obeisance to the mythologies of social class, a type of social determinism of cultural practices and arrangements. Equally significant are the discursive practices of these narratives. Previously, we have viewed these discursivities in isolation: as the narratives of literature, the narratives of science, the narratives of history, and so forth, without viewing them comparatively.[87] And however distinct these discursive narratives, we have usually viewed them from *within* rather than without: that is, as insiders rather than outsiders; as practitioners or critics of the distinct narratives of literature, of science, of history, and so forth, with the consequence that we have lost the gaze of the outsider, whose perspective can be the more acute by reason of distance, balance, objectivity and sturdiness. So that those who are most familiar with the discursive practices of literature – for example – might actually see deeply into the narratives we have called science, and historians into the narratives commonly reserved for the literary critic's eye. Nor have we comparatively studied allusions and tropes common to all these narratives, to establish, as it were, a grammar of allusions, common to all the various discursive narratives.

But it would require another essay to interpret the social processes I have been discussing, especially the dynamics of class formation, and the antagonisms of the upper and lower classes, just as another chapter would be needed to evaluate the theoretical models best suited to this material. But most of all, the discourses of the nerve are significant for embedding the deepest so-called scientific and metaphysical questions of the last hundred years: what is the universe, what is human life, what are mental processes? Can mental processes operate without a body? What is a computer? What is the body of a computer? Can a computer think for itself, feel, make love, use its imagination, grieve? What is artificial intelligence? What is sexual desire, sexual attraction, sexual orientation? All these questions have been asked in our century within the context of biology. But more recently, and as the new biology, psychology and sociobiology have shown themselves to rely on an even more fundamental neurophysiology, the deepest of these questions would all seem to have neurophysiological underpinnings. These developments demonstrate why the action, so to speak, and heat in science today (especially the trends in funding) lies at the juncture of neurophysiology and computer science, the points of reference where one can explore questions about mental processes operating *without* a body. And perhaps this is why the human body itself, and its various discourses, have been so predominantly privileged in the last few decades. Indeed, no set of discourses has been more emphasised than those of the body, isolating their tropes and metaphors as well as their cultural referents and comparative allusions.

But neurophysiology did not spring full-blown, as if from Athena's head, in the twentieth century. It has an older legacy, especially in the seventeenth and eighteenth centuries. In the face of these ancient but nevertheless somewhat contradictory discourses of the nerve I leave them with much wonder and awe, bewildered at the idea of our contemporary neurophysiologists that for us humans there is only nerve and brain, and nothing *but* nerve and brain, a notion the eighteenth century would not have believed itself all that uncomfortable with. To think that in the most abstract and non-referential discourse of all – classical music – that even there, and especially in its performance and interpretation, there should be brain and nerve primarily. Some would say only. Horowitz performing classical music – Scarlatti, Schumann, Chopin, Rachmaninoff – as nothing but an extraordinary nerve machine! Perhaps in the end the new positivistic neurophysiologists are right: nerves have perhaps been the greatest gift of all. It may be so, but it nevertheless remains an odd position when viewed *outside* Weisskopf's geological time and *inside* the parochial approach usually taken to cultural history.

Notes

This essay was originally delivered in a much shorter version in Aberdeen, Scotland in July 1987 under ordinary constraints of time, and the reader will soon see that my style abounds in ellipticisms (wherever possible two words instead of ten). I have allowed this version to stand rather than reconstruct the essay along somewhat artificial lines. Each style – the oral and the written – obviously has its advantages, and I trust that readers expecting a more formal and discursive version from this somewhat oral one will forgive whatever infelicities have crept into this version. Of course, the relation of orality (especially oral theory) to the nervous physiology developed here is interesting in itself. Unless noted otherwise, the place of publication is London.

1. Victor Weisskopf, *Knowledge and Wonder: The Natural World as Man Knows It*, Cambridge, 1979, p. 223 *et passim*.
2. Ibid., p. 264.
3. Ibid.
4. Ibid., p. 265.
5. Ibid.
6. Ibid.
7. For nerves and the ancients see: Friedrich Solmsen, 'Greek Philosophy and the Discovery of the Nerves', *Museum Helveticum* 18, 1961, pp. 150–67.
8. *Annual Register*, III, 1766, p. 234.
9. An extended note on the historiography of nerves provides some perspective here. As the seventeenth and eighteenth centuries progressed, nerves played an increasingly prominent role in all sorts of research agendas, including those of the laboratory as well as in more logocentric projects, as I tried to demonstrate over a decade ago in 'Nerves, Spirits and Fibres: Toward the Origins of Sensibility, *Studies in the Eighteenth Century*, ed. R. F. Brissenden, Canberra, 1975, pp. 137–57. My point, then and now, has been that there was a progression from nerves to the cults of sentiment and sensibility, and that European Romanticism could not have occurred without this sequence. It is, to be sure, a diachronic theory, and nowhere have I ever maintained that nerves and sensibility were *the* (superlative) cause of Romanticism. In this essay, I attempt to show more fully than previously the roles of the nerves in cultural history. But see also, for a medical historian's view of the subject, George Rosen, 'Emotion and Sensibility in Ages of Anxiety: A Comparative Historical Review,' *American Journal of Psychiatry* 6.124, 1967, pp. 771–83.

   Sensibility has had, of course, its own historiography, although it has been a ragged one, carved up by specialist disciplines, never gazed at sturdily or synoptically; see, for example, L. I. Brevold, *The Natural History of Sensibility*, Detroit, 1962; Caroline Thompson, 'Sensibility', *Psyche* XV, 1935, pp. 46–161; and for an astute lexical critique of the word that nevertheless limits its range to the narrow field, Raymond Williams, *Keywords: a vocabulary of culture and society*, 1976, pp. 235–8. There has not even been a bibliographical survey of *primary* works, such as Joanna Heywood's *Excessive Sensibility*, a tradition including such famous novels as Austen's *Sense and Sensibility*.

These traditions would require a book to adumbrate properly, especially in relation to the scientific movement, more specifically the rise and dissemination of mechanism then, but some important primary foci include: the works of Thomas Willis during the Restoration; later on, ca. 1705–20, the Boyle Lectures; in the 1750s, see Richard Barton, *Lectures in Natural Philosophy, Designed to be a Foundation*, 1751; Albrecht von Haller, 'Elementa Physiologiae Corporis Humanae', in *The Natural Philosophy of Albrecht von Haller*, ed. Shirley A. Roe, New York, 1981. Important secondary work includes: Edwin Clarke, 'The Doctrine of the Hollow Nerve in the Seventeenth and Eighteenth Centuries', *Medicine, Science, and Culture: Historical Essays in Honor of Owsei Temkin*, eds. Lloyd G. Stevenson and Robert P. Multhauf, Baltimore, 1968 pp. 123–41; Theodore M. Brown, 'From Mechanism to Vitalism in Eighteenth-Century English Physiology', *Journal of the History of Biology*, 7, 1974 pp. 179–216; Jacob Bronowski, 'A Sense of the Future: Essays in Natural Philosophy', in Rita Bronowski and Piero E. Ariotti (eds), *The Visionary Eye; essays in the arts, literature and science*, Cambridge, MA, 1978; and, in relation to the sciences of man, S. Moravia, 'From Homme Machine to Homme Sensible: Changing Eighteenth-Century Models of Man's Image', *Journal of the History of Ideas* 39, 1978 pp. 45–60, and *Filosofia e scienze umane nell 'eta dei lumi*, Florence, 1982. Even the Swiss professor of philosophy and philology at the University of Basel during this period, Samuel Werenfels, whose writings were influential for rhetoric and style in England, develops a programme that includes 'nervous science' in discussions of the rhetoric of sublimity; see, for example, his *Discourse of Logomachys, or Contraversys [sic] about Words . . .* (1711).

But the agenda was not limited to natural science; it was also evident in the arts (as in Daniel Webb's *An Inquiry into the Beauties of Painting*, London, 1760, where the nervous system is discussed in relation to art), and in music (as in Richard Browne's *Medicina musica: Or a Mechanical Essay on the Effects of Singing, Musick, and Dancing, on Human Bodies*, 1729). Guichard Duverney, a late seventeenth-century French anatomist who made the ear his organ of expertise, believed the sense of hearing among humans to be the most divine of all; the nerves in the ear constructed more subtly by the Deity and prompting 'Men and Birds to excite one another to sing'; see *A Treatise of the Organ of Hearing*, 1737; originally pub. 1683, p. 88. This agenda was also evident in theories of acting and dancing (as in John Hill's, *The Actor*, 1750), as well as theories about the branches of government and political economy (as evident in Adam Smith's discourses). For nerves and theories of acting, see George Taylor, ' "The Just Delineation of the Passions": Theories of Acting in the Age of Garrick,' *Essays on the Eighteenth-Century English Stage*, 1972. This widely disseminated agenda was a cultural development of the Enlightenment of whose ideology and influence we have yet to take stock. A sense of its breadth is discussed, most perceptively and intuitively, in Christopher Lawrence, 'The Nervous System and Society in the Scottish Enlightenment', *Natural Order: Historical Studies of Scientific Enlightenment*, ed. Barry Barnes and Steven Shapin, Beverly Hills, 1979 pp. 19–40; even so, Lawrence merely scratches the surface; much more remains to be done outside the local confines of eighteenth-century Edinburgh.

But even the novice reading through these diverse discourses of the nerve soon realises that the nerves were genderised throughout this period, especially as discourses about the nature of women's bodies developed; see, for example, Edward Shorter, *A History of Women's Bodies*, Harmondsworth, 1983. The means by which nerves became genderised – male nerves taking one set of attributes, female ones another, more delicate, if hysterical version – would also require another essay, if not a book in itself. Suffice it to comment here that, as theories developed separating the genders further and further than they had been, and accounted for all types of hermaphrodites and monsters, as well as sexual deviants (in our anachronistic language homosexuals and lesbians), the nerves were thoroughly implicated, and it is equally inconceivable to imagine Enlightenment discourses of these anatomical types without discussion of the nerves. By mid-century a physician such as J. Raulin could maintain in his *Traité des affections vaporeuses du sexe*, Paris, 1758, that nervous disorders were limited entirely to the *female* sex; over a century later, Albert Moll, a German doctor, claimed in his treatise, *Berühmte Homosexuelle: Grenzfragen des Nerven-und Seelenlebens*, Wiesbaden, 1910, 11, no. 75 that the etiology of homosexuality in both genders was caused entirely by defective, degenerating nerves. The nerves, then, were anything but delimited in their perceived ability to account for the generation of physiological and psychological types.

The recent neurophysiological critique of nerves is at once much more idealistic and positivistic; i.e., claiming that neurophysiology explains everything, thoroughly confident in the view that no other set of explanations, either mental or physical, can compete with it. See, for example: John Eccles, *The Neurophysiological Basis of Mind*, Oxford, 1953; E. Graham Howe, *Invisible Anatomy: A Study of Nerves, Hysteria and Sex*, 1955; Walther Riese, *A History of Neurology*, New York, 1959; I. H. Burn, *The Automatic Nervous System*, Oxford, 1963; J. Spillane, *The Doctrine of the Nerves*, 1981; *Historical Aspects of the Neurosciences: A Festschrift for Macdonald Critchley*, New York, 1982; E. Clarke and L. S. Jacyna (eds), *Nineteenth-Century Origins of Neuroscientific Concepts*, Lost Angeles, 1987.

10. Mind–body dualism was still in its post-Cartesian flowering: being massively attacked on all sides, but still far from overthrown. The most bewildering physical and metaphysical question debated then was whether the nerves were inherently a part of mind (soul) or body, or some substance in between, and there was no agreement about the proper method of asking and answering this question. There was also (ca. 1700–80) a vast developing discourse of the nerves in relation to mental faculties, passions and insanity, and to the specific role of the nerves played in pathological pleasure states. For the dualism of mind and body, see L. J. Rather, *Mind and Body in Eighteenth-Century Medicine*, 1965 and M. D. Wilson, 'Body and Mind from the Cartesian Point of View', *Body and Mind: Past, Present and Future*, ed. R. W. Rieber, New York, 1980. The most thorough treatment is found in G. S. Rousseau (ed.), *The Languages of Psyche: Mind and Body in Enlightenment Thought*, Berkeley and Los Angeles, 1990, whose bibliography of primary and secondary works should be consulted for nerves as well. For 'vital spirits' and vitalism, see particularly: Jacques Roger, *Les Sciences de la Vie*, Paris, 1963; A. A. Cournot, *Materialisme, Vitalisme, Rationalisme*, ed. Claire Salomon-Bayet, Paris, 1979; Richmond

Wheeler, *Vitalism – Its History and Validity, 1939; John W. Yolton, Thinking Matter: Materialism in Eighteenth-Century Britain*, Minneapolis; 1983.

11. The argument from nature, more locally from anatomy and physiology, was made many times, with reference to both genders and under virtually every type of ideological banner, conservative and radical, and in practically every political shade then available. A conventional example is found in William Derham's *Creation of the World*, 1712, one of the Boyle Lectures, and in the next generation in the physician-philosopher David Hartley's influential *Observations on Man, his frame* . . . , 1749, containing 'pt. 1: Observations on the frame of the human body and mind, and on their mutual connexions and influences,' but there were others. In general, the clergy were especially quick to claim that defective nerves, often arising from a diseased religious melancholy, would lead to morbid hypochondria and hysteria, about which more will be said later. As late as the 1760s, R. J. Boscovich, the philospher of science, was inquiring about the role played by nerves within the human organism's functioning (see his *Theory of Natural Philosophy*, Venice, 1763) and in the 1780s the Catholic apologist Laurent François Boursier was still explaining how those who had fallen from grace through sin also fell into convulsions through the explicit deterioration of their nerves; see his *Mémoire théologique sur ce qu'on appelle les secours violens dans les convulsions* (Paris, 1788).

12. For the generalised view, see Herbert Thursten, *The Physical Phenomenon of Mysticism*, 1950. There remains no study of what I label *the tradition of counter-nerve*, embodied in the critiques – especially through the trope of analogy (*analogia*) – of the mechanistic approaches to the nerves developed by Paracelsus and van Helmont through Swedenborg, Blake, Ebenezer Sibley (in *A Key to Physic, and the Occult Sciences: opening to mental view, the system and order of the interior and exterior heavens; the analogy betwixt angels, and spirits of men*, 1794) and eventually Carlyle; a post-Rabelaisian world of jumbled discourses rather than social structures, culled from competing disciplines, some anti-scientific, others not, which Bakhtin would have well understood if he had stumbled upon counter-nerve. For some sense of this tradition of counter-nerve before the seventeenth century, see Brian Vickers, *Scientific and Occult Mentalities in the Renaissance*, Cambridge, 1984. It is a tradition that flowered throughout Europe, as rich and diverse (if not empirically predictive) as that of the nerves. For example, de Valmont, a French speculator, produced a long *Dissertation sur les Maléfices et les Sorciers selon les principes de la théologie et de la physique, ou l'on examine en particulier l'état de la fille de Tourcoing*, Tourcoing, 1752, dealing with witchcraft in relation to nervous physiology, and there were others throughout the period who continued to believe that the nerves could be manipulated in supernatural ways: alchemically, zodiacally, nutritionally, demonically. Throughout the nineteenth century, there were attempts to retrieve this critique, as in Richard Robert Madden's (he was the author of a best-selling Victorian book about nightmares, gothic illusions, dreamlike monsters, male *couvade* – the entire spectral nightime world) 'Nervous States-Inspired Religious Vision', in *Phantasmata or illusions and Fanaticisms of Protean Forms*, 2 vols, 1857, 2. 517, and in Walter Cooper Dendy's (he was the Sussex surgeon who also wrote poetry and travel literature) 'Fantasy from Sympathy with the Brain', in *Psyche: A Discourse on the Birth and Pilgrimage*

of Thought, 1845, pp. 115–16. The most serious attempt to retrieve counter-nerve was made at the turn of this century by a French physician, Lucien Nass, who ransacked the closet of history to discover what extreme convulsive states had done to figures of the past; see his *Les Névroses de l'histoire*, Paris, 1908, a book as mystical as it is envious of the scientific.

13. Looking at its empirical dimensions we can now see, in hindsight, that this was nothing less than the history of the anatomy and physiology of the eighteenth century, as Jacques Roger showed long ago in *Les Sciences de la Vie*, Paris, 1963.

14. Cullen, *Vitalism*, 1750; see also n. 75. The role played by the nerves in the Enlightment debates over racism was considerable and should not be minimised by scholars interested in the eighteenth-century discourses of sex, race and gender. For one development, see G. S. Rousseau, 'Le Cat and the Physiology of Negroes', *Racism in the Eighteenth Century*, ed. Harold Pagliaro, Cleveland, 1973 pp. 369–87. The nerves were especially crucial in these debates; for example, R. C. Dallas, writing in a *History of the Maroons*, 2 vols, 1803, adjudged that negroes could never be integrated into 'cold climes' because their nerves and fibres could not withstand 'the pinching of frost' (I, pp. 200–1).

15. See Maureen McNeil, *Under the Banner of Science: Erasmus Darwin and his Age*, Manchester, 1986, and Peter Morton, *The Vital Science: Biology and The Literary Imagination*, 1984, for discussion of the nerves in Darwin's works.

16. Such prolific commentators on nerves in relation to health as Nicholas Robinson, George Cheyne and Robert Whytt had much to say on this subject. For Robinson, see: *A new system of the spleen, vapours, and hypochondriack melancholy; wherein all the decays of the nerves, and lownesses of the spirits are mechanically accounted for. To which is subjoined, a discourse upon the nature, cause, and cure of melancholy, madness, and lunacy*, 1729 and *A new theory of physick and diseases, founded on the principles of the Newtonian philosophy*, 1729.

17. Some idea of the range of this application in biology is found in Brian Easlea, *Witch Hunting, Magic and the New Philosophy*, Brighton, Sussex, 1980, esp. ch. 4; Shirley A. Roe, *Matter, Life and Generation: Eighteenth-Century Embryology and the Haller–Wolff Debate*, Cambridge, 1981, and 'John Turberville Needham and the Generation of Living Organisms', *Isis* 74, 1983, pp. 159–84; C. U. M. Smith, *The Problem of Life: An Essay in the Origins of Biological Thought*, London, 1976.

18. England and Scotland did not practise a tradition requiring medical students to produce a Latin medical dissertation as did the Continental universities, but many medical treatises dealing with the nerves were written in English, as, for example, David Bayne's *New Essay On the Nerves*, 1738, or, a generation later, John Hill's (the notorious Renaissance man of mid-Georgian England sometimes publicly known as the 'Inspector' or the quack Dr Hill) *Construction of the Nerves*, 1768. In Germany, a spectacular medical treatise about the nerves not originally written as a university medical dissertation was J. F. Isenflamm's *Versuch einiger praktischen Anmerkungen über die Nerven zur Erläuterung verschiedener Krankheiten derselben, vornehmlich hypochondrischer und hysterischer*, Autälle, 1774. The Dutch, who produced the largest number of medical dissertations after the Germans, poured forth thesis upon thesis dealing with the bile in relation to nervous disorders, as in T. W. Gartzwyler's *De bile atra*,

ejusque effectibus, Leyden, 1742. In Italy, many medical works on the nerves also appeared, such as Giovanni Giacinto Vogli's *Fluidi nervi in istoria*, Padua, 1720.

19. See Malcolm Flemyng, *Neuropathia: sive de morbis hypochondriacis et hystericis*, 1740. Flemyng, who was obsessed with mechanistic and vitalistic questions about the nerves, had also written 'A New Critical Exam of an Important Passage in Locke's "Essay On Human Understanding" to which is added an extract from the fifth book of anti-Lucretius, concerning the same subject...', 1751. John Armstrong, another physician who (like Flemyng) also wrote poetry, produced *The Art of Preserving Health*, 1744, a long didactic poem teaming with images of nerves, spirits and fibres.

20. Beckett's nervous laughter in his plays.

21. In James Makittrick Adair's *Medical Cautions for the Consideration of Invalids...Containing Essays on Fashionable Diseases*, Bath, 1786. Adair, MD, was a member of the Royal Medical Society, as well as Fellow of the College of Physicians in Edinburgh. The *Medical Cautions* was also perpetually sold by Dodsley in London, as it was a bestseller.

22. See John Evelyn, *The History of the Rebellion*, 2 vols., 1850, 2. 281. The 'nervous style' in English prose is discussed below, in n. 41. Suffice it to say here that it had roots in Port Royal Grammar and in Ben Jonson's 'full-blooded style', whose rhetorical components were correlated to the anatomical organs; but in Jonson style is skittish and fickle and can meander any sexual way, male or female, whereas the aesthetics of nervous prose during the Enlightenment always mandated its masculinity.

23. For the body politic as a phrase and concept in English, see David Armstrong, *Political Anatomy of the Body*, Cambridge, 1983; Martha Banta, 'Medical Therapies and the Body Politic', *Prospects, An Annual of American Cultural Studies*, ed. Jack Salzman, Cambridge, 1985; John O'Neill, *Five Bodies: The Human Shape of Modern Society*, Ithaca, 1985; and John Blacking, 'The Anthropology of the Body, *ASA Monographs*, ed. John Blacking, 1977, Monograph 15, pp. 19–21.

24. See John Lowde, *A Discourse concerning the nature of man...both in his natural and political capacity*, London, 1694.

25. An entire vocabulary of words originating as technical terms in anatomy that later lost their technical usage and became common household phrases begs for study. These include: tension, corruption, delicacy, irritation, sensibility. Indeed, much of the vocabulary of the School of Taste in the period from Reynolds to Wordsworth appropriated this technical anatomical language for its own aesthetic purposes.

26. This idea has been much discussed in our time in the works of Roland Barthes, Pierre Bourdieu, Nancy Cartwright, Paul Feyerabend, Michel Foucault, Ronald Giere, Jürgen Habermas, Ian Hacking, Mary Hesse, Karin Knorr-Cetina, T. S. Kuhn, Larry Laudan, Bruno Latour, Jean-François Lyotard, Michael Mulkav, Steven Shapin. Sharon Traweek.

27. The close tie between suicide and nerves was constantly noticed at that time, especially as applicable to the situation of persons in high rank and class. Later on, in the 1770s, as sentimental cults were more dispersed, and as increasingly more persons aped the habits of the great, suicide grew more common, its etiology and dynamic changing as well. In France, J. P. Falret

called suicide a class malaise in *De l'hypochondrie et du suicide. Considérations sur les causes, sur le siège et le traitement de ces maladies, sur les moyens d'en arreter les progrès et d'en prévenir le développement*, Paris, 1822. The social history of suicide in the Enlightenment remains to be written.

28. For whatever complex reasons, the feminists have not explored this aspect of Richardson's masterpiece, although they have understood so much else about it. A broad approach is found in Catherine Gallagher and Thomas Lacqueur, eds, *The Making of the Modern Body*, Berkeley and Los Angeles, 1986; see also Ann van Sant's forthcoming study of sensibility and the novel (Cambridge, 1991).

29. See Adam Smith, *Theory of Moral Sentiments*, Edinburgh, 1759.

30. James Makittrick Adair, *Essays on Fashionable Diseases*, 1786.

31. That is, in the theoretical sense that semiotics provides the deepest clue to the concept of the 'fibre' (here one wants to say nerve-centre, except for the obvious ineptitude) of a culture; see Tzvetan Todorov, *The Conquest of America*, New York, 1985. Good work in this semiotic vein that is also particularly germane to the cultural history of the Englightenment is found in Sylvain Auroux, *La Sémiotique des encyclopédistes: Essai d'épistémologie historique des sciences du language*, Paris, Payot, 1979.

32. Henrietta Knight [Lady Luxborough], *Letters of Lady Luxborough . . . to the poet William Shenstone*, London, 1775.

33. Tonic strength then denoted the degree of essential health of the nerves, its opposite being a state of morbid weakness, but there were many synonyms, common usages, and metaphoric abbreviations (that is, exquisite delicacy) applied linguistically as well. By the 1730s a dense metaphoric jungle of words describing this constellation, both healthy and diseased, had arisen. In his dialogical *Treatise of the hypochondriack and hysterick diseases*, 1711, Mandeville often enquires about the meaning of the 'tonic strength' of the nerves; see pp. 160, 172; later on John Armstrong versified some of the same ideas of tonic strength in his long didactic poem, *The Art of Preserving Health*, 1744. The late Raymond Williams has teased out some of the extended metaphors of sensibility in *Keywords: a vocabulary of culture and society*, 1976, pp. 235–8, but without referring to the great nervous underbelly – anatomically and metaphorically – of the development. If he had, he would have discovered a rich untapped vocabulary of phrases such as 'tonic strength', 'essential tension', 'exquisite tautness', and 'irritable', which were originally medical, but which by mid-century had passed into common parlance as part of the diverse cults of sensibility, about which more is said below. Moreover, tonic strength of the nerves was believed by many of the so-called 'nerve doctors' to be seasonal, producing the most 'tonic period' in the spring and summer, when nerves could enjoy the benefits of the six non-naturals (exercise, diet, good air, sleep, regular evacuation, passions); their worst as the leaves were falling under cold, grey, dark skies; see the medical exposition of this theory in Andrew Wilson, MD, *Short Remarks on Autumnal Disorders of the Bowels*, Newcastle upon Tyne, 1765.

34. The great nerve doctors of the day included, in England, such well-known authors as Thomas Willis, perhaps the first physician to elevate the nerves;

Thomas Sydenham, important for his theory of hysteria, as well as his generally empirical approach to nervous disorders; Bernard Mandeville; George Cheyne, whose *English Malady*, 1733, was one of the century's best-selling books; Robert Whytt, the so-called philosophical doctor, whose *Observations on the nature, causes, and cure of those disorders which have been commonly called nervous, hypochondriac, or hysteric*, Edinburgh, 1765, became the classic statement of his generation; Thomas Coe, Francis Adair, and many others. Their lives and practices beg for a proper narrative. Even John Fothergill, Johnson's friend, whose medical practice did not specialise in the nerves, wrote *An Account of a Painful Affection of the Nerves of the Face, Commonly Called Tic Douloureux*, 1804. Pain was a crucial domain for nerves.

35. New Historicism has further shown us why these doctors were bound to exert terrific power and sway in a society as stratified and hierarchical as that between 1660 and 1820 in England, the so-called long eighteenth century; see D. Veeser, *The New Historicism*, New York, 1989.

36. Henrietta Knight [Lady Luxborough], *Letters of Lady Luxborough . . . to the poet William Shenstone*, London, 1775.

37. Mrs Ralph Allen, an educated woman whose great parlour and reception rooms in Bath could be considered the English equivalent of a salon, would have known more than most mothers of the period about the intimacy between the fevers of their children and the inflammation of the nerves; it had been spelled out by many of the nerve doctors she herself had heard pronounce on these matters in the privacy of her own home; see Thomas Kirkland, MD, *A Treatise on Child-Bed Fevers . . . to which are prefixed two dissertations, the one on the Brain and Nerves; the Other on the Sympathy of the Nerves, and on Different Kinds of Irritability*, 1774, esp. pp. 168–72.

38. See C. F. Mullett, ed., *The Letters of Dr George Cheyne to the Countess of Huntingdon*, San Marino, CA, 1940. Richardson's prose is permeated with the language of the nerves which forms an intrinsic part of his version of sensibility; for his personal commentary about nerves, see Anna L. Barbauld ed., *The Correspondence of Samuel Richardson*, 1804, 4, pp. 30, 283–4, and Raymond Stephanson, 'Richardson's "Nerves": The Physiology of Sensibility in *Clarissa*,' *Journal of the History of Ideas* 49, 1988, pp. 267–85. At this time nervous mythology was intrinsically tied to myths about the English nation and their developing nationalism as the most melancholic people on earth: depressed by their perpetually foul weather, dispirited by new stresses of high living, even unusually suicidal; as the poet Thomas Gray would say in a letter dated 27 May 1742, a nation epidemically stricken by 'White Melancholy' and 'Leucocholy'.

39. Frances Burney, *Evelina*, ed. Edward A. Bloom, 1968, p. 286 (vol. 3, letter III), where Mr. Lovel tells Lady Louisa Larpent that 'Your Ladyship's constitution is infinitely delicate', to which Louisa replies: 'Indeed it is,' cried she, in a low voice, 'I am *nerve* all over!'

40. *Register of Folly*, 1773. See also Peter Wagner, ed., *Christopher Anstey: The New Bath Guide*, Hildesheim, 1989.

41. Johnson's 'nervous prose' has received some attention as 'masculine' and 'energetic'; see Cecil S. Emden, 'Rythmical Features in Dr Johnson's Prose', *RES* 25, 1949, pp. 38–54; John Arthos, *The Language of Natural Description in Eighteenth-Century Poetry*, Ann Arbor, 1949; W. V. Reynolds, 'Johnson's

Opinions on Prose Style', *RES* 9, 1933, pp. 433–46; and the classic study by
W. K. Wimsatt, *Philosophic Words*, New Haven, 1948. But the development of
an Enlightenment *nervous style* at large, crossing national boundaries and
different cultures, has not been viewed within the contexts of the semiotic
of the nerves. Considered synoptically, nervous style was an ellision for all
things masculine in language, tough, strong, assertive, taut, concise –
anything but feminine and soft, loose and spacious, weak and flaccid. It was
a much admired, if also phallocratic, style in the mid-eighteenth century,
whose cultural production and dynamic has not yet been explored. Fielding,
Johnson, Smollett, Goldsmith – all were decorated, so to speak, at one time
or another, by some *male* critic or commentator, as their critical heritages
show, for displaying this nervous *je ne sais quoi* in their prose, no one more
so than the second of these, the literary lion of the age; and when Charles
Churchill, the decadent satirist of the 1760s, commented in *The Apology* (line
164) on Smollett's 'nervous weakness', his ironic inversion merely under-
scored the opposite, that is, that an irritable temperament in real life had
produced such a 'nervous style' in art. Elsewhere, Dr James Drake, a medical
doctor whose tropes are ridiculed and whose metaphors are satirised in the
pages of Sterne's *Tristram Shandy*, and a man who had written the most
popular textbook of anatomy in a generation until Cheselden's replaced it in the
1730s, summed up 'nervous style' succinctly when he wrote of a colleague's
that 'it [his prose style], both *Latin*, and *English*, was Manly yet Easie;
Concise, yet Clear and Expressive'; see James Drake, *Anthropologia Novum*,
1707, p. ix. A much fuller discusion of nervous style is found in Rousseau in
The Social History of Language (see note 25).

42. Much has been written about Mesmer, of course, in many languages and
countries, but often without clear sight of the direct role he played in nerve
therapy. But what was animal magnetism if not the strongest stimulus the
nerves could receive from an artificial, external source?

43. The late eighteenth century was the era *par excellence* of developmental patent
medicine, the first patent medicines having been brought out in the 1770s
after the Patent Office had opened in England; it is not surprising that a
flurry of these quack therapies and remedies would be rushed to the public
before they could be scrutinised by the officers of the Patent Office. See
J. H. Young, *The Toadstool Millionaires*, Princeton, 1961.

44. Lewis Knapp, ed., *The Adventures of Humphry Clinker*, Oxford, 1966, 34.

45. Matthew Bramble, Smollett's last hero, serves as a perfect example of his
maker's (Smollett's) intuition. Bramble has spent much of his life trying to
understand his 'nerves', only to discover that they continue to elude him.

46. Sion House MSS.

47. Dr William Derry has compiled a still unpublished mss archive of the
eighteenth-century Bath doctors. As early as 1699 the soothing effects of these
spa waters specifically on the *nerves* had been commented upon (Benjamin
Allen, *The Natural History of the Chalybeat and Purging Waters of England*, 1699).
A generation later, Dr Thomas Guidott claimed that the restoration of the nerves
to health was the principal value of a visit to the pump rooms (Thomas Guidott,
An Apology for the Bath, 1724), indicating how well-understood stress already
was in developing urban sprawl. But British Enlightenment theory did not
merely generate abstract discussion of societal stress; it explicitly located that

stress in a particular part of the anatomy in an attempt to discover how the nerves could be repaired and strengthened after depletion through wear and tear. For nerves in relation to social rank and class in Bath society, see Georges Lamoine, *'La vie littéraire de Bath et de Bristol 1750–1800'* (University of Paris III doctoral dissertation, 1978). Comments on Bath doctors and their patients, sorted out by class and wealth, are mentioned in an anonymous tract in the British Library: *Two Letters from a Physician in London, to A Gentleman at Bath . . . with some Observations on the Present . . .*, 1744. Useful work on Dr George Cheyne at Bath is also being conducted at present by Dr Anita Guerrini. Most of all, though, it is necessary to demonstrate how this semiotics and mythology of nerves reflected views then held regarding gender, class and race.

48. Dr Richard Mead, thought by some to be the nation's leading physician and a former President of the Royal Society, also wrote about the nerves, especially within the terms of Newtonian aether; see his *Mechanical Account of Poisons*, 1702, pp. 9–21, reprinted in his *Collected Works* 1762, pp. 455–61 for nervous juices and fluids.

49. A useful study of Tissot's theory of nerves in relation to his medical practice and therapy is found in Heinrich Walther Bucher, 'Tissot und sein Traité des Nerfs,' *Zürcher Medizingeschichtlicher Abhandlungen*, ed. E. H. Ackerknecht, Zurich, 1958.

50. If cultural history crosses the lines of traditional disciplines and teaches us how to view these boundaries and borders with scepticism, it also allows us to retrieve lost discourses such as those of the crucial nerve, as Weisskopf in our opening section would say; and within this specific domain it demonstrates that Adair merits a full-length biography, as do Dr George Cheyne and the Bath eccentric and proflic commentator Phillip Thicknesse. Useful information about Cheyne's career as the doyen of 'nerve doctors' is found in William Falconer, *Remarks on Dr. Cheyne's Essay on Health and Long Life*, Bath: Leake, 1745.

51. James Makittrick Adair, 'Medical Cautions for the Consideration of Invalids', 1786, in *Essays on Fashionable Diseases*, 1786.

52. Ibid., 4–9.

53. Thomas Coe, *A Treatise on Biliary Concretions*, 1757. Coe, like other leading physicians (Cadogan, Hill, Robinson), claimed that the gout could be an entirely 'nervous affliction' whose first sign was the debilitation of the nerves and fibres.

54. George Cheyne, *The English Malady: Or a Treatise of Nervous Diseases of All Kinds*, Bath and London, 1733.

55. See Thomas Dover, *The Ancient Physician's Legacy*, 1733, which went through several editions in only a few years and was translated into French within twelve months as *Leys d'un ancien médicin àa sa Patrie*, The Hague, 1734.

56. This direct link between nerves and hysteria requires full treatment, and is being discussed in a book by G. S. Rousseau and Roy Porter called *Beyond Hysteria*, now in preparation at the University of California Press. Briefly, the theory of the eighteenth century (in so far as one can reduce its diversity and generalise about it) was that all mental and emotional states depend on this elasticity or tightness of the nerves. Mechanists and vitalists alike, indeed most others too, shared in the belief, especially in the notion of tension as

the key element. Through this doctrine the looser cultural concept of 'tension' between persons, and even more internally between one part of the psyche and another, arose, and was eventually metaphorised into popular culture at large as a psychological state. The precise mechanisms by which the nerves interacted with fluids in different degrees of tension, often causing melancholy and even the more extreme hypochondriasis (in men) and hysteria (in women), was the subject of many medical dissertations in Europe; see, the material presented in the Appendix below.

57. Adair, *Essays on Fashionable Diseases*, 1786, p. 12. As a medical theorist Adair had been much influenced by the Montpellier medical writer François Boissier de Sauvages, a Stahlian vitalist who believed that the soul (brain too?) activated the nervous mechanisms of the body, whose *Nosologia methodica*, 1768, represents an extreme Linnanean application to taxonomise all disease.

58. The imagination was becoming medicalised under the influence of the seventeenth-century mechanists, and became increasingly so in the eighteenth century; for this development, see G. S. Rousseau, 'Science and the Discovery of the Imagination in Enlightened England' for the Aristotelian tradition, see Michael V. Wedlin, *Mind and the Imagination in Aristotle*, New Haven, 1989. Some of this work, as found in C. G. Gross, *De morbis imaginariis hypochondriacorum*, 1755, specifically addressed the imagination in relation to somatic diseases generated in the locale of the hypochondrium. In 1691, Timothy Rogers, a sedentary MA from Oxford, published a confessional treatise, *A Discourse concerning Trouble of Mind, and the Disease of Melancholy*, linking mind and melancholy through the medium of the Nerves. Others, such as the German physician J. F. Mossdorff, writing *De valetudinariis imaginariis, von Menschen, die aus Einbildung kranck werden*, 1721, were more concerned with illnesses that had no detectable somatic manifestations (that is, what we would call psychological conditions). In Italy, Lodovico Antonio Muratori, the empirical philosopher-poet whose book on imagination and dreams (1747) was widely discussed, suggested that the imagination played a central role in the formation of illness. In England, J. Richardson (of Newent) wrote *Thoughts upon thinking, or, a new theory of the human mind; wherein a physical rationale of the formation of our ideas, the passions, dreaming, and every faculty of the soul is attempted upon principles entirely new*, 1755, and suggested that the nerves mediate between ideas and illness. In all these discussions, and others, the nerves played a central role.

59. I have found only one copy of this obscure work, in the BL.

60. Richard Kuhn has surveyed the long tradition in *The Demon of Noontide: Ennui in Western Literature*, Princeton, 1976, but is rather inadequate on the Enlightment, leaving out such obvious candidates as melancholic Boswell – perhaps the greatest sufferer of the century, as his Dutch journals show – who even titled his most sustained work of periodical journalism *The Hypochondriack*; see the edition by Marjorie Bailey called *Boswell's Column*, 1951.

61. See Anna L. Barbauld, ed., *The Correspondence of Samuel Richardson*, 1804, 4, p. 30.

62. Ian Watt, *The Rise of the Novel*, Berkeley and Los Angeles, 1957, p. 184.

63. George Sherburn, ed., *The Correspondence of Alexander Pope*, Oxford, 1956, IV, p. 526. In his *Essay on the Genius and Writings of Pope*, 1762, Joseph Warton attempts to show that much of Pope's genius was tied to a delicate

sensibility founded on a nervous personality. Warton also considered 'nervous' composition as one of 'three different species,' which he adumbrates in the *Essay*, I, p. 170. Adam Smith commented in his *Theory of Moral Sentiments*, Edinburgh, 1759, on the stylistic (i.e. couplet) 'nervous precision of Mr. Pope'.

64. Misaurus Philander, *The Honour of the Gout*, 1720, pp. 18–19.

65. The full title deserves a place in the history of stress-related illnesses associated with global economic depression, such as the worldwide crash of 1929; see John Midriff, *Observations on the Spleen and Vapours; Containing Remarkable Cases of Persons of both Sexes, and all Ranks, from the aspiring Directors to the Humble Bubbler, who have been miserably afflicted with these Melancholy Disorders since the Fall of the South-sea, and other publick Stocks; with the proper Method for their Recovery, according to the new and uncommon Circumstances of each Case*, 1721.

66. William Smith, *A Dissertation upon the Nervous System*, 1768, whose purpose was to show its influence upon the soul. In France, Le Camus, a physician, also constructed a *Médecine de l'Esprit*, 1769, showing the link between the nerves and the soul.

67. Boswell's melancholy has been studied by Alan Ingram in *Boswell's Gloom*, London, 1984, but without paying attention to the scientific, medical, or cultural semiotics of the matter.

68. See Elizabeth Carter, *The Correspondence of Elizabeth Carter and Catherine Talbot*, 4 vols, 1809, 2, p. 156. More generally the passions, reason, morality and insanity were linked together specifically by the nervous apparatus, as suggested here and in dozens of other similar passages in different kinds of writing by both sexes. Two generations after Elizabeth Carter wrote, the prolific (if also prolix) Reverend Trusler, the moraliser of Hogarth who made his fortune by combining alleged medical expertise with clerical eccentricity, claimed to have penetrated to the truth about *cowardice* – in his view the most *feminine* of all moral defects, implying just the kind of *genderised nerves* we have seen gradually developing thoughout the century. Trusler wrote in his *Memoirs*, Bath, 1806, p. 46:

> What then is cowardice? – It is the effect of weak nerves – Who would not be brave if he could? *Acquired* courage may be the result of strong reasoning, refined courage, and a sense of duty, as in the simple case of the officer: *mechanical* courage, is often the effect of example, as in the soldier: – one man keeps the line, because another does, they consider themselves merely as parts of a great machine; but *natural* courage is the effect of strong nerves, which every man is not blessed with [compare Mrs Donnellan's advice to Samuel Richardson and my synoptic paradigm in note 61 above]. I might pity a coward, but I would not condemn him for want of resolution, more than I would condemn a weak man for want of strength. They are, like *nerves*, gifts of Providence bestowed on particular men.

69. Ibid., 2, p. 156.

70. Robert Halsband, ed., *The Complete Letters of Lady Mary Wortley Montagu*, Oxford, 1965, II, 63.

71. See Frank H. Ellis, ed., *Poems on Affairs of State: Augustan Satirical Verse, 1660–1714, Volume VI: 1697–1704*, New Haven, 1970, 64, lines 35–6.

72. *European Magazine*, 1812. William Cowper, the poet, almost blindly subscribed to a genderised version of nerves, but the myth was so broadly disseminated throughout his culture that one can hardly fault him for being less vigilant than he was to its sexual resonances. See his letter to the Reverend Unwin in *Collected Letters*, ed. Thomas Wright, 1780, for 2 July, 1780:

> . . . I like your epitaph, except that I doubt the propriety of the word immaturus; which I think, is rather applicable to fruits than flowers; and except the last pentameter, the assertion it contains being rather too obvious a thought to finish with: not that I think an epitaph should be pointed like an epigram. But still there is a closeness of thought and expression necessary in the conclusion of all these little things, that they may leave an agreeable flavour upon the plate. What ever is short should be nervous, masculine, and compact. Little men are so; and little poems should be so; because, where the work is short the author has no right to the plea of weariness; and laziness is never admitted as an available excuse in anything.

73. 'Of the Hypp', is to be found in the *Universal Spectator*, November 18, no. 214; *Gentleman's Magazine*, 2, 1732, 1062–3.

74. See Pierre Fedida, 'Les Exercises de l'imagination et la commotion sur la masse des nerfs: un érotisme de tête,' in *Oeuvres complètes du Marquis de Sade*, 16 vols, Paris, 1967, 9, pp. 613–25. Perceptive discussion of the nerves in Sade's prose is found in David Morris, 'The Discourses of Pain in Revolutionary France,' in G. S. Rousseau (ed.), *The Languages of Psyche: Mind and Body in Enlightenment Thought*, Berkeley and Los Angeles, 1990, pp. 291–300. While Sade was generating his fictional version of moral and revolutionary hedonism, Cabanis, among the most philosophical of physicians of the post-revolutionary period, correlated the nerves to specific stages of human perfection in an almost Lamarckean and pre-Darwinian sense; see Pierre-Jean-Georges Cabanis, *On the Relations Between the Physical and Moral Aspects of Man*, ed. George Mora, 2 vols, Baltimore, 1981.

75. See William Cullen, *Nosologia*; translated as *Nosology; or, a Systematic arrangement of diseases*, Edinburgh, 1768; 2nd edn, 1800, p. 238. J. M. Lopez Pinero has traced the tradition from Willis and Cullen down to current time in his *Historical Origins of the Concept of Neurosis*, Cambridge, 1983.

76. *The Works of John Dryden, The California Dryden*, ed. H. T. Swedenberg *et al.*, 18 vols, Berkeley and Los Angeles, 1961–2, p. 10.

77. In the same Paris milieu in which Gautier and his fellow decadents flourished, Pierre Jules Descot, a physician, published *Dissertation sur les affections locales des nerfs*, Paris, 1882, showing how all pleasure and pain was situated in the nerves at the tip of the genitals and why these erotogeneous zones were consequently the most crucial part of human anatomy, for pleasure as well as reproduction.

78. *The Works of Charles Lamb*, 7 vols, 1903–5. Across the English Channel, the literary-medical milieu in England was also interwoven with figures having an impact in each realm. In Lamb's world, the physician-poet Thomas Trotter could publish a significant study of *A View of the Nervous Temperament*, 1807, which correlated personality types according to their anatomical-physiological

constitutions, as well as a volume called *Sea Weeds: Poems written on various occasions, [written] chiefly during a naval life*, Newcastle and London, 1829.

79. Taking her cue from Cheyne's *English Malady* of 1733 (see note 34 above), Professor Elaine Showalter has studied the conditions under which the nerves were invoked in analyses of female somatic and psychogenic disorders in the nineteenth century; see *The Female Malady: Women, Madness and English Culture, 1830–1980*, 1987. During the peak of high Victorianism, countless doctors wrote about the nerves; in Germany, discourses of the nerve were as important as they were in England, and treatises such as the German physician M. H. Romberg's (*Nervous Diseases of Man*, 1853) were quickly translated into English.

80. This link remains the one to be explored among various discourses, and it is a glaring shortcoming of this essay that I do not undertake it here. My excuse (such as it is) that I have not had the space will perhaps not stand up, but I nevertheless wish to acknowledge how crucial it seems to me to establish these networks of connection.

81. The extraordinary matter to be grasped here is not that Austen should allow common parlance about nerves to invade her highly eclectic prose vocabulary, but rather that she permits it without more irony, as any systematic lexical study of her use of nervous language would show (I have collected over two dozen of these passages but do not include them here for reasons of space). Whole ranges of scientific vocabularies and their metaphors are denied entry to her fictive discourse; but the nerves enter with little if any resistance; the question is why. Perhaps Austen knew more about neural medicine than has been credited to her. As she was composing her novels of sentiment and delicacy, her *Emma* and *Mansfield Park* (which has the largest number of uses of nervous vocabulary and which is drenched in the female sensibility of Richardson and Frances Burney), English doctors in her geographical locale, such as Dr M. Hall, were writing *On the mimoses: or, A descriptive, diagnostic and practical essay on the affections usually denominated dyspeptic, hypochondriac, bilious, nervous, chlorotic, hysteric*, 1818.

82. For this tradition, see note 9 above and Robert Brissenden, *Virtue in Distress: Studies in the Novel of Sentiment from Richardson to Sade* London 1974.

83. Coleridge's anatomy and medicine have not received the attention they deserve, despite Trevor H. Levere's study of his science, *Poetry Realized in Nature: Samuel Tayor Coleridge and Early Nineteenth-Century Science*, New York, 1981; without understanding these two realms one cannot comprehend Coleridge's contribution to the discourse of the nerves. One would have thought that his philosophical lineage as a Hartleyan and his aesthetics of immediacy would automatically privilege the nerves; see Wallace Jackson, *Immediacy: the Development of a Critical Concept from Addison to Coleridge*, Amsterdam, 1973; but although Coleridge writes abundantly about immediacy there is less material about nerves in his prose than one would have imagined; nevertheless see his pronouncements 'On Sensibility,' in *Complete Works*, Princeton: Bollingen, 1962–. At the same time Coleridge was pronouncing, such scientific associates of his as Drs Thomas Young (the prolific naturalist who wrote about colour theory) and John cooke (the president of the

Medico-Chirurgical Society) were also writing about the nerves: Young especially in a *Treatise on Phthisis*, 1822, and Cooke in a two-volume *Treatise on Nervous Diseases*, 1823; Boston edn 1824.

84. Virginia Woolf's relation to this tradition merits some attention, not least because she herself was among the most 'nervous' of writers and connected the creative act to the state of the nerves during creation. Like Mary Shelley's mad scientist, whose 'nervous agony' was renewed at the mere sight of a 'chemical instrument' ever since the appearance of his vision, she often composed under nervous duress and extreme agitation. The passage in *To the Lighthouse* in which Lily is struggling with her painting captures this essential belief about the relation of physiology, desire and artistic creation: 'Phrases came. Visions came. Beautiful pictures. Beautiful phrases. But what she wished to get hold of was that very jar on the nerves–the thing itself before it has been made anything.' Sterne was Woolf's favourite prose writer of the eighteenth century, though she also read and delighted in the witty prose of Addison and Steele (as one would know from reading *Orlando*); yet it is inconceivable that in her constant reading of Sterne she had overlooked his own fascination with nervous prose (note 41 above) and the various ways in which Sterne and his contemporaries had transformed the discourse of animal spirits, nerves and fibres – literally from the first paragraph of *Tristram Shandy* – turning it upside down and metaphorising and satirising medical dissertations about nerves of just the type discussed in this essay. For further discussion of Sterne in this sense, see G. S. Rousseau, 'Smollett and Sterne: A Revaluation', *Archiv für das Studium der neuren sprachen und Litteraturen*, CCVII, 1972, pp. 286–97. Yet Woolf should be viewed in a wider, and more medical, context than this. She was growing up in a late Victorian England that had heard (for example) medically trained lecturers like Andrew Wilson speaking on *The Origin of Nerves. A Lecture Delivered Before the Sunday Lecture Society on 24 December 1878*, London, 1879. And she herself was profoundly interested in theories of nervous disorders and neurosis, in part as a result of her own mental states when she composed, but also in view of her own mental states when she composed, but also in view of her oblique sexuality. Her diaries continually exude wrenching remarks about the effort her writing wrung from her: 'it calls upon every nerve to hold itself taut'. Later on, in her mature years during the 1920s, there was talk everywhere – Havelock Ellis, Edward Carpenter, D. H. Lawrence, Walt Whitman, Otto Weininger, even within the Bloomsbury circle – about the relation of nervous mechanism and physiology in sex and love. In passing, one suspects that Woolf might have sympathised with Schreber's analysis of his own nervous conditions; see *Daniel Paul Schreber: Memoirs of My Nervous Illness*, 1955.

85. A. Alvarez, *The Savage God: A Study of Suicide*, New York, 1970, p. 19. J. Babin-ski and J. Froment, *Hysteria or Pithiatism and Reflex Nervous Disorders in the Neurology of War*, 1918, p. 311.

86. By the turn of this century a whole school of thought had developed in the belief that nervous ailments could be treated successfully by moral therapies; see Arnold Stocker, *Le traitement moral des nerveux*, Geneva, 1945. Most have been discredited by now.

87. But now, in our post-disciplinary age, the discourse of postmodernism, as Habermas has suggested, brings them together.

Appendix

These dissertations were written largely by medical students and their professors, although some (indeed the first work, by Albert, and the cultural historical work by Dubois d'Amiens) were composed by non-medical writers. The reader will soon see that their number dramatically increased in the nineteenth century, although the practice was already institutionally fixed by the eighteenth century, where hardly a year goes by without several such tracts appearing. The list below is highly selective. An exhaustive bibliography would be many times larger, and might even be impossible to compile, given the large number of such works that have disappeared and the poor handwritten catalogues of most Continental libraries during the eighteenth century.

J. Albert, *Essai sur l'hypochondrie* (Paris, 1813).

M. Albertus, *De hypochondriaco-hysterico malo* (Halle, 1703).

T. H. Arens, *De mali hypochondriaci symptomatis et causis* (Berlin, 1844).

F. Arnisaeus, *De malo hypochondriaco* (Copenhagen, 1654).

G. J. A. Baltz, *De malo hypochondriaco* (Berlin, 1845).

J. F. Becker, *De morbo hypochondriaco* (Berlin, 1820).

J. C. Below, *Dissertatio casum matronae hypochondriacae exhibens* (Erfurt, 1685).

J. Ben, *De suffocatione hypochondriaca* (Leiden, 1683).

C. A. Berthelen, *De hypochondriasis origine* (Leipzig, 1846).

A. L. Birotheau, *Sur l'hypochondrie* (Paris, 1830).

Sir Richard Blackmore. *A treatise of the Spleen and Vapours: or hypochondriacal and hysterical affections, with three discourses in the nature and cure of the cholick, melancholy, and palsies. Never before published* (London, 1725).

E. Blum, *De dolore hypochondriaco, vulgo sed falso putato splenetico* (Leipzig, 1671).

J. G. Boemer, *De hypochondria* (Berlin, 1817).

E. J. F. Bourrelly, *Sur l'hypochondrie* (Paris, 1819).

F. Bouteiller, *Essai sur l'hypochondrie* (Paris, 1820).

J. Bouwer, *De affectione hypochondriaca* (Utrecht, 1688).

J. L. Brachet, *Recherches sur la nature et le siège de l'hystérie et de l'hypochondrie* (Paris, 1832).

P. Brand, *De malo hypochondriaco* (Copenhagen, 1676).

J. C. C. Brandt, *De malo hypochondriaco rite cognoscendo* (Wittenberg, 1811).

J. H. Brechtfeld, *De morbo hypochondriaco* (Helmstedt, 1662).

H. L. J. Brequin, *Sur l'hypochondrie* (Paris, 1831).

P. A. Brunereau, *Du siège, de la nature, des causes de l'hypochondrie* (Paris, 1857).

C. G. Burghart, *De malo sic dicto hypochondriaco* (Wittenberg, 1703).

J. Cahen, *De natura atque causis hypochondriae* (Berlin, 1843).

J. H. Calestroupat, *Sur l'hypochondrie* (Paris, 1823).

A. F. V. Carilian, *Sur l'hystérie et l'hypochondrie* (Paris, 1818).

J. B. L. P. Castagnon, *Essai sur l'hypochondrie* (Paris, 1858).

H. Cellarius, *De affectu hypochondriaco* (Jena, 1671).

M. Chabert, *De hypochondria* (Paris, 1805).

J. P. Champagne, *Sur l'hypochondrie* (Paris, 1827).

A. E. C. A. Chauvin, *Parallèle de l'hypochondrie avec la mélancholie* (Strasbourg, 1824).

L. H. Chevalier, *Sur l'hypochondrie* (Paris, 1820).

G. Clasius, *De therapia passionis hypochondriacae* (Halle, 1713).

C. C. Colnot, *Etude sur le délire hypochondriaque* (Paris, 1878).

M. A. Colohri, *De passione hypochondriaca* (Frankfurt, 1751).

D. Corbet, *De hypochondriasi* (Edinburgh, 1821).

G. S. Cotta, *Dissertatio aegrum chylificațione laesa hypochondriaca laborantem exhibens* (Jena, 1689).

J. Cowling, *De hypochondriasi* (Edinburgh, 1768).

T. Cupples, *De hypochondriasis causis* (Edinburgh, 1777).

A. L. Dejoye, *Essai sur l'hypochondrie* (Paris, 1866).

H. G. Delagrye, *De l'hypochondrie* (Paris, 1817).

Nicholas François Dellehe, *Tentamen medicum de affectione hypochondriaca seu hysterica* (Avignon, 1788).

J. B. Derivaux, *Essai sur l'hypochondrie* (Strasbourg, 1836).

E. F. Dubois, *Ueber das Wesen und die gründliche Heilung der Hypochondrie und Hysterie. Herausgegeben und mit einer Einleitung versehen von Dr. K. W. Ideler* (Berlin, 1840).

E. F. Dubois d'Amiens, *Histoire philosophique de l'hypochondrie et de l'hystérie* (Paris, 1837).

J. M. D. Duc, *Sur l'hypochondrie* (Paris, 1827).

G. Duché, *De la nature essentielle de hypochondrie* (Paris, 1833).

A. Dufour, *Etude sur l'hypochondrie et de délire hypochondriaque* (Paris, 1860).

J. C. Dupont, *Recherches sur l'affection hypochondriaque* (Montpellier, 1798).

G. G. Dynnebier, *De morbo hypochondriaco* (Berlin, 1820).

J. F. Entlicher, *De hypochondriasi* (Prague, 1813).

G. Erdmann, *De hypochondriasi* (1847).

J. B. Ernoul-Provoté, *Essai sur l'hypochondrie* (Paris, 1816).

F. F. Ettling, *De hypochondriasi* (Berlin, 1850).

A. A. Etzel, *De morbo hypochondriaco* (Vienna, 1789).

G. B. Faber, *Ulterior expositio novae methodi Kaempfianae curandi morbos chronicos inveteratos, praecipue malum hypochondriacum* (Tübingen, 1755).

J. P. Falret, *De l'hypochondrie et du suicide. Considérations sur les causes, sur le siège et la traitement de ces maladies, sur les moyens d'en arrêter les progrès . . . le développement* (Paris, 1822).

J. Faure, *Sur l'hypochondrie* (Paris, 1823).

J. Feist, *Morbi hypochondriaci cum hysterico comparatio* (Berlin, 1819).

J. Fellner, *De hysteria et hypochondria* (Würzburg, 1837).

J. C. Fischer, *De malo hypochondriaco* (Erfurt, 1713).

Malcolm Flemyng, *Neuropathia; sive de morbis hypochondriacis et hystericis, etc.* (1740).

A. Fracassini, *Naturae morbi hypochondriaci ejusque curationis mechanica investigatio* (Verona, 1754).

L. Fraser, *De morbo hysterico sive hypochondriaco* (Edinburgh, 1750).

G. Fruth, *De hypochondria* (Munich, 1842).

M. Fuker, *Disquisitiones nonnullae circa hypochondriam* (1833).

S. V. Gadebusch, *De affectione hypochondriaca* (1685).

P. S. Garboe, *Experimenta quaedam circa malum hypochondriacum* (Halle, 1762).

G. M. Gattenhof and F. Zuccarinus, *Hypochondriasis* (Heidelberg, 1769).

F. P. Gauné, *Sur l'hypochondrie* (Paris, 1826).

M. Geiger, *Microcosmus hypochondriacus, sive de melancholia hypochondriaca tractatus* (Munich, 1652).

A. B. de La Geneste, *De morbo hypochondriaco* (Leiden, 1763).

J. A. Genser, *Dissertatio pathologiam mali hypochondriaci inquirens* (Wittenburg 1797).

Oscar Giacchi, *L'isterismo e l'ipochondria avvero il malo nervosa . . . Giudizii fisio-clinici-sociali* (Milan, 1875).

W. Gibbons, *On hypochondriasis* (Philadelphia, 1805).

M. Giraldus, *De singulari sensibilitate hypochondriacorum ejusque causis* (1749).

P. M. B. Goullin, *Sur l'hypochondrie* (Paris, 1821).

A. A. Grenet, *Sur l'hypochondrie* (Paris, 1840).

C. G. Gross, *De morbis imaginariis hypochondriacorum* (1755).

J. F. Haack, *De affectione hypochondriaca* (1678).

A. Hafner, *De hypochondriasi ut morbo coenaesthesis* (1808).

M. Hall, *On the mimoses: or, A descriptive, diagnostic and practical essay on the affections usually denomianted dyspeptic, hypochondriac, bilious, nervous, chlorotic, hysteric, etc.* (London, 1818).

A. Haro, *Considérations générales sur l'hypochondrie* (Strasbourg, 1834).

C. V. Hareaux, *Essai sur une variété d'hypocondrie particulière aux femmes de l'âge critique* (Paris, 1837).

A. Hay, *De affectionibus hystericis et hypochondriacis* (Leiden, 1765).

F. L. Hedenaberg, *De differentia et similiandinibus hypochondriae et hysteriae* (1815).

G. Heideman, *De hypochondriae caussis* (Berlin, 1838).

H. G. Herfelt, *De affectione hypochondriaca* (Duisberg, 1678).

J. G. Heyman, *De praecipuo literatorum morbo affectione hypochondriaco* (Leiden, 1732).

Nathaniel Highmore, *Exercitationes duae, quarum prior de passione hysterica, altera de affectione hypochondriaca.* 2d ed (Amsterdam, 1660).

Nathaniel Highmore, *De hysterica et hypochondriaca passione. Responsio epistolaris ad Doctorem Willis* (London, 1670).

J. Hill, *Hypochondriasis. A practical treatise on the nature and cure of that disorder; commonly called the hyp and hypo* (London, 1766).

J. F. Isenflamm, *Versuch einiger praktischen Anmerkungen Ueber die Nerven zur Erläuterung verschiedener Krankheiten derselben, vornehmlich hypochondrisch und hysterischer* (Autälle, 1774).

J. M. Israel, *De hypochondriaco malo monita quaedam* (1798).

J. N. Jessenwanger, *Dissertatio sistens morbum hypochondriacum et hystericum* (1778).

D. A. Koch, *De infarctibus vasorum in infimo ventre ceu caussa plurium pathematum chronicorum, speciatim eorum, quae sub mali hypochondriaci nomine veniunt* (Strasbourg, 1752).

L. I. Kohen, *De morbo hypochondriaco* (1729).

V. Lebas, *Observation de mélancholie, et quelques propositions sur cette maladie* (Paris, 1820).

J. Le Blanc, *Sur l'hypochondrie* (Paris, 1826).

A. A. Lecadre, *Sur le siège et la nature de l'hypochondrie* (Paris, 1827).

J. G. Lehmann, *De duumviratu hypochondriorum* (Leipzig, 1689).

P. G. Léhu, *Sur la pathidie, vulgairement nommée hypochondrie, considérée en général, et particulièrement sous le rapport de son siège* (Paris, 1831).

J. G. Leidenfrost, *De mali hypochondriaci ad minimum sextuplici specie* (Duisberg, 1797).

J. G. Leisner, *De malo hypochondriaco-hysterico* (1749).

L. M. Le Siner, *De l'hypochondrie* (Paris, 1841).

Louyer-Villemary, *Sur l'hypochondrie* (Paris, 1802).

L. Löwenberg, *De hypochondria* (Berlin, 1841).

B. Mandeville, *A treatise of the hypochondriack and hysterick diseases* (London, 1730).

C. T. Matschke, *De stabilienda hypochondriae et hysteriae notione* (1806).

C. H. Matthiae, *Morbi hypochondriaci cum hysterico comparatio* (Würzburg, 1845).

A. Mecklenburg, *De hypochondria* (Berlin, 1851).

A. C. Meineke, *De vera morbi hypochondriaci sede indole, ac curatione* (1719).

J. Meyer, *De natura morbi hypochondriaci* (Berlin, 1867).

C. F. Michéa, *Traité pratique, dogmatique et critique de l'hypochondrie* (1845).

A. F. F. Mohring, *Dissertatio sistens cogitata quaedam de malo hypochondriaco atque hysterico* (1798).

C. Mongin-Montrol, *Sur l'hypochondrie* (Paris, 1823).

De Montallegry, *Hypochondrie-spleen ou névroses trisplanchniques. Observations relative à ces maladies, et leur traitement radical* (1841).

G. H. Morin, *De hypochondrie* (Paris, 1831).

L. Müller, *Dissertation sur le spasme et l'affection vaporeuse* (Strasbourg, 1813).

F. W. Nolte, *Die Hypochondrie* (Utrecht, 1840).

S. Ochlitius, *De passione hypochondriaca* (Jena, 1666).

J. J. Otto, *De malo hypochondriaco* (1722).

M. Pallier-Lapeyrière, *Coup d'œil philosophique sur l'hypochondrie* (Paris, 1837).

A. F. Pelgrom, *De morbo hypochondriaco* (Leiden, 1759).

William Perfect, *Cases of Insanity . . . Hypochondriacal Affection . . .* (London, 1781).

Joannes Fridericus de Pre, *Dissertatio inauguralis medica de melancholia hysterica . . .* (Erfunt, 1728).

F. Private, *Coup d'œil sur l'hypochondrie* (Paris, 1827).

J. T. Rauch, *De affectu hypochondriaco* (Jena, 1755).

E. H. Reichel, *De hypochondria et hysteria* (Jena, 1803).

J. Reid, *Essays on hypochondriacal and other nervous affections* (Philadelphia, 1817).

T. Remmets, *Ueber die Hypochondrie* (Bonn, 1872).

C. Retenbacher, *De mali hypochondriaci causa proxima* (1838).

J. E. Riemer, *De affectu hypochondriaco* (1728).

H. Rigius, *De affectione hypochondriaca* (1649).

C. Ringelmann, *Ueber die Natur, das Wesen und die Behandlung der Hypochondrie und Hysterie* (1824).

Nicholas Robinson, *A new system of the spleen, vapours, and hypochondriack melancholy; wherein all the decays of the nerves, and lowness of the spirits are mechanically accounted for. To which is subjoined, a discourse upon the nature, cause, and cure of melancholy, madness and lunacy* (London, 1729).

A. Rossi, *De hypochondriasi* (1842).

G. Roth, *De hypochondriasi* (1833).

William Rowley, *A treatise on female, nervous, hysterical, hypochondriacial, bilious, convulsive disease; apoplexy & palsy with thoughts on madness & suicide, etc.* (London, 1788).

J. L. Rudiger, *De variabili hypochondriacorum mente* (1746).

H. O. Schacht, *De melancholia hypochondriaca* (1693).

E. Schiller, *De hypochondria* (Prague, 1841).

J. B. Schlosser, *Ueber die Hypochondrie* (Munich, 1838).

F. G. Schroeerus, *De morbo ex hypochondriis* (1760).

G. Schultz, *Dissertatione sistens aegrum laborantem malo hypochondriaco scorbutico* (1670).

W. G. Schuyt, *De differentia inter hypochondriacum et hysteriam* (Amsterdam, 1847).

S. Schwartz, *Nonnulla ad malum hypochondriacum spectantia* (1757).

C. D. Seboldt, *Mali hypochondriaci, veri ac nervose, seu morbi sine materie aucta, notio et natura* (1796).

J. F. Seyffert, *De hypochondriasi* (Leipzig, 1824).

J. Siess, *Dissertatio sistens ideam pathematum hypochondriaco-hystericorum cum singulari huc faciente historia morbi* (1780).

J. C. Sommer, *De melancholia imprimis hypochondriaca* (1706).

J. G. Sonnenmayer, *De vero ortu mali hypochondriaci et hysterici* (1769).

J. P. Spring, *De malo hypochondriaco* (Munich, 1758).

J. M. Starckloff, *De sputatione hypochondriacorum* (Halle, 1730).

J. Stark, *De malo hypochondriaco* (Edinburgh, 1783).

A. Staub, *Allgemeiner Leitfaden zur Bearbeitung der Hypochondrie und Hysterie* (Würzburg, 1826).

J. A. Steininger, *Centuria positionum medicarum de melancholia hypochondriaca* (Wittenberg, 1625).

J. C. Storck, *De malo hypochondriaco* (Altdorf, 1685).

L. Storr, *Untersuchungen über den Begriff, die Natur und die Heilbedingungen der Hypochondrie* (Stuttgart, 1805).

J. T. Strassburg, *De affectu hypochondriaco* (1696).

J. D. Strauss, *Dissertatio aegrum affectu hypochondriaco, capitisque steatomate laborantem exhibens* (Giessen, 1683).

J. L. Sustermann, *De valetudine ex hypochondriis* (Göttingen, 1752).

Thomas Sydenham, 'Processes Integri: Chap. 1: On the Affection called Hysteria in Women; and Hypochondriasis in Men', *The Works of Thomas Sydenham*, trans. R. G. Latham, 2 vols (London, 1848).

F. P. A. Tartivel, *De l'hypochondrie* (Paris, 1852).

L. Theill, *De malo hypochondriaco* (Jena, 1668).

P. Thewalt, *Ueber die Ursachen der Hypochondrie nebst einigen begleitenden Bemerkungen* (Würzburg, 1846).

J. C. Tode, *Nödig underwisning für hypochondrister som wilja rätt lära känna sitt tillstränd och förbättra det. Ofversättning fränstränd och förbättra det.* (Stockholm, 1809).

J. C. Troppinniger, *De malo hypochondriaco* (Leipzig, 1676).

W. Turner, *De morbo hypochondriaco* (Edinburgh, 1756).

D. van Buren, *De affectione hypochondriaco* (Leiden, 1711).

C. van der Haghen, *De melancholia hypochondriaca* (1715).

D. J. van der Meersch, *De hypochondria* (1817).

P. van Suchtelen, *De melancholia hypochondriaca* (1718).

P. Venables, *De hypochondriasi* (Edinburgh, 1803).

E. Vielloehner, *De hypochondriae et hysteriae differentiis* (Berlin, 1841).

J. C. I. Voigt, *Tractatus medicus Galeno-chymicus, de passione seu affectione hypochondriaca, authoritatibus Galeni et Hippocratis suffulsus* (Prague, 1678).

L. de Wahl, *De causa hypochondriae proxima* (Berlin, 1832).

G. Walther, *De mali hypochondriaci natura et causis* (Berlin, 1845).

G. Wehrmeister, *De hypochondriasi* (1846).

B. N. Weigelius, *De malo hypochondriaco* (1745).

Robert Whytt, *Observations on the nature, causes, and cure of those disorders which have been commonly called nervous, hypochondriac, or hysteric, to which are prefixed some remarks on the sympathy of the nerves*, 2nd edn, 8 vols (Edinburgh, 1765).

Thomas Willis, *Affectionum quae dicuntur hystericae et hypochondriacae, pathologia spasmodica vindicata, contra responsionem epistolarem Nathanael Highmori* (Leiden 1671).

C. F. Winneke, *De morbo hypochondriaco e plethora oriundo* (1792).

J. M. Wirtz, *De sede et causa proxima hypochondriae* (Bonn, 1830).

C. L. Wischke, *Mali hypochondriaci veri ac nervosi signa et diagnosis* (1795).

C. Witter, *De hypochondria* (Berlin, 1837).

T. Wittmaack, *Die Hypochondrie (Hyperaesthesia physica, Romberg) in pathologischer und therapeutischer Beziehung, nebst einigen vorgängigen Bemerkungen über die Bedeutung der psychischen Heilmittel* (Leipzig, 1857).

P. Zacchia, *De' mali hipochondriaci Libri tre* (Rome, 1644).

P. Zacchia, *De affectionibus hypochondriacis libri tres. Nunc in Latinum sermonem translati ab Alphonso Khonn* (1671).

G. V. Zeviani, *Del flato a favore degli ipocondriaci* (Verona, 1794).

K. J. Zimmermann, *Versuch über Hypochondrie und Hysterie* (Bamberg, 1816).

J. C. Zopef, *De malo hypochondriaco* (Jena, 1676).

C. Zurborn, *De hypochondriasi* (Berlin, 1845).

G. Zwingenberg, *De malo hypochondriaco* (Berlin, 1856).

9

'Strange Pathology': Nerves and the Hysteria Diagnosis in Early Modern Europe (1993)

This excerpt is taken from the middle of a long chapter about the hysteria diagnosis three centuries before Freud's. It appeared in a five-handed book – together with Sander Gilman, Helen King, Roy Porter and Elaine Showalter – called Hysteria Beyond Freud *in which each of us was responsible to 'frame hysteria' in a particular epoch, mine bounded by hysteria's middle-ground during 1650–1820. The title, a 'strange pathology,' takes its cue from Thomas Sydenham and recognizes how the chain of nerves, fibers and spirits had paved the way to a new hysteria diagnosis at that transformative moment in the generation of Willis and Sydenham before the Paris doctors altered it yet again while retaining its nervous base.*

The doctors I discuss in this essay also assumed a primarily nervous base for human life, one whose sediment deepened over the 170 years surveyed here. The value of their diagnoses for us is that they were closer to the source of hysteria's nervous heritage than we are. They never questioned this fundamental etiology of the condition, or category, they explored. Our brain-dominated modern era acknowledges that all sensation and thought starts and finishes in the brain, but (except for the neurologists among us) we understandably overlook that the brain would cease to exist without the synapses and rest of the nervous system. The plasticity of this system, presiding over all aspects of human life, also constituted the root of consciousness, including dark despair and lofty sublimity; as well as sublime thoughts and ecstatic sensations. These and others were enabled by particular types of nervous systems. The task of describing these nuanced bodily states, in an epoch when they were barely understood, posed hurdles in retrieval as well as description.

This reconstruction of the hysteria diagnosis also persuaded me of the modernity of Enlightenment sensibility. Hysteria was, after all, much older. The intensely 'nervous culture' of the eighteenth century was similar, in many respects, to ours, with the proviso that our technology has made it possible to pinpoint discrete functions (thoughts, experiences, memories, mental states) with increasing precision. Moreover, the parallels between an evolving nervous system and the hysteria diagnosis summoning up that system was noteworthy. Whether I paused over the imaginative literature of sensibility, or – again – the medical discourse about it in Sydenham's observations on hysteria in the 1680s, in Dutch medicine of the 1690s, Boerhaave's applications around 1700, or the nerve doctors at the time of the Queen Anne Wits (some of whom were referred to as 'wit doctors' for applying Newton's laws of mechanics so adroitly to the body) – what struck me was the sense of an evolving view of man as exquisitely malleable in his responses and nuanced in his movements, precisely because of his nervous system. Throughout the eighteenth century he continued to be configured as a binary creature: composed of mind and body, reason and passion, virtue and vice, and others as well, in part because such a coordinated set of antitheses (or near-opposites) enabled creatures to interpret themselves and be cured from malady, especially the inexplicable hysteria.

The main argument of our book, reflected in its title, Hysteria Beyond Freud, was that Freud represented the 'end' rather than the beginning of a process regarding the hysteria diagnosis. An ancillary point was that hysteria had steadily been evolving: in a complex arc from the Ancients to the Moderns. More locally it seemed that the evolution of the nervous system, from the time of Galen, had been concurrent. Oliver Sacks notes with awe the consequence of this development: 'A single conscious visual percept may entail the parallel and mutually influencing activities of billions of nerve cells.' Neuroscientist Gerald Edelman extends the argument in the books cited here to describe how consciousness itself evolved over billions of years, and elsewhere that 'whether richly structured or simple, nervous systems evolved to generate individual behaviour that is adaptive within a specific econiche in relatively short periods of time.' How else could all this neuronal activity, fusing perception and memory, past and present, together into dynamic consciousness and reflection, occur if the nerve themselves had not evolved? And if nerves were evolving, then why should not new nervous states, or at least nervous maladies like the protean hysteria?

The point of this nervous evolution, complex for a non-neurobiologist, was to permit creatures to survive in the wild and adjust to their changing environments. Hence the brain itself had to be plastic – in our language, soft- rather than hard-wired. The brain must alter and change itself through experience and learning to reach goals and satisfy the needs of survival. Hence the 'neurological chaos in the brain,' to which I referred in this chapter (p. 318), was itself a product of adaptation to demanding forces required for survival, albeit social rather than protein in recent man. Hysteria – even the early hysteria of *the early modern period surveyed here – was a response to the ardours of such adaptation. It assumed, first, the existence of a gendered nervous system enabling women to calibrate themselves to adaptation, and then, secondly, a cultural set of values and codes typified as 'sensibility' which guided her through the complex signs required for socioeconomic adaptation. The signposts of sensibility had thus grown up in a milieu that prized nerves and practically made a fetish out of them. No wonder then that cultural icons of sensibility were thrown up. Or that – for the first time in modern culture – some people suddenly became alert to the double entendre of words such as sense, sensitive, sensible, sensuous, spirited, spiritual, and the now equally semantically charged 'nervous' whose domain was greatly expanded in the eighteenth century.*

'A Strange Pathology' (1993)

I

In the progression from Willis and Sydenham to Cheyne and Bernard Mandeville – the satirist of *The Fable of the Bees* – and their successors later in the eighteenth century, it was Sydenham who took the largest strides. Willis made free use of the hysteria diagnosis in managing sick women, saw hysteria as a somatic disturbance, treated patients with drug-based therapeutics, and considered the probability that men could be afflicted too. Inasmuch as women of all ages and ranks could suffer from it, he prudently dismissed the notion of Dr. Nathaniel Highmore, his contemporary, that hysteria was due to bad blood.[1] He doubted that it was owing to any specific uterine pathology and identified the central nervous system, spanning the brain and the spinal cord, as the true site. Being 'chiefly and primarily convulsive,' he argued, 'hysteria flared on the brain, and the nervous stock being affected.'[2] The animal spirits were specially vulnerable: 'The Passions commonly called Hysterical... arise most often [when]... the animal spirits, possessing the beginning of the Nerves within the head, are infected with some Taint.' So he, like Sydenham, concluded that hysteria could not, technically speaking, be solely a female complaint; he offered the weaker nervous constitution as the reason why women were worse afflicted.[3] The obvious conclusion, although both Willis and Sydenham were too cautious to proffer it, was that men with clear symptoms of hysteria were effeminate.[4]

These schematizations shifted the ground to the nervous system as the key through which to understand and interpret hysteria as a category as well as human illness, and the paradigmatic shift is important for Enlightenment medicine.[5] But if hysteria, as both Sydenham and Willis claimed, was the Proteus of maladies – the elusive medical condition par excellence – then we should expect the medical theory of the period to view the nervous system as the key to practically *all* illness, not merely hysteria.

This it did. The best theory of the day did not, naturally, endow the nerves with the key to every disease, but once the mechanical philosophy had completed its work and the paradigmatic shift was absorbed (roughly by 1700), there were few if any diseases without nervous implications. Eventually this monolithic attribution would be seen for the foreshadowing of modern nervousness that it is. At the time, it was viewed as the only respectable medical course possible. Dealing with affluent clienteles, the highly influential Italian physician Georgio Baglivi and satirist Bernard Mandeville carved out comparable concepts of hysteria to encompass the protean ailments of the polite, whose sensibilities to pain were as extensive as their vocabularies, and who may have been adroit at manipulating the protective potential of sickness. Mandeville, a brilliant writer of prose, was sensitive to the languages of hysteria, especially their jumbled vocabularies and dense metaphors. He had commented profusely on the metaphoric kingdoms of 'the animal spirits' – commenting pejoratively most of the time and demonstrating how little he believed that medical writers had followed the pious credos of the Royal Society espousing *nullius in verba*, loosely 'nothing in the word.' In his dialogic *Treatise of the Hypochondriack and Hysterick Passions*, Mandeville makes a character proclaim: 'You Gentlemen of Learning make use of very comprehensive Expressions; the Word *Hysterick* must be of a prodigious Latitude, to signify so many different Evils,' suggesting that a type of 'madness' would arise from nomenclature itself, a form of illness every bit as real as the genuine 'hysteric's affliction.'

Drawing upon his extensive clinical experience, Baglivi demonstrated how patients commonly presented symptom clusters resistant to rigid disease categories, though responsive to the personal tact and guile of the physician.[6] Mandeville, for his part a profound social commentator as well as a sought-after medical practitioner, made much of the fashionable life-style pressures disposing women to hysteria while their husbands sank into hypochondriasis.[7] Was there a *determinant* anatomico-physiological etiology for the disorder? Mandeville, like Sydenham, deflected the question, concentrating instead upon those behavioral facets – languor, low spirits, mood swings, depression, anxiety – integral to the presentation of the self in everyday sickness. Mandeville's substantial contribution to the theory of hysteria was revisionary more than anything else. He ridiculed the elaborate speculative models of mechanico-corporeal machinery floated by Willis, especially the idea that erratic mood shifts were literally due to 'explosions' in the animal spirits, and derogated the highly analogical language Willis used to capture the iatromathematical motion of these nervous eruptions. Mandeville was less troubled by Willis's theory of *sympathy* than with his version of *idiopathy*: the idea that the 'explosion' could convey its neuroanatomic effects throughout the body by sympathy. Idiopathy and 'detonation' were Mandeville's unrelenting gripe, especially the unpredictable onset of the 'detonations,' not a theory of medical sympathy that had historically antedated Willis nor neurophysiological disagreement about the manner of conveyance through the nervous pathways. Furthermore, the metaphoric dangers of 'detonation in the human body' struck the satiric Mandeville as comic, even hilarious. Anatomic detonations, nervous explosions, sudden eruptions: what reason did nature have for infusing the human microcosm called 'the body' with these sudden 'detonations,' especially if they could 'explode' at any moment and throw the organism into a paroxysm of hysterical illness?[8]

Subsequent theorists of hysteria took up Mandeville's caveat, favoring the sympathetic transmission over the idiopathic. But by now – the eighteenth century – the neural transmission of hysteria had almost completely replaced the 'bloody' and uterine, 'explosions' or not. The old dualistic categories of spirit and body, rational and physical dimensions, were replaced by a more or less integral 'nervous system' (however poorly defined and ill understood) transmitting all manner of 'nervous disorders,' of which hysteria was indubitably the supreme. As the discourse on hysteria made its way through the world of the Enlightenment, at least three of its most cherished beliefs were quashed. Set the dials roughly to the first quarter of the eighteenth century and hysteria is now a rampantly spreading malady that clearly afflicts *both* genders, women primarily because of their *weaker nervous systems*, and while stress and daily routine are crucial in its genesis, nothing is more important than *the state of the nerves* and *the animal spirits* that govern them.

When Baglivi wrote in *The Practice of Physick, reduc'd to the ancient Way of Observations, containing a just Parallel between the Wisdom of the Ancients and the Hypothesis's of Modern Physicians* (1704) that 'Women are more subject than Men to Diseases arising from the Passions of the Mind, and more violently affected with them, by Reason of the Timorousness and Weakness of their Sex,' he meant weakness in the *nerves*. Baglivi was widely read throughout Europe, from north to south, from the avant-garde medical schools of Holland to those in Spain and Salerno. His theory of 'Diseases arising from the Passions of the Mind' as diseases of gender took hold almost instantly. This eighteenth-century view represented a narrow conception of a disease that had puzzled doctors for long, even if men and women then invested in the ideologies of the animal spirits in ways now almost irretrievable. It was a narrow conception, and it demonstrates that the paradigmatic shift from a uterine to a nervous model for hysteria was the most significant shift the conception of hysteria experienced since its medicalization in the sixteenth century and until its genuine psychogenic formulation in the nineteenth.

II

I hope I have explain'd the Nature and Causes of Nervous Distempers *(which have hitherto been rockon'd Witchcraft, Enchantment, Sorcery and Possession, and have been the constant Resource of Ignorance) from Principles easy, natural and intelligible, deduc'd from the best and soundest* Natural Philosophy.

– GEORGE CHEYNE, *The English Malady*

The paradigmatic shift is, of course, self-evident to the careful reader of these discourses, especially as former 'hysterical' complaints now become monolithically 'nervous.' Sydenham died in 1689, almost at the moment that Newton's *Principia* (1687) was being interpreted and Locke's *Essay Concerning Human Understanding* (1690) printed, works providing evidence that paradigmatic shifts were then taking place in other fields as well as in medical theory.[9] Within a generation, to be hysterical was to be *nervous*: the two became synonymous, the latter eventually a shorthand, a metonymy, almost a code word, for the broad class of hysterical and hypochondriacal illnesses. Another feature of the theory of hysteria (not merely the fact of its existence as a medical condition) affords a clue to this transformation into nervous illnesses: the sense that nervous disease *permeates* society. This pervasiveness had never been a primary dimension of the older theories of hysteria.[10] For generations, at least since the time of Weyer and Jorden, it had been thought that hysteria was present and could be found in segments here and there but that it was not omnipresent or pervasive in European society. Now, in the generation between the death of Sydenham and the succession of the Hanoverians (1689–1714), the pervasiveness of nervous disease became as entrenched as the mechanical revolution in science more widely.[11] Was it for that reason, perhaps, that a large number of cases began to surface in the eighteenth century in comparison to previous periods? Even more puzzling, why should diagnoses of hysteria suddenly reach such epidemic proportions? Were there the cases to support the diagnoses, or were doctors on some type of crusade to hystericize (i.e., neuralize) medical illness and encourage the perception that disease was now fundamentally nervous?

The answers must be sought in the discourses themselves as well as in the views of women then and in social transformations then occurring. We tend to think of the nineteenth century as the golden age of hysterical women in part because – we think – the eighteenth century refused to problematize the female sex[12] – that is, to see women in all their biologic and social complexity. Yet authoritative social history reveals the opposite: for example, Sydenham's remarkable social construction of women and their chief disease. The degree to which an epoch problematizes women varies of course; it is perfectly true that *all* epochs problematize their women; nevertheless, in the period of the Enlightenment it was high. Throughout the Restoration and eighteenth century, at least in the British Isles and France, even the healthy woman was still seen as a walking womb. Several dozen rebels – the Bluestockings, the Aphra Behns and Charlotte Charkes, the Lady Mary Wortley Montagus and Madame de Staëls, and other sophisticates in the leading courts and capital cities of Europe – challenged this characterization, but they and their cohorts were unable to put a significant dent in the armor of that social world.[13]

For some, spleen and vapors, often used interchangeably, were still proofs of demonic possession rather than somatic ailment; this is not surprising since witches were still being tried in the early eighteenth century (until the 1730s), even if not so vigorously as they had been previously.[14] But for most, 'the vapors' was the colloquial cousin of hysteria, as Dr. John Purcell, a self-professed 'nerve doctor,' insisted.[15] Dr. John Radcliffe, for whom Oxford's Radcliffe camera is named, was dismissed from Queen Anne's service after telling Her Majesty that she suffered only from the vapors, thereby implying that hers was an imaginary and doubtful malady. This was nothing Her Majesty wished to hear; the Queen wanted a diagnosis indicating *real illness* that could be treated with acceptable therapy, not some imaginary delusion, like 'the vapors,' for which her character could be impugned and to which no attention would be paid.[16] We glimpse a different view in the poet Pope's treatment of Belinda when she descends into 'The Cave of Spleen' in canto 4 of the famous mock-epic poem *The Rape of the Lock* (1714). Belinda's sudden hysterical seizure embodies the older connotation of the medical doctors, and becomes the sign of the unstable postpubescent and nubile nymph burdened with her essential uterine stigmata:[17]

> Safe past the *Gnome* thro' this fantastic Band,
> A branch of healing *Spleenwort* in his hand.
> Then thus address the Pow'r – Hail wayward Queen

Who rule the Sex from Fifty to Fifteen,
Parent of Vapours and of Female Wit,
Who give th' *Hysteric* or *Poetic Fit*,
On various Tempers act by various ways,
Make some take Physick, others scribble Plays.

(lines 55–60)[18]

The poetry succeeds brilliantly here because of a sustained ambivalence between real and imaginary delusion: *'Hysteric'* and *'Poetic'* fits: that never-never land capturing genuine dementia versus imagined, even feigned, vapors. Pope thereby enables Belinda to enjoy a status unavailable in actual life had she been the historical, precocious, upper-class Arabella Fermor suffering from medically diagnosed hysteria.[19] Unlike Belinda, real patients craved diagnoses that did not brand them as possessed or deluded by imaginary or pretended illnesses. They wanted to be told by their physicians and apothecaries that they were suffering from genuine nervous afflictions that had attacked specific parts of their nervous systems for which there existed pharmacological remedies and other tonic nostrums.[20] Alternatively, in medical theory as distinct from the diagnostic and therapeutic spheres, nothing persuaded doctors and patients alike so well as numbers and mathematics. So long as the physician could quantify the malfunction of the diseased animal spirits and apply arithmetic and even Newtonian fluxions to the motions (i.e., the contractions and expansions) of the nervous system, both diagnosis and therapy seemed possible. Specialized 'nerve doctors' were well served by iatromechanical training. For the rest, quantification and numbers had proceeded so far in the mechanical imagination of the day that nothing therapeutic succeeded so well as pills and potions designed to normalize the mechanical motions of the animal spirits within the nerves that had caused the hysteria in the first place.

The path ahead for the theory of hysteria lay then in its iatromechanical applications, i.e., its mathematical charting.[21] The followers of Sydenham, especially Baglivi and Mandeville, and of their counterparts Archibald Pitcairne (a Scot who became an important professor of medicine in Leyden and Edinburgh) and Herman Boerhaave in Holland,[22] avowed a medical Newtonianism aspiring to establish the laws – static, dynamic, hydraulic – governing the mechanics of the organism and preferably couching their findings in these mathematical expressions. Anatomical attention to the body's solids would provide, they contended, surer foundations for medical laws than the traditional Galenic preoccupation with the humors and fluctuations of the fluids. Dr. George Cheyne in particular had nothing but scorn for talk of humors and those 'fugitive fictions,' the animal spirits.[23] Mechanist physicians, treading lightly in Willis's footsteps, pointed to the experimentally demonstrable role of the nervous system – a sensory skeleton variously imagined as comprising nerves, fibers and spirits, strings, pipes, or cords – in mediating between brain and body, anatomy and activity. As I have described elsewhere, Cheyne and his medical peers in Enlightenment England launched an aggressively somaticizing drive to modernize medicine in a Newtonian mode. 'Physic,' Cheyne advised his brethren, must aspire to the condition of physics. The possibility of diseases, especially hysteria, springing primarily from the mind was discounted – no longer, in the main, because such disorders would be deemed diabolically insinuated, but because they would thereby be rendered empirically unintelligible. For the theory of hysteria this represented an invigorating somaticizing that totally undid Sydenham's *cultural* unraveling.[24]

The Newtonian mechanics of cause and effect meant that no reflex, no disturbance of consciousness, no sensation or motor response, was to be admitted without presuming some prior organic disturbance communicated via the senses and the nerves. 'Every change of the Mind,' pronounced the enthusiastic Newtonian Dr. Nicholas Robinson in 1729, 'indicates a change in the Bodily Organs,'[25] a view Cheyne endorsed in *The English Malady* by adumbrating its workings in the intimate interplay between the digestive organs and healthy nerves' tonicity:

> I never saw a person labour under severe, obstinate, and strong nervous complaints, but I always found at last, the stomach, guts, liver, spleen, mesentery [i.e., thick membranes enfolding internal organs], or some of the great and necessary organs or glands of the belly were obstructed, knotted, schirrous, spoiled or perhaps all these together.[26]

Cheyne subsumed hysteria – which in his fashionable medical practice covered a multitude of symptoms ranging 'from Yawning and Stretching up to a mortal Fit of Apoplexy' – under the umbrella of nervous diseases, its being due to 'a Relaxation and the Want of a sufficient Force and Elasticity in the Solids in general and the *Nerves* in particular.'[27] Cheyne's 'nerves' thereby endorsed the Sydenham/Willis exoneration of the womb, relocating the distemper as the neighbor of the spleen and vapors, and closely situated next to melancholy. Time elapsed, however, before the educated public caught up with Cheyne's reforms, and even someone as knowledgeable of Cheyne's theory of hysteria as the novelist Samuel Richardson, Cheyne's great friend, conflated his version of hysteria with the vapors and spleen. In Richardson's last novel, *Sir Charles Grandison* (1753), the willowy heroine Clementina endures the three stages of 'vapours' Cheyne described in *The English Malady*, proceeding from fits, fainting, lethargy, or restlessness to hallucinations, loss of memory, and despondency (Cheyne recommended bleeding and blistering at this stage), with a final decline toward consumption. To cure her, Sir Charles follows Cheyne, prescribing diet and medicine, exercise, diversion, and rest, and the story is considerably affected when Clementina's parents adopt unquestioningly Dr. Robert James's further recommendation that 'in Virgins arrived at Maturity, and rendered mad by Love, Marriage is the most efficacious Remedy.'[28]

In the perceptions and practice of early Georgian medicine, these nervous complaints constituted a block of relatively nonspecific ailments and behavioral disorders. One need merely think of the letters and diaries of the period to see what resonance spleen and vapors emitted.[29] They are even more frequently referred to in the poetry and drama of the period, where virtually no author is exempt. From the mad hack's attacks of spleen in Jonathan Swift's *Tale of a Tub* to Clarissa Harlowe's persistent bouts with vapors in the Richardson novel of that name, the nervous ailment exists as mundane reality as well as cliché and complex trope.[30] Gender proves no discriminating factor, as men and women alike, and in almost equal numbers, fall prey to its sudden attacks. But diagnosed inaccurately, the same symptoms could denote lunacy, insanity, dementia: the same madness Swift's hack clearly suffers from in the Rabelaisian *Tale of a Tub*.[31] To our way of thinking, the broad category melancholy would not seem to fit under this conception of hysteria. Yet it then did, one evidence of which is the consistent interchange of the two words in even the most technical medical literature. Furthermore, the line between melancholy and madness was delicate and thus greatly feared. Melancholy, madness, hysteria, hypochondria, dementia, spleen, vapors, nerves: by 1720 or 1730 all were jumbled and confused with one another as they had never been before. Anne Finch, the Countess of Winchelsea and a poet much admired by Pope and Wordsworth, turned this confusion about the status of hysteria to her advantage in *The Spleen: A Pindarique Ode by a Lady* (1709). This is her most ambitious work: a phantasmagoria about life, death, and the nocturnal reverie world – all conceived and executed by pondering reality through the gaze of the splenetic poet.[32]

The leading 'nerve doctors' – the Mandevilles and Cheynes and their group of lesser epigoni – grounded these hysterical symptoms entirely in somatic origins: to make certain through tact and expertise that patients understood that virtually all hysterical complaints were worlds apart from gross lunacy. Thus Dr. Purcell, mentioned earlier as a fashionable nerve doctor, claimed that 'the vapours' – a condition colloquially synonymous with hysteria – consisted entirely of an organic obstruction located 'in the Stomach and Guts; whereof the Grumbling of the one and the Heaviness and uneasiness of the other generally preceding the Paroxysm, are no small Proofs.'[33] Noting that one of Hippocrates's noblest contributions to medicine lay in recognizing that epilepsy was not a divine affliction ('the sacred disease') but entirely natural, Purcell insisted that the vapors (what the French would call the 'petit mal') were akin to epilepsy (the 'grand mal'); indeed that 'an epilepsie, is Vapours arriv'd to a more violent degree.'

What had become of Sydenham's revolutionary insights – the social conditions, daily stresses, nocturnal excesses, wasting away of women in a patriarchal world, all of which he had believed were important in the genesis of hysteria? Where was the view that the new Enlightenment codes of politeness and refinement, and the encroachment of unwanted foreign customs on civilized English and French life (coffee, tea, chocolate, snuff, etc.) played a part in creating these hysterical complaints? In England and later in Western Europe they had gone underground, subservient to, or overwhelmed by, a scientific milieu bristling with vigorous Newtonianism.[34] It is not easy to imagine that a wave of Newtonianism diverted the nerve doctors to such a preponderant degree despite theories such as Robinson's; nevertheless, the fact is that it did. Mental illness in our time has been construed so completely within the light of socioeconomic determinants, when it is not considered a genetic or hormonal disorder requiring chemical correction, that we find it hard to imagine an approach to hysteria so monolithically iatromathematical as the Newtonian one of Cheyne's world. Yet for a generation at least, extending well beyond the second quarter of the eighteenth century, personal and social stress were discounted as uninteresting to the theories of hysteria, while the limelight fell on the application of the new 'mathematical medicine' to existing cases.

Indeed, inquiry into the etiology of hysteria as a valid form of exploration regressed: all cases were deemed to result from deviant physiologies of the nervous system that could be understood only by Newtonian or other mechanical analyses. As the century evolved, it became clear that lunacy, insanity, and madness represented the great fears – the *grand peur* – of these early Georgians, not the chronic hysteria that doctors like Mandeville and Cheyne claimed they could *always* cure now that it was somaticized and released from its previous diabolical moorings. Lunacy was feared as the great hangman because even the best of the Newtonian doctors had no clue to its genesis and cure.[35] In cases of hysteria there was at least hope for the patient. Its onset, as the doctors assuaged their patients, had not even been mi'lady's or his lordship's fault. Madness, on the other hand, represented an unequivocal failing in the popular imagination: a fatal lapse of the soul, a disjunction of mind and body; the stigma *ne plus ultra*; in the brave new world of the Enlightenment it was a final, irrevocable state, usually ending in incarceration. It was not until late in the century that a new class of humane physicians – the Batties, Monros, Chiarugis, Crichtons, Pinels – demonstrated the same humanitarian attitude to madness that the Willises, Sydenhams, and Cheynes had for hysteria and other nervous disorders.[36]

Medical science thus led early Enlightenment physicians to make a great play of the organic rootings of problematic disorders. But so too did bedside diplomacy. Confronted with indeterminate ailments, Cheyne, for example, pondered the problem of negotiating diagnoses acceptable to doctor and patient alike. In his remarkable autobiography and tantalizingly ambiguous self 'case history,' he claimed to empathize with these victims because he himself suffered from such disorders.[37] Physicians were commonly put on the spot by 'nervous cases,' he noted, because such conditions were easily dismissed by the 'vulgar' as marks of 'peevishness,' or, when ladies were afflicted, of 'fantasticalness' or 'coquetry.'[38] But his own somaticizing categories were pure music to his patients' ears, for they craved diagnoses that rendered their hysterical disorders *real*. The uninformed might suppose that hysteria, the spleen, and all that class of disorders were 'nothing but the effect of Fancy, and a delusive Imagination': such a charge was ill-founded, Cheyne assured them, because 'the consequent Sufferings are without doubt real and unfeigned.'[39] Even so, finding *le mot juste* required tact. 'Often when I have been consulted in a Case,' Cheyne mused, 'and found it to be what is commonly call'd Nervous, I have been in the utmost Difficulty, when desir'd to define or name the Distemper.'[40] His reason was the predictable desire not to offend, 'for fear of affronting them or fixing a Reproach on a Family or Person.' For, 'if I said it was Vapours, hysterick or Hypochondriacal Disorders, they thought I call'd them Mad or Fantastical.'

What precisely was the sociology and linguistics of this annotated disgust? Did the patients disown their hysteria and the similar maladies because they reflected a perverse life-style? Some moral or religious failing? Or was it that somehow centuries of uterine stigma could not be wiped away so quickly, not even by the reforms of Willis and Sydenham? Throughout his prolific medical writings, commenting on the recoil of his patients in the face of a diagnosis of nerves or spleen, even when he gave the complaints a somatic basis, Cheyne recognized the degree to which he would have to educate them. Sir Richard Blackmore, another fashionable 'nerve doctor,' experienced similar difficulties, to the point of admitting that his hysterical patients were often viewed as freaks suffering from 'an imaginary and fantastick sickness of the Brain.'[41] The freaks thus became 'Objects of Derision and Contempt,' and naturally were 'unwilling to own a Disease that will expose them to Dishonour and Reproach.'

While Enlightenment doctors ignored what we would call the sociology of hysteria, they did accept the lack of gender distinctions. Blackmore was as mechanical and Newtonian a physician as one could find in the early eighteenth century, certainly as 'mechanical' as Robinson, his colleague, but he lost no opportunity to show that hysterical symptoms in women were identical to those in hypochondriacal men. Ridiculing uterine theories of hysteria as so much anatomical jibberish, Blackmore concluded, as Cheyne did, that 'the Symptoms that disturb the Operations of the Mind and Imagination in hysterick Women' – by which he meant 'Fluctuations of Judgment, and swift Turns in forming and reversing of Opinions and Resolutions, Inconstancy, Timidity, Absence of Mind, want of self-determining power, Inattention, Incogitancy, Diffidence, Suspicion, and an Aptness to take well-meant Things amiss' – 'are the same with those in Hypochondriacal Men.'[42] The condition, he maintained, was common to both sexes, and the many names given to it – melancholy, spleen, vapors, hysteria, nerves, among dozens of others – all amounted to the same thing: a genuine malady with somatic pathology requiring a new understanding between doctor and patient. The sensitive physician demonstrated his expertise by ridiculing theories that these nervous complaints were the result of a diseased womb, and he recommended identical therapy for hysteric male and female patients.

To gain acceptance for the term *hysteric* and its symptoms, these physicians proposed to yoke them with more common organic illnesses, investing them with labels and copper-bottomed organic connotations, for example, by speaking of 'hysterick colic' or 'hysterick gout.' The tendency persisted for sixty or seventy years at least. Thus one woman Cheyne treated had a 'hysterick lowness,' another 'frequent hysterick fits'; eventually the word *hysteric* was so flattened and became so neutral in its connotations as to mean almost nothing at all. The physician thereby spared himself the accusation of merely trading in words – which he was consciously doing anyway in view of the number of conditions that had come under the umbrella of 'nervous' – and imputations of shamming also were avoided. Robinson, already mentioned, insisted that such nervous disorders were not 'imaginary Whims and Fancies, but real Affections of the Mind, arising from the real, mechanical Affections of Matter and Motion.'[43] His reason was that 'neither the Fancy, nor Imagination, nor even Reason itself...can feign...a Disease that has no Foundation in Nature,' a position that hurls down the gauntlet to Sigmund Freud.[44] Organic agencies, such as stone, tumor, fistula, and so on, thus had to initiate the chain of reactions, no matter what the conversion process entailed: 'The affected Nerves...must strike the Imagination with the Sense of Pain, before the Mind can conceive the Idea of Pain in that Part.' Here then was the all-important role of the nerves in sensation, as well as all human pleasure and pain.

Cheyne, Blackmore, Robinson, and their contemporaries did not seek to deny the contribution of consciousness to the genesis of nervous disease nor reduce mind to body (Baglivi, so influential in southern Europe, went the other way, reducing all body to mind – a mind whose passions had been shaped exclusively by the state of the nerves). But their aspirations as 'scientific' doctors treating 'enlightened' patients (usually the elite of the population) disposed them to insist upon the priority of physical stimuli as part of their two-pronged strategy to win the confidence of their patients and the esteem of their medical peers. They relied on their academic-medical credentials to enforce this approach as being both objective and true. Credentials were, after all, one of the main factors in determining authority, popularity, and fashionability.[45] The most sought-after doctors in London and Edinburgh, Oxford and Cambridge, as well as at the spas and in the major cities of other countries, had been decorated, so to speak, for their academic achievements. If this approach rendered the species man – in a world increasingly explained by new theories about the sciences of man – l'homme machine, its philosophical materialism also had beneficial effects. Thus the establishment of nervous conditions as valid medical diseases helped to secure the credit of medicine itself in an era of rampant quacks and proliferating mountebanks, when doubts about its validity as a science were at an all-time high.[46]

More locally, within the realm of medical theory, this state of affairs amounted to a neurological approach to hysteria, which Veith has claimed was 'sterile' in a 'controversial century.'[47] Oddly, it was the dominance of this neurological approach to hysteria and the triumph of the nerve doctors with their patients (physicians such as Cheyne) that led Veith to this disastrous conclusion. Countering her judgment, we might note (without adopting any Victorian or Darwinian notions about the evolution of medicine or medical conditions) that late twentieth-century medicine has vindicated the neurological approach and returned to it the primacy of neurobiology.[48] This may prove nothing in itself but at least demonstrates the longevity of the neurological approach. Furthermore, the Enlightenment nerve doctors were immensely sympathetic to their patients. Even in an age, such as ours, when hysteria has become so politically and academically charged, this fact within the history of hysteria cannot be lightly dismissed. In the case histories detailed in the final section of *The English Malady*, Cheyne drew attention to the real woes of sufferers burdened with misery, depression, *taedium vitae*, ennui, hysteria, and melancholy – not least, to his own nervous misery.[49] His patients, unlike Sydenham's, shared one common thread: they uniformly came from the ranks of the rich and the famous.

III

Hysteria thus came of age in the openness of the Enlightenment, more specifically in the sunlight of the Newtonian Enlightenment. Virtually no important doctor in the first half of the eighteenth century placed the root of hysteria in the uterus, and this fact tells us as much about the patients of the epoch as its mostly male physicians. The modernization proved anatomically liberating, while also helping to discredit the theory based on the misogynistic sexual stigma of the voracious womb.[50] The new emplacement of hysteria in the world of Cheyne and his 'nerve doctor' colleagues moreover skirted vulgar reductionism. Its unmistakable language of the nerves – amounting to the heart of its linguistic discourse – pointed toward the mutual interplay of consciousness and body through the brain and the (often) still perplexing animal spirits as the primary nervous medium.[51]

This new linguistic footing, which had been developing since the days of Willis and Mandeville, had profound cultural and gender-based implications: cultural because society itself was growing 'nervous' in ways no one had anticipated, and gender-based as a consequence of this new nervous model of mankind mandating a weaker nervous constitution for women than men. The desexualization of hysteria was, of course, one part of a movement during the Enlightenment that demystified the entire body.[52] This process included the reproductive organs and the newly privileged mind over matter, as in Hume's examples and (especially under the weight of Linnaean taxonomy) the rule of species over gender. With demystification also came the shedding of much of the shame of hysteria. Its sufferers at mid-century were now seen as the victims of an interestingly delicate nervous system buckling under the pressures of civilization, typically the thorn in the flesh of elites moving in flashy, fast-lane society.[53] This was the essence of Cheyne's message in his best-selling book, *The English Malady*.

But the cultural reasons for this 'delicate nervous constitution' were to remain hidden and elusive for some time. Its personal effects, especially for patients, were described ad infinitem; the other effects, the larger images of those living an affluent life, could be seen in the new image the emerging Georgians held of themselves. At home, in the bedroom, this might entail paralysis, fear of the dark, as well as dread of the incubus and succubus, as evidenced by sleepwalking and amnesia.[54] (If the weekly and monthly magazines can be considered reliable, amnesia was more common than we might think.) These were the standard images of the somnambulant melancholic or insomniac hysteric in the caricatures of the time, as the accompanying plate demonstrates. More locally still, within the context of a now desexualized female hysteria, the suggestion was that coquetry verged on hysteria.[55] To the vulgar, as Pope had suggested in *The Rape of the Lock*, hysteria might signify nothing more than coquetry itself. But these examples, medical and literary, signified something more deeply ingrained in the world of the Georgians than has been thought: namely, the nervous self-fashioning of Augustan society.

Stephen Greenblatt and others among the New Historicists have written about such self-fashioning in the Renaissance.[56] Yet the latter period of the Enlightenment is even more revealing of the great personal tensions it raised between the sexes in a milieu of increasing desexualization in which women continued to enjoy greater freedom and equality than they had before. The Augustan wits – the Addisons and Swifts, virtually all the Scriblerians – encouraged us to believe that logic, wit and intelligence – all part of the realm of the mind – were the sine qua nons of polite society then. But the tension between men and women revolved around more than matching wits, competing intellects, wit and wit-would-be, even in a 'republic of letters' governed by an obsessive commitment to refinement and politeness, manners and etiquette. In addition, and most important, there was the unrelenting search for personal identity and self-fulfillment. This need is what the novel and drama of the period capture par excellence, and nothing reflects the mood of the epoch better than its great imaginative literature.[57]

All these cults of sensibility – as I have called them elsewhere[58] – demanded rising standards of behavioral achievement and necessarily called attention to their opposites: the realms of pathology and abnormality. This is why the medicine of the day, especially its theory based on bodily signs and symptoms, the semiology and pathology of illness, cannot be dismissed as so much esoterica.[59] We have devoted two generations of study to the literary language of the Georgians; their ideas of body would well repay half that attention. The Lady Marys and Duchess of Portlands were hardly norms capable of emulation, yet in their bodily motions were codified the brilliant new urbanity of the age. Their sophisticated postures swirled round in rarefied atmospheres of courtliness and polite town society, abiding by a code of language and gesture in which the body was always required to be disciplined and drilled, coy and controlled; always mannered, as we see everywhere from the roles of dancing masters, acting teachers, tutors, governesses, and gymnasts of the age.[60] Even so, new inner sensibilities had to find expression through refined and often subtly veiled bodily codes: one's bearing around the tea table, in the salon, at the assembly and pumproom, in town and country, at home and abroad, paradoxically revealing yet concealing at the same time, in actions, gestures, and movements that spoke louder than words.[61]

This was the source of tension now superimposed on the gender pressures spawned in the Restoration under the weight of urban sprawl and new sociopolitical arrangements. In England at least, the gender rearrangements of the Restoration were elevated to exponential highs in the ages of Anne and the Georges. Isn't this a principal reason why the drama from Etherege and Congreve to Gay and Goldsmith assumes its particular trajectory vis-à-vis the sexes and gender arrangements? Urban sprawl, new forms of consumer consumption, gender rearrangements, interpersonal tensions, crime and violence, class mobility, the transfer of money and goods into a process of unprecedented consumption: the phrases appear to describe our vexed world. This was, however, the eighteenth century, consuming itself in newly found nationalism and wealth and basking in its accompanying leisure time, especially in food and drink.[62] The lingua franca of such expression-repression-expression lay in the refined codes of nervousness: a new body language, ultraflexible, nuanced yet thoroughly poised within ambivalence. The essence of the code lay in these bodily gestures of recognition – whether blushing or weeping, fainting or swooning – which could act as sorting-out devices in times of doubt, certainly when love and marriage were involved. The comic drama from approximately 1730 onward demonstrates what heightened requirements the code placed on actors who tried to reflect it; our lack of recognition of the code itself results, in part, from the rarity with which any of these plays is now performed. Words were also tokens of recognition for the sensible and sensitive: sorting-out devices too.

Under duress and at great expense, the language (of gestures *and* words) could be learned, but even among the rich and great, the smart and chic, it was acquired at the cost of great personal risk and self-doubt. Risk lay everywhere in the new social arrangements represented – almost mimetically – in the proliferating idioms of nervous sensibility. The sheer number of the idioms then available has prevented us from seeing deeply (and some might say darkly) into the risks involved. Upon occasion we have even denied that the idioms existed. Readers may well wonder: What cults of nervous sensibility? And why *nervous*?[63] Want of *nerve*, for example, betrayed a clear effeminacy, unacceptable in all classes from the highest rakes and fops to the lowest laborers. Paradoxically, want of *nerves*, exposed a rustic dullness, a latent tedium, a resulting boredom odious to the British for all sorts of reasons and feared among the highest ranking of both genders. Yet florid, volatile nervousness – in both men and women – betrayed excess and confusion: symptoms that could result in hysterical crisis. And hysteria, no matter what appellation it was given and no matter how culturally positive in the popular semiotics of that world, was a refuge of last resort. It was the cry of the person (usually female) unable to cope with the sharp cultural dislocations and social norms that had occurred in such a relatively short time. Within this taxonomy of disease, then, hysteria was the final limit beyond which no condition was more baffling, none capable of producing stranger somatic consequences. The semiotics of the nerves, leading to understanding of hysteria, is therefore a way of knowing, and thereby decoding, the infirmity of excess, in much the same way that Foucault's hysteria is an understanding derived through comprehension of the female's inner spaces. And it was through this semiotics of the nerves that Foucault made the grandest claim of all: 'It was in these diseases of the nerves and in those hysterias [of the period 1680–1780], which would soon provoke its irony, that psychiatry took its origin.'[64]

The quest was rather for a golden mean filtered by decorum – the same variegated decorum extolled by the age. But decorum had its snares too; it was easier to conceptualize or verbalize than to put into practice, as weepy heroine upon heroine lamented, usually to her detriment, in the fictions of the age. The snare was the retention of one's individuality within this bodily and verbal control. In practice, the act resembled treading on a tightrope, the walker forever balancing over the abyss. This was the beginning of a way of life – as Cheyne above all others in his age seems to have recognized – where the participants lived on the edge and in the fast lane. Richard Sennett, the American sociologist, has located the origins of modern individualism within this fast-paced eighteenth-century culture.[65] More precisely, we might counterargue, individualism was created out of nervous tension and ambivalence over the self: the accommodation between the hyper-visible, narcissistic individual and a society that had craved it (i.e., the individualism), while at the same time demanding conformity to the civilizing process. This was the self-fashioning of the urbane Augustans, the codes on which the sexual politics of the new hysteria of the eighteenth century depended, and it would not have come about without the prior hypostases of the great nerve doctors – the Sydenhams and Willises, the Mandevilles and Cheynes – which resulted in the nervous codes that elevated sensibility to a new pinnacle.[66]

Here then was a different route to the golden age of hysteria, a different dualism than the old Cartesian saw about mind and body. This Georgian self was less a divided Cartesian self – the now unisex woman or man riveted by conventional mind and body – than a creature part public, part private, often hidden behind a mask (sometimes a literal vizard) that curtailed self-expression as well as permitted it to flourish. Here, in this passionate sexual ambivalence, was the heart (one might as well claim the stomach and liver for the visceral effect it had on lives then) of the cults of nervous sensibility. It imbued Augustan and Georgian culture; eventually it made inroads in Holland, France, Italy, all Europe. And it left its mark on the best philosophers: the Voltaires and Hallers and Humes without whom an eighteenth-century 'Enlightenment' is unthinkable.[67] It energized the Diderots and Sternes, the Casanovas and Rousseaus, as well as the fictional Clarissas and Evelinas, the Tristram Shandys and other noted 'gentlemen' – and gentlewomen – of feeling. How then could nervous sensibility have been born without a medical agenda that demystified the body and a subsequent Newtonian revolution that concretized its best hypotheses?[68]

In the intellectual domain, this nervous tension surfaced as a Sphinxian riddle of psyche–soma affinities, and spurred, in part, the literally hundreds of works on mind and body we have heard about for so long.[69] But in more familiar corners – at home and in church, in the theater and public garden, everywhere in polite society – it also appeared in subtle ways: in bodily motion, gait, affectation, gesture, even in the simple blush or tear, and in the most private thought that now could be read by another. Nervous tension was thus domesticized for the first time in modern history. Viewed from another perspective, it was also being mechanized for the first time, as manners themselves coagulated into an abstract code-language of mechanical philosophy: on the surface a loose application of Newtonian mechanics to the body's gait and gestures, but an application nevertheless.[70] The self-fashioning of nerves was thus significantly expanded: from mechanical philosophy it was medicalized, familiarized, domesticated, and eventually transformed into the métier of polite self-fashioning and even world-fashioning, in the sense that its code was eventually adopted as a universal *sine qua non* for those aspiring to succeed in the beau monde. The consequences for human sexuality and social intercourse were incalculable because passion and the imagination were implicated to such an extraordinary degree, as were the links between hysteria and the imagination. As soon as the imagination was aroused or disturbed, even in the most imperceptible way, somatic change was indicated. Of this sequence, the physicians had been certain from the mid-eighteenth century, if not earlier. 'It appears almost incredible,' Peter Shaw, His Majesty George II's Physician Extraordinary and the English champion of chemical applications in medicine, wrote in *The Reflector: Representing Human Affairs, As They Are: and may be improved* (1750, number 228), 'what great Effects the Imagination has upon Patients.' Later on the point was reiterated by William Heberden, another noted clinician in the tradition of Boerhaave whose life spanned nearly the whole of the eighteenth century and of whom Samuel Johnson said that he was '*ultimus Romanorum*, the last of our great physicians.' Heberden was as much a product of this 'nerve culture' as anyone else. After years of clinical experience he found that the indication of hysteria usually began 'with some uneasiness of the stomach or bowels.'[71] He listed the symptoms: 'Hypochondriac men and hysteric women suffer accidities, wind, choking, leading to giddiness, confusion, stupidity, inattention, forgetfulness, and irresolution.' The symptoms were diverse, perhaps too diverse; a powerful and wild imagination lay at their base. But when Heberden pronounced on the root cause of hysteria, he could only say that the condition was fundamentally *nervous*, that is, fundamentally real or nonimaginary; in his words, 'for I doubt not their arising from as real a cause as any other distemper.'[72]

Such nervous self-fashioning lay at the base of the social cults and linguistic idioms of Enlightenment sensibility, and were as influential as any other force in generating the theory of hysteria that we see reflected in the writings of the nerve doctors and their students.[73] The process would not be reversible. The doctors did not impose their vision of society on their culture; it was life with its tensions that drew even the doctors into its orbit and caused their theories utterly to reflect this new society.

Just as important, nerves in the new culture precluded moral blame, because there could be no censure in a social, almost *Zeitgeist*, disease. Enlightenment swoons and their subsequent numbness in both women and men came from the act of buckling under the pressures of civilization, especially for the elite who moved within the fast lane of society. The new violence and the threat of its omnipresence enhanced the panic, as John Gay and the early novelists observed. Amelia's strange disorder is described by Captain Booth in Fielding's *Amelia* in terms that make clear the price she has paid for living in the new fast lane. Booth knows not what to call her 'disease,' but eventually lands on 'the hysterics,' which seems as accurate to him as any other appellations. Fielding's case history is not very different from the one Jane Austen will narrate with laser precision in *Sense and Sensibility*; its Marianne Dashwood, with her swoons and sighs, is another 'hysteric' whose case has not yet been discussed in the detail it deserves, meticulously recounted as it is in that novel from the first onset of fits and starts to the patient's near demise and eventual recovery. In all these cases, real and imagined, panic stemmed not merely from male violence but from a new type of female as well, and society's fears were substantiated almost daily by the culprits and vagabonds apprehended and brought into the courts of law.[74] Life in the fast lane then, at least for the new urban rich, entailed high living, conspicuous consumption, reckless spending, more travel than previously (especially to the developing seaside resorts), late nights, and new gender arrangements, all combining to set off the beau monde from the other ranks of society. Neurological chaos in the body merely mirrored the social disorder of the time. Though the comparison may not have struck the average aristocrat, these forms of disorder never stood apart, nor did the hysteria of its women and men.

But did a delicate nervous organization predispose one to the buckling under, or did the buckling under alter the body's nervous organization? The question is hard but cannot be overlooked or swept away. The approach to the answers taken by the nerve doctors was not, as Veith has suggested, sterile; they recognized the psychogenic burdens of their patients and the role played by mind and imagination, even though the doctors grounded virtually all their diseases in nervous structures. This monolithic attribution remains the difficult aspect of their 'hysteria diagnosis' for us. Even so, the doctors often failed (almost always) to see the sociological roots of numbness and its radical enmeshment in language and its representations.[75]

This is a revelatory indication of the degree to which the new nervous culture of the eighteenth century had made inroads into the philosophy, psychology, and medicine of the time. In brief, Cheyne and his colleagues scientized hysteria by radically neuralizing it. They did not invalidate consciousness in human life or reduce mind to body. Theirs was rather a crusade against duplicitous disease, campaigned for in the sunny light and quasi-blind optimism of high Enlightenment science. Not even hysteria could hide from them or prove elusive. If the Enlightenment nerve doctors came back today – *Cheyne recidivus* – they could not agree with our contemporary Dr. Alan Krohn about hysteria as 'the elusive neurosis.' To them, hysteria was fundamentally knowable: a neurology of solids, an iatromathematics of forces, a neural web of nerves, spirits, and fibers.

IV

By the mid-eighteenth century, nerves seem to have run wild; the resulting hysteria was chronic among all those living in the fast lane and endemic, for different reasons, among the nation at large. Some women knew they had it, others did not: the inconsistency was less a defect of medical theory than the extreme fluidity of the diagnosis. For hysteria was not poured into a rigid mold by either the doctors or their patients. The diagnosis was usually made to fit the sufferer: a nonreductive expression of disorder. Linguistically speaking, hysteria profited from a new and very malleable vocabulary of the nerves as flexible and adjustable to the particular situation as the patient's symptoms themselves. In formal writing, by mid-century this vocabulary had been expressed in new nervous discourses: of poets, novelists, critics, didactic writers, in narratives of all sorts. An aesthetic of 'nervous style' began to emerge, endorsed by male writers, found suspect by female, which was unabashed in calling itself, after its patriarchal affinities, masculine, strong, taut – anything but feminine or epicene. And if style was then genderized to this degree, why should medicine not have been, especially the *maladia summa* hysteria – the genderized condition par excellence? Cheyne, above all, exploited this protean nervous idiom and procrustean vocabulary in his best-seller *The English Malady*, the real reason for its instant success. So too did his followers and disciples.

One of these, representative of these disciples in several ways, was Dr. James Makittrick Adair. Like Cheyne and William Cullen, Adair was also a Scot who had been deeply influenced by the Scottish Enlightenment. But Adair was also a Cheyne follower who saw what benefits could accrue to his career by worshiping, so to speak, within the 'Temple of the English Malady.' Adair had been taught in Edinburgh by Robert

Whytt, the 'philosophic doctor' who related 'nervous sensibility' to every aspect of modern life, and he never forgot the great medical precept of his teacher, which resounded in the lecture theaters Adair attended: 'The shapes of *Proteus*, or the colours of the *chameleon*, are not more numerous and inconstant, than the variations of the hypochondriac and hysteric diseases.'

But it was Cheyne's thought that lay in the deepest regions of Adair's imagination throughout his professional medical career.[76] Always acknowledging his teacher's famous essay of 1764–65 on nervous diseases (Whytt's *Observations on the nature, causes, and cure of those disorders which have been commonly called nervous, hypochondriac, or hysteric, to which are prefixed some remarks on the sympathy of the nerves*), Adair served up explanations his readers wanted to hear about hysteria. He also provided them with a natural history of nerves in the linguistic and cultural domain:

> Upwards of thirty years ago, a treatise on nervous diseases was published by my quondam learned and ingenious preceptor DR. WHYTT, professor of physick, at Edinburgh. Before the publication of this book, people of fashion had not the least idea that they had nerves; but a fashionable apothecary of my acquaintance, having cast his eye over the book, and having been often puzzled by the enquiries of his patients concerning the nature and causes of their complaints, derived from thence a hint, by which he readily cut the gordian knot — "Madam, you are nervous"; the solution was quite satisfactory, the term [nervous] became quite fashionable, and spleen, vapours, and hyp, were forgotten.[77]

It is an extraordinary explanation, showing the continuity of eighteenth-century nervous self-fashioning. It not only casts light on the aftermath of Cheyne's career following his death in 1743 and on Whytt's much-discussed treatise of 1764 but resonates with class filiation. Adair saw how shrewd his medical brethren had been to classify as 'nervous' those behavioral disorders free of determinate organic lesions: that is, vapors, spleen, hysteria, hypochondria, melancholy, and the dozens of subcategories spawned from these. Adair also recognized that naming and labeling played a large role in the hysteric's conceptualization. The Gordian knot was unraveled when words were deciphered. Likewise, in the previous generation, when Dr. Nicholas Robinson published a 'Newtonian dissertation on hysteria' and wrote that every maiden had become so nervous that coining new words to describe its minute grades was necessary, he knew whereof he spoke. He himself compiled a whole vocabulary of remarkable neologisms that had been coined in his time: hypp, hyppos, hyppocons, markambles, moonpalls, strong fiacs, hockogrogles – all jocularly describing hysteria's grades of severity. Still, it was the great male poet, the dwarf of Twickenham, who used the vernacular of nerves to describe the living consequences of male hysteria. As he lay dying at fifty-five, Alexander Pope claimed to those gathered around him that he 'had never been hyppish in his life.' There was no need to gloss the phrase. Presumably all knew what he meant.

The very sturdy and nonhysterical Lady Mary, already mentioned, may have considered the 'little poet of Twickenham' to be, like his fierce enemy Lord Hervey, a member of the 'third sex.' But even Lady Mary would have had to admit that Pope was essentially 'male.' How came it to pass that Pope, whose 'long Disease, my Life' had paved the way for him to become more intimate with medical literature than he would otherwise have been, assumed male hysteria to be in the normal course of affairs?[78] One can demonstrate, as I have tried, that as far back as the Elizabethan era, and probably earlier, males were assumed to be natural targets for 'the mother,' this despite their obviously not having the requisite anatomical apparatus. The progress of medical theory in the aftermath of Sydenham and outside the Cheyne–Adair circle also needs to be consulted if we are to understand how male hysteria shaped up in the eighteenth century.

For the fact is that virtually every serious medical author who wrote about hysteria after Sydenham's death in 1689, even the skeptics among the medical fraternity, included *men* among their lists of those *naturally* afflicted: in England, for example, these authors included some of the best-known doctors of the age, including Nathaniel Highmore, Richard Blackmore, Bernard Mandeville (the physician-satirist), John Purcell, and Nicholas Robinson; in Scotland, Thomas Cupples, Lawrence Fraser, William Turner, and nearly the whole of the Edinburgh medical school; in Holland, the 'Eurocentric' Boerhaave and his far-flung students, including Jan Esgers, C. van de Haghen, Lucas van Stevenick, as can be gleaned from dozens of medical dissertations written on hysteria at Leiden and Utrecht; in Denmark, Johannes Tode; in Switzerland and Bohemia, a certain number; in France, Jean Astruc, Nicholas Dellehe, J. C. Dupont, Pierre Pomme, and even the so-called father of psychiatry and transformer of therapies for the suffering insane, the great Philippe Pinel;[79] in Germany, Gustavus Becker, C. G. Burghart, Georg Clasius, C. G. Gross, J. F. Isenflamm, Johann Christoph Stock; in Italy, A. Fracassini, P. Virard, G. V. Zeviani. These names suggest little if anything now, but in their time these figures constituted something of an international gallery of medical stars.[80]

The treatment of *males* among the hysterically afflicted, and especially males of the upper classes, was a veritable industry in the eighteenth century. Whether the doctors were persuaded that males were clinically afflicted in the same way as women (*sans* 'the mother' and the rest of the female reproductive apparatus) we may never know, and Mark Micale's biographical researches do not extend far enough back to offer a clue.[81] Yet the medical literature from Sydenham forward speaks for itself and is unequivocal on the matter. Moreover, there seems to have been no major opponent to Sydenham's view about male hysteria to challenge his theory in the long course of the eighteenth century, neither in England nor elsewhere. Once the notion of *male hysteria* took root as a clinically observed phenomenon, which it had not done a hundred years earlier, its existence appears to have been guaranteed. The huge annals of eighteenth-century medical literature corroborate this position, and examples citing Sydenham as their fount are replete in the record. It is more difficult, however, to discover examples roughly contemporary with Sydenham, perhaps suggesting to what degree the notion of male hysteria had been absorbed into the medical imagination.[82]

For example, consider the curious but still far from clear relationship between Thomas Guidott and John Maplet. Both were English physicians practicing in the Restoration and early eighteenth century in and around Bath. Guidott owed his entire Bath practice to Maplet, who helped him acquire it. After Guidott lost his practice in Bath through imprudence, libel, and squandering, he moved to London, remained loyal to his former patron, and continued to diagnose and treat his (Maplet's) ailments until the end of his life.[83] This would seem to be a case of professional patronage larded over with friendship, but it also had its profound medical side useful in these explorations of male hysteria. What survives are Guidott's accounts (not Maplet's), and considering Guidott's colorful character, his record may not be entirely reliable or complete. But it does provide enough information to comprehend what it was about Maplet's 'male hysteria' that so attracted and excited Guidott, who wrote many years after Maplet's death:

> [He] was of a tender, brittle Constitution, including to Feminine, clear Skin'd, and of a very fair Complexion, and though very temperate . . . yet inclinable to *Hysterical* Distempers, chiefly Gouts and Catarrhs, which would oftentimes confuse his Body, but not his Mind [mind and body construed as separate entities], which was then more at

Liberty to expatiate, and give some Invitation to his Poetick Genius…
to descant on the Tormentor, and transmit his Sorrow into a Scene
of Mirth.[84]

Multiple aspects of this analysis give us pause: Guidott's strange linking
of hysteria to gout and catarrh and in other writings his subclassification of
'hysterical gout'; his post-Cartesian version of the mind/body split; the
assumption that creativity and hysteria ('Poetick Genius' and 'the
Tormentor') are cousins; above all, the presumption that in educated
and intelligent males like Maplet 'hysterical mania' is merely the outward
sign (again a semiotics of the malady) of an almost 'Feminine' nervous
'Constitution.' Here, in nervous anatomy and 'Tender Constitution,' lies
the origin of temperamental sensitivity in men. Later, Guidott discusses
Maplet's delicate nerves, metaphorically isolating them as 'suspects' in
this quasi-criminal hysterical disorder.[85] 'Suspects' in both the positive
and pejorative dimension: positive in that they virtually breed sensitivity
and creativity; negative in their pathological predisposing toward the
condition. All this is what we would expect after unraveling and decoding
the complex medical theory of the time.

Much less expected is Guidott's leap to friendship. He claims to be
'attracted' to the nervous, brittle, delicate, tender, frail, white-skinned
Maplet – not attracted sexually, certainly, not primarily as a consequence
of Maplet's professional generosity, although one would presumably be
interested in the arm and leg of patronage, but attracted intellectually
and humanly. Guidott's life is not sufficiently understood to hazard any
guesses about his sexuality, but his case history of Maplet suggests the
existence by approximately 1700 of a new Sydenhamian paradigm
about *male hysteria* that yokes anatomy, physiology, and psychology to
culture, gender formation, and society.[86]

What better evidence could there be of *gender* basis in this account?
Maplet is the 'tender, nervous, brittle' male who has become afflicted
and requires diagnosing and treating by Guidott; he is also the soft,
creative, nervous male predisposed to hysteria and friendship. Guidott's
language does not yet reveal the developed jungle of nerves and fibers
that will flourish in Cheyne and Richardson, and later even more
metaphorically and densely in the fictions of Sterne and the Scottish
doctors. But it remains one of the earliest and most interesting accounts
of *male* hysteria in English, certainly a prototype of sorts. Guidott himself
was somewhat 'poetically inspired,' though he is not known to have been
'hysterical.' He had composed poetry at Oxford and wrote poetic satire
when he quarreled with the London physicians.[87] And he had matured

in a world overrun with male enthusiasts of all sorts – the broad spectrum that permeates the great satires of the age, such as Swift's *Tale of a Tub*. Guidott's London, like that of Sydenham, his contemporary, displayed ranting enthusiasts on every corner, often said by the 'doctors' to be male hysterics let loose on the Town. Though their numbers increased and decreased according to the luck of the time, decade by decade, their presence was commonly explained, as Swift had suggested in the *Tale*, in the language of the vapors and spleen, nerves and fibers, all their raving and madness attributable to 'hysterical affections.'

This was a motif – the connection between religious inspiration and *male* hysteria – that would extend throughout the course of the eighteenth century. As newly inspired sects became more visible, so too the varieties of their *male* hysterics, and in almost every case where documentation survives there lingers the implication of a 'hysterical affection' of one or another variety. If epilepsies and convulsions were the signs of secular distraction, they also afflicted men crazed in groups by their religious enthusiasm; Philippe Hecquet, a French physician of the ancien régime, claimed in *Le naturalisme des convulsions dans les maladies de l'épidémie convulsionnaire* (1733) that convulsions among the mob were anatomically experienced no differently than among individuals.[88] Charles Revillon, another French physician, supported this view in *Recherches sur la cause des affections hypochrondriaques* (Paris: Hérissant, 1786), explaining that sudden and unexpected catastrophic events trigger hysteria in the 'mob's body' exactly as they do in the individual body. Historically there were – to browse through the century cursorily – the strolling French prophets, or Camizards, in the first two decades; the new alchemists and preachers of the mid-century; the melancholic visionary poets (the Grays, Smarts, Collinses, Cowpers), all of whom suffered some type of religious melancholy and were either incarcerated in their colleges, like Gray, or in madhouses) to say nothing of the non-religious sects and the spate ranging from Hogarth's comic varieties to Dame Edith Sitwell's gallery of rogues.[89] Male hysteria coursed down through the century. Whole books could be written about it, deriving much of their information from the pages of popular reviews like the *Gentleman's Magazine*, one of the most widely circulated outlets of the Enlightenment, British or non-British. For example, the November issue of 1734 recounts a story embellished by the twist of cross dressing. Both the husband and wife have been 'hysterically affected,' she more acutely than he. More familiar than she with the medical profession, the husband persuades a friend to impersonate a physician, who treats his hysterical wife by prescribing 'the simple life.' The wife is duped, follows her therapy, and recovers. More common cases reveal afflicted males, prescribed to by bona fide doctors, who do *not* recover quickly.

By 1775, Hugh Farmer, the dissenting minister who was the friend of Dr. Philip Doddridge and enemy of Joseph Priestley, persuaded his publishers that there was sufficient interest in contemporary male hysteria to resuscitate it in the oldest extant texts. Farmer did so himself in *An Essay on the [male] Demoniacs of the New Testament*, a work aimed to show how ancient the lineage of inspiration was.[90] Farmer, like Christopher Smart and William Cowper, had himself been afflicted with a variety of religious melancholies that left him as debilitated as many chronic male hysterics. As a dissenting minister with a parish to look after and duties to attend to, Farmer was utterly uninterested in male license and liberty and, like Smart and Cowper, had maintained a queasy fear of women, especially older, sisterly women who forever rescued him and looked after him. The mindsets of all these figures lie far from the medical theory I discussed earlier, but not so far as to escape its effects. As I continue to suggest here, culture is a large mosaic whose individual pieces do fit together if the historian can only relate them. The English lyric poets, those of the ilk of William Collins and Smart, who were diagnosed male hysterics and melancholics, glimpsed the solipsism of their condition. All they discovered was an omniscient God whose powers of insight they could worship and emulate through their own visionary capabilities.[91] More broadly though, the greater the *resistance* to hysteria among men (in that century there was a surfeit of resistance), the more it revealed about their male sexuality in an era growing increasingly patriarchal and fastidious about its sexual mores. All these conditions and individual cases, far-flung and disparate as they are, some more anecdotal than others, presaged the scenario for male hysterics in the nineteenth century.

Still, the preeminent matter of gender in cases more or less hysterical hardly vanished in the second half of the eighteenth century. Granting that both sexes could become afflicted, perhaps in equal degree, profound questions about hysteria's anatomical prefigurements lingered. This is not surprising after centuries in which the feminine gender base had been strengthened by *men* exorcising hysterical *women* in need of help. No one to my knowledge has ever attempted to compile a list of eighteenth-century cases by gender.[92] If it were tried, even on a limited basis, it would be evident that women were *said* to have become afflicted in far greater numbers. The trend is even reflected in the lamp of imaginative literature. One and only one clearly delineated hysterical figure, for example, appears in Fielding's mock-epic novel *Tom Jones*: the young Nancy Miller, steeped in love sickness. Given the care with which Fielding is known to have constructed his symmetrical work of heroic proportions, the fact is not insignificant and can be demonstrated with similar results for other writers of the epoch. In Tobias Smollett there are many more: even the male hysteric Launcelot Greaves, a modern British version of Don Quixote, whose 'nerves' become damaged from his circulation in a crime-ridden, dangerous environment. Smollett was morbidly fascinated with crime in an almost sociological way. He eventually concluded that it had perpetrated the most heinous attack against the society of his day and formed the bedrock on which chronic diseases like hysteria flourished.[93]

Provided that medical and nonmedical discourses are gazed at in tandem, and without undue concern for validity in evidence, it becomes apparent that for most of the eighteenth century the nerves, not gender, were the burning issue for hysteria; that is, the nerves in their variegated anatomical, physiological, vivisectional, linguistic, ideologic, and even political senses. In the first published treatise on nymphomania, J. D. T. Bienville's curious work of 1775, there is no distinction whatever in regard to gender, no sense that the irritation or excitation of the genital area specifically is the cause of his new nymphomania.[94] 'Nymphomania,' Bienville wrote, arises from 'diseased imagination' taking root on the nervous stock, and it could afflict men as readily as women. Perhaps this occurred, in Bienville's view, because both genders had the potential for a 'diseased imagination.' It is an odd position to maintain, considering that his mind was formed in a world in which the close connection between sex and hysteria was taken for granted. Cases of 'erotomania,' a fierce and heightened form of erotic melancholy caused by love sickness, were regularly chronicled in the newspapers of the day. Erasmus Darwin, the poet and scientist, had mentioned one severe case (James Hackman's shooting of Martha Ray), but others were also written up. In all of them, the nervous system had flared out of control as the result of passion. The nerves were the zone Bienville was trying to penetrate in his discourse; the healthy or unhealthy state of the nerves, as well as the anatomic condition of the genital area (morbid, tonic, flaccid, put to use or not, aroused), the determinants. Bienville, a French mechanist about whom surprisingly little is known, ultimately wanted little truck with and underlying mental malady.

Turn the page, so to speak, to more literary annals, and hysteria blends in with other conditions from which its commentators barely differentiate it. Hysteria, hypochondria, melancholy – all are nervous maladies of one grade or another. Sterne's eternally melancholic Tristram may have been, in just this sense, the greatest and most self-reflective male hypochondriac of all the fictional characters of the century. He calls his confessional book 'a treatise writ against the spleen,' and knows, as his opening paragraph makes plain, that his animal spirits and nervous fibers have been irrevocably mutilated, rendering him a type of male hysteric. This is why he (like so many male patients in the next century) must be 'taken out of himself' as it were, through his own hobbies and the hobbyhorses of others. The nervous 'tracks' on which 'his little gentlemen' traveled during conception have been damaged. But a visit from Tristram to the great 'nerve doctors' – the Cheynes, Cullens, and Adairs – would have proved futile: he might as well have sent his manuscript, which is as good a case history of a 'male hysteric' as has ever been compiled. Yet Tristram himself might have been shocked to have been tendered this diagnosis. What Sydenham and his medical followers opined about male hysteria and gender at the end of the seventeenth century took decades to filter down to the ordinary person in any sophisticated way. Popular culture was indeed permeated with notions of hysteria, as I have been suggesting throughout this chapter, but Sydenham's views required decades to filter through to other doctors, let alone the lay public. A generation after Laurence Sterne's death in 1768, Edward Jenner, the Gloucestershire doctor and medical researcher into smallpox, was astonished to find himself a member of this filtered class. 'In a female,' Jenner wrote, 'I should call it Hysterical – but in myself I know not what to call it, but by the old sweeping term nervous.'[95] The difference was extraordinarily significant for him.

One of hysteria's other paradoxes was that it was alleged both to afflict males and to safeguard them *against* it. This was a curious double take seemingly reserved for hysteria, although traces of the incongruity are also found in the theory of gout and consumption at the time. The double bind rendered men safe and vulnerable at the same time. How are these theoretical 'doubles' explained? Under what framing? If run through the gamut of possibilities, it is seen that gender and patriarchy, power and marginalization alone can explain the double status of hysteria. The nerves have merely been the convenient pawns of a grander landlord. For the professional medical world of the eighteenth century was still preponderantly – as it would be in the nineteenth century and much of our own – a male-centered universe.[96] William Hogarth's male doctors, 'consulting' as they often do in his prints, could not see to what degree they were monolithically set against the few females who appeared in them and were an indirect cause of the very hysterical suffering they claimed they sought to relieve. It is hardly surprising then that the theory of *male* hysteria between Sydenham and the Victorians revealed what it genuinely was by describing its Other, its Counter, its Double: *female* hysteria.

Hordes of male doctors, exclusively generating medical theory, now – for the first time – institutionalized female hysteria by claiming that men could be afflicted by it but in actuality rarely were. Whether in Scotland or the West Country, in France or Germany, the results of these gender debates were more or less identical, often derived from one another.[97] The task then was to demonstrate precisely *why* women were more prone. But as the uterine debility hypothesis had been overthrown, the most persuasive mode was to argue from so-called incontrovertible universals: women's innate propensity to nervousness; their domestic situation in a private world conducive to hysterical excess; their insatiable sexual voracity granted from time immemorial – these as God-given, inevitable, unchangeable conditions. But all the while it was acknowledged that men were also prone, and proving theoretical consistency by occasionally diagnosing male hysterias and documenting them in the published literature.

We understand the complexity of Enlightenment hysteria only if we are willing to view its paradoxes, its double binds, within large social and cultural contexts, and only if we are capable of conceding that medical theory then was consistent and internally logical so long as doctors were not asked to be held accountable for the cultural conditions in which hysteria flourished. The state of laboratory verifiability and clinical observation of patients in a condition such as hysteria was still small compared to other maladies. A hundred years later, in Freud's Vienna, there would still be debate about the objectivity of the clinician's gaze. What counted for more than objective gaze in the world of Whytt, Cullen, and Jenner was a view of 'women' that naturally – almost preternaturally – seemed to lend itself to the hysteria diagnosis.

Notes

1. Highmore espoused his theory of hysteria in three works primarily: *Excercitationes duae, quarum prior de passione hysterica, altera de affectione hypochondriaca* (Amsterdam: C. Commelin, 1660); *Hysteria* (Oxford: A. Lichfield and R. Davis, 1660); *De hysterica et hypochondriaca passione: Responsio epistolaris ad Doctorem Willis* (London, 1670).
2. Quoted in Richard Hunter and Ida Macalpine, *Three Hundred Years of Psychiatry, 1535–1860* (London: Oxford University Press, 1963). See also Isler, *Thomas Willis*.
3. T. Willis, *An Essay on the Pathology of the Brain* (London, 1684), 71.

4. Beliefs about the effeminacy of men antedate the Restoration, of course, but the idea acquired altogether different currency then. For some of the reasons see Trumbach, 'The Birth of the Queen'; J. Turner, 'The School of Men: Libertine Texts in the Subculture of Restoration London' (a talk given at UCLA, 1989); for a remarkably detailed case history of male effeminacy of the playwright Richard Cumberland in the eighteenth century, see K. C. Balderston, ed., *Thraliana: The Diary of Mrs. Thrale* 1776–1809, 2 vols (Oxford: Clarendon Press, 1942; rev. edn, 1951), 2: 436–40.

5. The term *category* as I have been using it in this chapter should not suggest philosophical so much as medical category. Disease was then understood almost entirely within the terms of categories and classifications, as the wide taxonomic tendencies of the era had doctors compiling and classifying every disease in terms of its major symptoms, anatomic presentations, organic involvements, and so forth. See D. Knight, *Ordering the World: A History of Classifying the World* (London: Macmillan, 1980).

6. Baglivi held a chair of medical theory in the collegio della Sapienza in Rome, having been elected to it by Pope Clement XI. His book *De praxi medicina* (1699; English trans. 1723) was written with a knowledge of Sydenham's theories. He believed that hysteria was a mental disease caused by passions of the troubled mind; in this sense, he is less accurate and intuitive than Sydenham but nevertheless important. For Italian hysteria and hypochondria see Oscar Giacchi, *L'isterismo e l'ipochondria avvero il malo nervosa . . . Giudizii fisio-clinici-sociali* (Milan, 1875).

7. See B. Mandeville, *A Treatise of the Hypochondriack and Hysterick Passions* (London, 1711; reprinted 1715; 3rd edn, 1730).

8. Willis's anatomical 'explosions' are discussed by R. G. Frank, 'Thomas Willis and His Circle: Brain and Mind in Seventeenth-Century Medicine,' in *The Languages of Psyche*, ed. Rousseau, 107–147; Sacks, *Migraine*, 26–27; for Willis's rhetoric and language see D. Davie, *Science and Literature* 1700–1740 (London: Sheed & Ward, 1964).

9. For these shifts in knowledge at large see Thomas S. Kuhn, *The Structure of the Scientific Revolution* (Chicago: University of Chicago Press, 1970; rev. ed.); Rom Harré, 'Philosophy and Ideas: Knowledge,' in *Ferment of Knowledge*, ed. Rousseau and Porter, 11–55.

10. Even Veith's survey in *Hysteria* makes this fact abundantly clear.

11. The evidence for entrenchment is provided in the remaining portion of this chapter and remains a central theme of this essay, as it does in J. Wright, 'Hysteria and Mechanical Man,' *Journal History of Ideas* 41 (1980): 233–47, and for numbers of medical historians such as A. Luyendijk.

12. For some of the evidence of the opposite view see P. Hoffmann, *La femme dans la pensée des Lumières* (Paris: Ophrys, 1977); Hill, *Women and Work*.

13. So much has now been written about this relatively small group that one hardly knows where to direct the curious reader; a good place is J. Todd, *Sign of Angellica: Women, Writing, and Fiction*, 1660–1800 (London: Virago, 1988), and for one case history, written in depth, R. Perry, *The Celebrated Mary Astell: An Early English Feminist* (Chicago: University of Chicago Press, 1986).

14. A thorough linguistic study of these words ('spleen,' 'vapors,' 'hysterics') reconstructed in their local contexts would reveal shades of difference, but there are an equal number of examples of overlap and interchangeability; see

also section XIII. For the witch trials, see K. Thomas, *Religion and the Decline of Magic* (Harmondsworth, Middlesex: Penguin, 1973); for the famous 1736 case of the witch of Endor, B. Stock, *The Holy and the Demonic* (Princeton, NJ: Princeton University Press, 1983).

15. John Purcell, *A Treatise of Vapours, or, Hysterick Fits* (London: J. Johnson, 1707), 91.

16. Radcliffe had a large and established practice of wealthy aristocratic clients, many of whom suffered from hysteria, but he wrote little; his famed repertoire of remedies continued to be published during and after his lifetime and was edited by apothecary Edward Strother; see J. radcliffe, *Pharmacopoeia Radcliffeana* (London, 1716).

17. A good discussion of the scene is found in John Sena, 'Belinda's Hysteria: The Medical Context of *The Rape of the Lock*,' *Eighteenth-Century Life* 5, no. 4 (1979): 29–42.

18. J. Butt, ed., *The Twickenham Edition of the Works of Alexander Pope* (New Haven, Conn.: Yale University Press), 234.

19. For the post-Popean iconography of Belinda as hysteric see C. Tracy, *The Rape Observ'd* (Toronto: University of Toronto Press, 1974), 81, especially D. Guernier's illustration of Belinda swooning.

20. An interesting pharmaceutical study could be written compiling these remedies in the eighteenth century. For example, the *Gentlemen's Magazine* regularly printed 'receipts' for female hysteria and 'male lovesickness'; see the June 1733 issue, p. 321, prescribing the tying of a woman's head in a noose next to a cricket allegedly stung by the noise! Domestic *vade mecums* such as W. Buchan, *Domestic Medicine* (London, 1776), and standard pharmacopeias such as J. Quincy, *The Dispensatory of the Royal College of Physicians* (London, 1721, many editions), also prescribed. Hysteria was a virtual industry for apothecaries for the entire period, especially in cordials to prevent miscarrying.

21. For the all-important iatromechanism of the period at large see T. M. Brown, 'From Mechanism to Vitalism in Eighteenth-Century English Physiology,' *Journal of the History of Biology* 7 (1974): 179–216; Rousseau, 'Nerves, Spirits and Fibres'; G. Bowles, 'Physical, Human and Divine Attraction in the Life and Thought of George Cheyne,' *Annals of Science* 41 (1974): 473–88; H. Metzger, *Attraction Universelle et Religion Naturelle chez quelques Commentateurs Anglais de Newton* (Paris: Nizet, 1938); for iatromechanism in the work of Dr. Cheyne, see G. S. Rousseau, 'Medicine and Millenarianism: "Immortal Doctor Cheyne,"' in *Hermeticism and the Renaissance: Intellectual History and the Occult in Early Modern Europe*, ed. Ingrid Merkel and Allen Debus (Washington, DC: Folger Shakespeare Library, 1988), 192–230, and for the roles of rhetoric and language in Cheyne's writings, see Rousseau, 'Language of the Nerves,' in *Social History of Language*, ed. Burke and Porter. I consider Cheyne's *Essay of the True Nature and Due Method of Treating the Gout* (London: G. Strahan, 1722) among his most important works for laying out his theory of iatromechanism and post-Newtonian application.

22. The Dutch were important in the development of a *mechanical* theory of hysteria, the great and influential Dr. Boerhaave himself having identified hysteria as the most baffling of all female maladies. Boerhaave's writings set hysteria on a firm mechanical basis on the continent; for his theory of hysteria and its adoption by his followers, especially Anton de Haen in

Holland, Gerard van Swieten in Austria, and Robert Whytt in Scotland, see A. M. Luyendijk, 'Het hysterie-begrip in de 18de eeuw,' in *Ongeregeld zenuwleven*, ed. L. de Goei (Utrecht: NcGv, 1989), 30–41, a volume rich in the bibliography of hysteria and dealing exclusively with the modern history of female uterine maladies. Luyendijk is right to claim that throughout the eighteenth century every aspect of 'the sick woman' was sexually charged and sexually liminal; see A. M. Luyendijk, 'De Zieke Vrouw in de Achttiende Eeuw,' *Natuurkundige Voordrachten* 66 (1988): 129–136.

23. See Rousseau on Cheyne ('Medicine and Millinarianism'). By 1750 hysteria had become 'nationalized' (i.e., Dutch hysteria, Scottish hysteria, etc.) and a study of its nationalistic idiosyncrasies would make for fascinating reading.

24. It undid his psychologizing and cultural determination, neglected his primary point about hysteria as a disease of imitation, and replaced it with a radical anatomizing and mechanizing of the nervous system capable of accounting for rises and falls of hysteria in both genders. Indeed, after Sydenham the theory of imitation virtually went under, finding no place in Cheyne's system, where the word never appears. It may be more than coincidental that Sydenhamian hysteria as a *disease of imitation* declines concomitantly with the larger aesthetic and philosophical theory of imitation in the same period; see F. Boyd, *Mimesis: The Decline of a Doctrine* (Cambridge, Mass.: Harvard University Press, 1973).

25. For Robinson see *A New System of the Spleen* (London, 1729), quoted in Richard Hunter and Ida Macalpine, *Three Hundred Years of Psychiatry*; Klibansky et al., *Saturn and Melancholy*; Jackson, *Melancholia and Depression*, 291–4; T. H. Jobe, 'Medical Theories of Melancholia in the Seventeenth and Early Eighteenth Centuries,' *Clio Medica* 19 (1976): 217–31.

26. G. Cheyne, *The English Malady: or, a Treatise of Nervous Diseases of All Kinds* (London: Strahan & Leake, 1733), 184; see also O. Doughty, 'The English Malady of the Eighteenth Century,' *Review of English Studies* 2 (1926): 257–69; E. Fischer-Homberger, 'On the Medical History of the Doctrine of the Imagination,' *Psychological Medicine* 4 (1979): 619–28, which discusses the medicalization of the imagination in relation to the hysteric affection, and, most important, R. Porter, 'The Rage of Party: A Glorious Revolution in English Psychiatry,' *Medical History* 27 (1983): 35–50.

27. Cheyne, *English Malady*, 14. Samuel Richardson, the novelist and printer, had printed the book for his friend and claimed that Cheyne chose the title ('English') because he held the squalor and polluted air responsible for London's being 'the greatest, most capacious, close and populous City of the Globe' – and also called it the '*English* malady' because hysteria was so called in derision by continental writers (*English Malady*, 55; C. F. Mullett, ed., *The Letters of Doctor George Cheyne to Samuel Richardson* 1733–1743 [Columbia: University of Missouri Press, 1943, 15]).

28. R. James, *A Medicinal Dictionary; Including Physics, Surgery, Anatomy, Chemistry, and Botany, in All Their Branches. Together with a History of Drugs . . .* (London: T. Osborne, 1743–1745), article entitled 'hysteria.'

29. Curiously, no systematic study has been undertaken despite the large amount of recent feminist scholarship in the field of eighteenth-century studies; it awaits its avid student, for whom the sheer amount of material between 1700 and 1800 will make for a field day of scholarship. Some material for the

nineteenth century is found in Y. Ripa, *La ronde des folles: Femme folie et enfermement au XIXe siecle* (Paris: Aubier, 1986). Müller, who became a leading anthropologist in Germany, wrote his medical thesis at the University of Paris in 1813 on 'le spasme et l'affection vaporeuse'; as late as the 1840s some French doctors still considered 'spleen' a valid category of the hysteria–hypochondria syndrome; see D, Montallegry, *Hypochondrie-spleen ou névroses trisplanchniques. Observations relative à ces maladies et leur traitement radical* (Paris, 1841).

30. For Swift and hysteria see Christopher Fox, ed., *Psychology and Literature in the Eighteenth Century* (New York: AMS Press, 1988), 236–7.

31. M. DePorte, *Nightmares and Hobbyhorses: Swift, Sterne, and Augustan Ideas of Madness* (San Marino, Calif.: Huntington Library Press, 1974), 125 ff.

32. For evidence of the linguistic confusion in the primary medical literature, see W. Stukeley, *Of the Spleen* (London, 1723); J. Midriff, *Observations on the Spleen and Vapours; Containing Remarkable Cases of Persons of both Sexes, and all Ranks, from the aspiring Directors to the Humble Bubbler, who have been miserably afflicted with these Melancholy Disorders since the Fall of the South-see, and other publick Stocks; with the proper Method for their Recovery, according to the new and uncommon Circumstances of each Case* (London, 1720); J. Raulin, *Traité des affections vaporeuses du sexe* (Paris, 1758). There is also a wide literature of spleen and vapors, as in Matthew Green, *The Spleen, and Other Poems . . . with a Prefatory Essay by John Aikin, M. D.* (London: Cadell, 1796). For comparison of this early eighteenth-century, outbreak of spleen with outbursts in America at the end of the nineteenth century, see T. Lutz, *American Nervousness, 1903: An Anecdotal History* (Ithaca, NY: Cornell University Press, 1991), a study of the 'neurasthenia plague' of 1903 that gave rise to hundreds of cures and potions. Midriff wondered if certain types of 'spleen' appeared in particular types of wars and not others.

33. See Purcell, *Treatise of Vapours*; some discussion of these matters is found in O. Temkin, *The Falling Sickness* (Baltimore: Johns Hopkins University Press, 1974).

34. For the extensiveness of this Newtonianism in medical theory, see N. Robinson, MD, *A new theory of physick and diseases, founded on the principles of the Newtonian philosophy* (London, 1725), with much emphasis on hysteria; in theology and cosmic thought, J. Craig, *Theologia . . . Mathematica* (London, 1699); more generally, I. Prigogine, *Order Out of Chaos: Man's New Dialogue with Nature* (New York: Bantam Books, 1984). James Thomson the poet and author of *The Seasons*, the most widely read English poem of the eighteenth century, also reflects this pervasiveness; see A. D. McKillop, *The Background of Thomson's Seasons* (Minneapolis: University of Minnesota Press, 1942). For Newtonianism and the popular imagination, M. H. Nicolson, *Newton Demands the Muse* (Princeton, NJ: Princeton University Press, 1946).

35. Roy Porter has chronicled aspects of this development in *Mind-Forged Manacles: A History of Madness in England from the Restoration to the Regency* (London: Penguin, 1987); see also for madness in this period and its relation to current scientific movements: V. Skultans, *English Madness: Ideas on Insanity 1580–1890* (London: Routledge, 1979); M. Foucault, *Madness and Civilization: A History of Insanity in the Age of Reason* (New York: Pantheon Books, 1965), 120–32. Dr. Charles Perry, a mechanist and contemporary of Cheyne, Robinson, and

Purcell, makes perceptive points about madness in relation to hysteria in his treatise *On the Causes and Nature of Madness* (London, 1723).
36. For the humanitarianism of madness, see D. Weiner, 'Mind and Body in the Clinic: Philippe Pinel, Alexander Crichton, Dominique Esquirol, and the Birth of Psychiatry,' in Rousseau, *Languages of Psyche*, 332–40.
37. See Rousseau, 'Medicine and Millenarianism'.
38. G. Cheyne, quoted in L. Feder, *Madness in Literature* (Princeton, NJ: Princeton University Press, 1980), 170. Cheyne's prose abounds with weird syntax, ungrammatical constructions, and neologisms; 'fantastical' rather than the simpler word *strange* is just the sort of word found in his vocabulary.
39. *The English Malady* (1733), 353.
40. Ibid., 354.
41. Sir Richard Blackmore, *A Treatise of the Spleen and Vapours* (London, 1725), 320. In a rather similar prose, William Buchan in his *Domestic Medicine* (Edinburgh, 1769), 561, discussing 'hysteric and hypochondriacal affections,' noted that these nervous disorders were 'diseases which nobody chuses to own.' It is important to insist on the yoking of hysteria and hypochondria *ever since Sydenham undercut (except in name) hysteria as a gendered disease.* Blackmore argued from the perspective of one who had lived through the revolution in nomenclature as well as gender: 'Most Physicians have looked upon Hysteric Affections as a distinct Disease from Hypochondriacal, and therefore have treated some of them under different Heads; but though in Conformity to that Custom I do the same, yet...I take them to be the same Malady.' Blackmore admitted that women suffered worse, 'the Reason of which is, a more volatile, dissipable [sic], and weak Constitution of the Spirits, and a more soft, tender, and delicate Texture of the Nerves.' Yet, he insisted, 'this proves no Difference in their Nature and essential Properties, but only a higher or lower Degree of the Symptoms common to both.' This more 'delicate Texture of the Nerves' was the fulcrum on which the theory of nervous diseases, including hysteria, was to be pegged for the next century and remains a crucial development in the history of medicine in the Enlightenment. For some of its cultural resonances, see Rousseau, 'Cultural History in a New Key,' in *Interpretation and Cultural History*, ed. Pittock and Wear, 25–81.
42. Blackmore, *Treatise of the Spleen and Vapours*, 319. It is important to reiterate Sydenham's consistent use of this nomenclature for males, which fell under his gender collapse of the disease and which was generally adopted by his students and followers into the time of Blackmore and Robinson: men were always 'hypochondriacal,' while women remained 'hysterical,' and no amount of anatomical similitude between the genders could account for the linguist disparity; for some discussion, see E. Fischer-Homberger, 'Hypochondriasis of the Eighteenth Century – Neurosis of the Present Century,' *Bulletin of the History of Medicine* 46 (1972): 391–401.
43. Nicholas Robinson, *A new system of the spleen, vapours, and hypochondriack melancholy; wherein all the decays of the nerves, and lownesses of the spirits are mechanically accounted for. To which is subjoined, a discourse upon the nature, cause, and cure of melancholy, madness, and lunacy* (London, 1729), 144.
44. Ibid., 345. More generally for this 'physiological psychology' see DePorte, *Nightmares and Hobbyhorses*; Rather, 'Old and New Views of the Emotions and Bodily Changes'; Jobe, 'Melancholia in the Seventeenth and Eighteenth Centuries'.

45. Looking ahead, these factors will coalesce later on in the century, in the world of Adair, Heberden and Cullen – Cheyne's followers. For the medical profession in the eighteenth century in relation to the development of other professions, see Geoffrey S. Holmes, *The Professions and Social Change in England 1680–1730* (Oxford: Oxford University Press, 1981), and idem, *Augustan England: Professions, State and Society, 1680–1730* (London: Allen & Unwin, 1982).

46. For the role of quacks in this milieu see R. Porter, *Health for Sale: Quackery in England 1650–1850* (Manchester: Manchester University Press, 1989), and 'Female Quacks in the Consumer Society,' *The History of Nursing Society Journal* 3 (1990): 1–25.

47. Veith, *Hysteria*, 155.

48. I.e., the essentially anti-vitalistic principle that all is brain and body, nothing mind. Twentieth-century science has spelled the death knell of scientific vitalism despite its many vestiges in the biological and neurological realms. For the anti-vitalistic strains and what I am calling the triumph of the neurophysiological approach of contemporary twentieth-century science, see J. D. Spillane, *The Doctrine of the Nerves: Chapters in the History of Neurology* (Oxford, New York: Oxford University Press, 1981); W. Riese, *A History of Neurology* (New York: MD Publications, 1959); for the linguistic implications, M. Jeannerod, *The Brain Machine: The Development of Neurophysiological Thought* (Cambridge, Mass.: Harvard University Press, 1985); H. A. Whitaker, *On the Representation of Language in the Human Brain: Problems in the Neurology of Language* (Los Angeles: UCLA Working Papers in Linguistics, 1969).

49. Cheyne, *English Malady*, 271 ff.

50. For nymphomania, see chapter 6 in this volume.

51. The animal spirits continued to prove troublesome for experimenters and theorists until the middle of the eighteenth century; for this complicated chapter in the history of science and medicine, see E. Clarke, 'The Doctrine of the Hollow Nerve in the Seventeenth and Eighteenth Centuries,' in *Medicine, Science, and Culture: Historical Essays in Honor of Owsei Temkin*, ed. L. G. Stevenson and Robert P. Multhauf (Baltimore: Johns Hopkins University Press, 1968), 123–41; for its linguistic representations and diverse metaphorical uses, Rousseau, 'Discovery of the Imagination'; the interchanges between the rhetorical and empirical (or scientific) domains here would make a fascinating study that has not been undertaken on a broad canvas.

52. See Laqueur, *Making Sex*; Feher, *History of the Human Body*.

53. Cheyne, *English Malady*, ii (preface).

54. For nightmares and hysteria, see A. M. Luyendijk-Elshout, 'Mechanism contra vitalisme: De school van Herman Boerhaave en de beginselen van het leven,' *T. Gesch. Geneesk. Natuurw. Wisk. Techn.* 5 (1982): 16–26; idem, 'Of Masks and Mills'.

55. Two generations after Pope, Hannah Webster Foster (1759–1840) thought that the nerves of the coquette distinguished her from other types; see *The Coquette; or, The history of Eliza Wharton. Reproduced from the original edition of 1797* (New York: Columbia University Press, 1939), as did David Garrick in his play of the same name, but a century earlier there was no such notion in Philippe Quinault's *La mère coquette* (written as Sydenham was composing his essay on hysteria) or in the *State Poems* on court coquettes written during Swift's period.

56. S. Greenblatt, *Renaissance Self-fashioning: From More to Shakespeare* (Chicago: University of Chicago Press, 1980), whose use of self-fashioning must be credited.

57. P. M. Spacks, *The Female Imagination* (London: Methuen, 1976); idem., *Imagining a Self: Autobiography and Novel in Eighteenth-Century England* (London: Routledge, 1976); K. O. Lyons, *The Invention of the Self* (Carbondale: Southern Illinois University Press, 1978); J. Mullan, *Sentiment and Sociability* (Oxford: Oxford University Press, 1988); and literary criticism dealing with the literature of sensibility.

58. G. S. Rousseau, 'Nerves, Spirits and Fibres'; for the scientific dimension in the mid-eighteenth century, see Haller's physiological revolution; for the popular cults, see an anonymous 'Descant on Sensibility,' *London Magazine* (May 1776); for the literary dimension, Hagstrum, *Sex and Sensibility*; and L. I. Bredvold, *The Natural History of Sensibility* (Detroit: Wayne State University Press, 1962).

59. I tried to document this point about the semiology of disease then in ' "Sowing the Wind and Reaping the Whirlwind": Aspects of Change in Eighteenth-Century Medicine,' in *Studies in Change and Revolution: Aspects of English Intellectual History* 1640–1800, ed. Paul J. Korshin (London: Scholar Press, 1972), 129–59.

60. The new code is not evident in John Playford's seventeenth-century treatises, but begins to be apparent in the drama (Wycherly's *Love in a Wood*) and in treatises by dancing masters written after ca. 1740.

61. This complex and largely nonverbal code remains to be deciphered; it is something as yet not understood about the Augustan 'self-fashioning' (to invoke Greenblatt's fine term) of the nerves.

62. The new role of consumption of every type cannot be minimized in this period: see N. J. McKendrick et al., *The Birth of Consumer Society: The Commercialization of Eighteenth-Century England* (London: Europa Publications Limited, 1982); J. Brewer, *The Sinews of Power: War, Money, and the English State*, 1688–1783 (Cambridge, Mass.: Harvard University Press, 1990); for the reaction, M. Caldwell, *The Last Crusade: The War on Consumption* (New York: Atheneum Publishers, 1988); for the medical diagnosis and its economic implications see such contemporary medical works as C. Bennet, *Treatise of Consumptions* (London, 1720); for drink and its relation to nervous sensibility, compare T. Trotter, *An Essay, Medical, Philosophical and Chemical on Drunkenness* (London: Longmans, 1804).

63. I tried to explain the chain of reasons from medical and philosophical, to social and popular, in 'Nerves, Spirits and Fibres' and 'Cultural History in a New Key', but much work remains to be done – I have barely scratched the surface of the Enlightenment cults of sensibility.

64. See M. Foucault, *Madness and Civilization: A History of Insanity in the Age of Reason* (New York: Pantheon Books, 1965), 132. My own thought has been influenced as much on the semiotic domain by Tzvetan Todorov in *The Conquest of America: The Question of the Other Translated from the French by Richard Howard* (New York: Harper & Row, 1985).

65. Richard Sennett, *The Fall of Public Man* (New York: Alfred A. Knopf, 1979).

66. It could not have elevated sensibility and the conditions (hysteria) that depended on it, *without* a prior theory of the 'sciences of man.' There are fine studies of this subject, but they usually omit the medical dimension entirely;

for the best, see Sergio Moravia, *Filosofia e scienze umane nell'eta dei lumi* (Florence: Sansoni, 1982). The point needs to be related to the development of the science of man; Moravia saw much but did not make the important connections; he saw narrowly only the new science of man but not its implication for self-fashioning.

67. Even Peter Gay had made this seminal point about Haller in the opening pages of *The Enlightenment: An Interpretation*, 2 vols. (New York: Alfred A. Knopf, 1966–69), 1:30, in 'The Spirit of the Age' and 'The Recovery of Nerve,' as did Henry Steele Commager in *The Empire of Reason* (New York: Anchor Doubleday, 1977), 8–10, in the famous paean to Haller who 'took all knowledge for his province' (p. 8) and who 'in the breadth and depth of his knowledge was perhaps unique' (p. 10). However, Haller's shrewd *fusion* of a medical *and* literary language of sexual sensibility (*sensibilität*) has been less well understood by historians forever bent on merely assessing his contribution to the history of European science, the Swiss Enlightenment, or the intellectual development of Göttingen.

68. Elsewhere I have tried to make the argument that the medical and scientific revolutions of the Enlightenment have still not been integrated into the culture at large, nor into the developing medical profession; Goldstein's *Console and Classify* is an exemplary book for this type of work carried out for the next century. For the legacy of the 'nervous revolution' in medicine in the next century see also Oppenheim, 'Shattered Nerves'.

69. See Rousseau, *Languages of Psyche*.

70. Mechanical philosophy had been applied to every other domain, including painting, diet, health, government, so why not to manners? For a list of applications, see Rousseau, 'Language of the Nerves', 60–1; for an example in music, R. Browne, *Medicina musica: Or a Mechanical Essay on the Effects of Singing, Musick, and Dancing, on Human Bodies* (London, 1729). As late as 1757, manners are still being described in mechanical metaphors; see J. Brown, *An Estimate of the Manners and Principles of the Times* (London: L. Davis & C. Reymers, 1757).

71. William Heberden, *Medical Commentaries* (London: T. Payne, 1802), 227.

72. Ibid., 235. Heberden did insist, however, that 'their force will be very different, according to the patient's choosing to indulge and give way to them.'

73. The role of medical schools was also great in this.

74. See F. J. McLynn, *Crime and Punishment in Eighteenth-Century England* (London: Routledge, 1989).

75. For the evidence see Wright, 'Hysteria and Mechanical Man'. Servants often aped these affectations of spleen and vapors to other servants, but rarely would they do so with their mistresses, who usually saw through the pretense. In Gay's *The Beggar's Opera*, Lucy explains her unacceptable behavior to the rivalrous Polly in terms of the vapors, but without recalling (if she ever knew it) that 'Affectation' had been one of the handmaidens in Pope's 'Cave of Spleen' in *The Rape of the Lock*.

76. For his life and works, see Rousseau, 'Cultural History in a New Key'; Philip Gosse, *Dr. Viper: The Querulous Life of Philip Thicknesse* (London: Cassell, 1952); A. Brunschwig, *Enlightenment and Romanticism in Eighteenth-Century Prussia* (Chicago: University of Chicago Press, 1974). For Whytt, see R. K. French, *Robert Whytt, the Soul, and Medicine* (London: Wellcome Institute for the History of Medicine, 1969).

77. James Makittrick Adair, *Essays on Fashionable Diseases* (N.P., 1786), 4–7.
78. The phrase is usually quoted from *An Epistle to Dr. Arbuthnot*, line 132; see also Marjorie Hope Nicolson and G. S. Rousseau, *This Long Disease My Life: Alexander Pope and the Sciences* (Princeton, NJ: Princeton University Press, 1968). But Pope had used it earlier in a letter to Aaron Hill, March 14, 1731 (*Correspondence*, III. 182), commenting on his chronic infirmities, which he thought had predisposed his 'manly temperament' to certain 'softer activities.'
79. For Pinel and hysteria see D. Weiner, 'Mind and Body in the Clinic,' in Rousseau, *The Languages of Psyche*, 391–5.
80. For a list of many of these medical dissertations see G. S. Rousseau, 'Discourses of the Nerve,' in *Literature and Science as Modes of Expression*, ed. F. Amrine (Dordrecht: Kluwer Academic Publishers, 1989), 56–60.
81. M. Micale, 'A Review Essay of Male Hysteria,' *Medical History* (1988).
82. For some examples see Boss, 'Transformation of the Hysteric Affection'.
83. Biographical material is found in Thomas Guidott, *The Lives and Characters of the Physicians of Bath* (London, 1676–77; reprint of 1724–25 is edition referred to here) and *Some Particulars of the Author's [i.e., Guidott] Life* in Guidott's ed. of Edward Jorden's *Discourse of Natural Bathes and Mineral Waters* (London, 1669, 3d ed.). Guidott dedicated his books to Maplet and in 1694 saw through the press Maplet's treatise on the effects of bathing.
84. Guidott, *Lives and Characters of Physicians of Bath*, 128–42. Subsequent passages are found on these pages.
85. Throughout my reading I wondered if Guidott had read Sydenham on hysteria, but have been unable to make a case for or against. The larger point, however, is that one would not have to read a particular text to know, and even espouse, the fundamental aspects of the paradigm.
86. Elsewhere I shall demonstrate that it was this paradigm that informed, in part, theoretical explanations of all-male friendship (on grounds that sensitivity gravitated to like sensitivity), and that became the substratum of later discussions about effeminacy and sodomy.
87. For example, *Gideon's Fleece; or the Sieur de Frisk. An Heroic Poem . . . by Philo-Musus, a Friend to the Muses* (London, 1684).
88. Philippe Hecquet, *Le naturalisme des convulsions dans les maladies de l'épidémie convulsionnaire* (Soleure, 1733); Hillel Schwartz, *The French Prophets: The History of a Millenarian Group in Eighteenth-Century England* (Berkeley, Los Angeles, London: University of California Press, 1980); idem, *Knaves, Fools, Madmen, and that Subtile Effluvium: A Study of the Opposition to the French Prophets in England, 1706–1710* (Gainesville: University Presses of Florida, 1978).
89. Edith Sitwell, *The English Eccentrics* (Boston: Houghton Mifflin Co., 1933).
90. Hugh Farmer, *An Essay on the Demoniacs of the New Testament* (London: G. Robinson, 1775). For Farmer's interest in miracles, demons, spirits, and hysterics, as well as his medical case history and life, see Michael Dodson, *Memoirs of the Life and Writings of the Late Reverend and Learned Hugh Farmer* (London: Longman & Rees, 1804). This work differs from physician Richard Mead's *Treatise concerning the Influence of the Sun and the Moon upon Human Bodies, and the Diseases Thereby Produced* (London, 1748). In Mead, male hysteria is explained according to *external* phenomena (for example moon, waves, tides) acting through Hartleyan vibrations and magnetism upon the human Nerves and then the imagination. In this sense Mead, like Farmer, different

though their professions were, should both be considered kindred in the mindset of counter-nerve. For counter-nerve see Rousseau, 'Cultural History in a New Key', 70–75, and Richard Kuhn, *The Demon of Noontide: Ennui in Western Literature* (Princeton, NJ: Princeton University Press, 1976).

91. Loneliness was an element of their alienation as securely as any other factors, as has been noticed by John Sitter in his *Literary Loneliness in Mid-Eighteenth-Century England* (Ithaca and London: Cornell University Press, 1982).

92. For a very limited study in one hospital during the 1780s see G. B. Risse, 'Hysteria at the Edinburgh Infirmary: The Construction and Treatment of a Disease, 1770–1800,' *Medical History* 32 (1988): 1–22. Risse has suggested that the organic diagnosis rather than any remotely psychogenic etiology enhanced the bedside discourse shared between these Edinburgh professors and their pupils. Men were not taken in at Edinburgh, but they were in Paris and Vienna. Highborn and low, female and male: all were treated and eventually admitted without regard to gender.

93. *The Adventures of Tom Jones, a Foundling* (1749), Bk. XVI. Smollett, a physician-novelist who knew medical theory more intimately than Fielding, portrays many more hysterics, male as well as female, especially in his 'psychiatric novel' *The Adventures of Sir Launcelot Greaves* (1762). Karl Miller believes that Greaves's 'weakness of the nerves,' the malady his quack doctor assigns, is a foreshadowing of modern, almost Beckettian, 'nervousness,' and 'the more nervous people there are, the more we may need spitting images, a comedy of hurt.' See his provocative chapter entitled 'Andante Capriccioso,' in his *Authors* (Oxford: Clarendon Press, 1989).

94. Bienville, *Nymphomania*. Works had been written before 1775 on the behavior or activity we would now, anachronistically, call nymphomania, but Bienville was the first to write an entire treatise using the word and concept.

95. G. Miller, ed., *Letters of Edward Jenner* (Baltimore: Johns Hopkins University Press, 1983).

96. This fact surfaces repeatedly in the study of female maladies in Barbara Duden, *The Woman beneath the Skin: A Doctor's Patients in Eighteenth-Century Germany* (Cambridge, Mass.: Harvard University Press, 1991).

97. Those who think 'hordes' is excessive to describe the proliferation of hysteria theory should consult the bibliographical evidence; see J. Sena, *A Bibliography of Melancholy* (London: Nether Press, 1970).

Part III
Epilogue

Epilogue 2004

As I correct the galley proofs for this collection, I cannot imagine how these 'acts' might be construed as devoid of presentist concerns, i.e., the notion that they recapture discourses of the past that have now disappeared. Although most of us do not usually think about it, the nerves continue to be prominent everywhere in modern life, aesthetically as well as corporeally, intellectually and practically. The ancient mind–body split, and its attendant binaries, are so profoundly ingrained that we conceive of our waking states – in the morning, during the coffee-break, after lunch, just before bedtime – as pre-eminently 'psychological' and therefore as existing *apart* from our bodies; yet if pressed to the wall about what we really believe, and who we inherently are, most of us will honestly profess to being monists who cannot separate how we 'think' from how we 'feel' or our cognitive attitudes from our emotional states of being.

Besides, even if this ingrained dualism did not rule us, the stress of contemporary life impels us to believe it should. It is much easier to live as if 'mind' and 'body' *were* separate, especially when we are well: each category has its functions and when things go wrong you fix the part rather than the whole. Illness, especially chronic and complex malady, provides a different attitude and forces us to recognize that every attack on the body is also one on the mind and vice-versa. Yet dualism is breaking down as the result of our swift pace – the sheer breakneck speed of modern life. Technology and the bombardment of information have created the sense that time is now accelerating more quickly than ever before. Multifaceted stress and widespread depression, now doubtlessly our Western nervous conditions par excellence, have become especially fixed in the public consciousness; and they continue to frighten us more than most of us can say, not least in the new terrorist contexts. Apart from cancer and cardiac failure, they are our most chronic ailments and ever on the rise without the slightest intimation that they will wane over the coming decades. Consciousness may continue to be contested and culturally constructed by neuroscientists and social philosophers on the boundary points where body and mind meet – as if to provide a grammar of hope that the old philosophical dualism has finally lost out – yet our conscious states still get construed by most *non*-scientists as an attribute of mind *apart* from body. Nor has the tyranny of the mind, or will, given way to the needs of the body to any degree that makes a difference to the way consciousness is configured.

Francis Crick, the pioneer molecular biologist and discoverer (in collaboration with J. D. Watson) of DNA, has just died,[1] believing that some day his deterministic 'awareness neurons' would be found, offering the key to a biological explanation for consciousness.[2] The revolution in contemporary brain theory recently explained by Joseph LeDoux, Mike Gazanigga and Antonio Damasio, affirming that our brains are soft- rather than hard-wired and that personal experience can therefore change them neuroanatomically through the synapses, is slowly seeping into the popular imagination and exerting some influence. In literature and the arts, the nerves loom as vividly, though we rarely pause to reflect how or why, even when we see 'nervous art' in museums of modern art. 'It's me nerves,' a Cockney figure desperately exclaims to another in a current play in the London West End and set in the 1950s, the coinage still apt today to pinpoint the spot where life's largest crises will be waged: not just in the 'mind' but also in the nervous system, because each of us is an indivisible biochemical unit. Life's hurdles and impediments continue to multiply: sooner or later their toll on the domain of nerves – that kingdom on which consciousness, even the conscious will, depends – is apparent; that convenient linguistic shorthand, the 'nerves,' the catch-all for everything mysterious about our identities even if we know that our brains are the one organ most of us would never trade in (some incurable neurotics and psychotics doubtlessly would if it could delete their memories and give them a new lease on life). If only we could unravel the whole of the brain, we sometimes think, we could make genuine progress about the roots of consciousness and finally decide whether we are monists or dualists – it does make a difference. Yet if it took centuries to understand the circulation of the blood and to listen percussively to the heart, imagine the eons of time that will be required to penetrate the brain's darkest secrets. Another ice age may have arisen by then.

The 'nervous acts' implied in the essays collected in this volume primarily entail the constructions of adequate historical contexts for similar anxieties – cognitive, emotional, racial, sexual – in the past. The acts have grown all the more 'nervous' as the consequence of a revolution in brain theory that has occurred since many of them were written in the 1970s. Today the view is that neuronal circuits are modified by what we learn and remember. The brain systems that underlie thinking, emotion, and motivation develop, interact with, and influence each other to make us who we are. This hypothesis (not yet proved) is not so different from the eighteenth-century paradigm that nervous bodies produce literary selves; or – at the very least – that nervous physical bodies produce the environmental conditions in which poetic selves can begin to deconstruct their identities. The Enlightenment anticipation – with conceded differences – of our contemporary position was diffused in huge hulks of scientific and medical literature written over the course of 150 years (1650–1800). The idea that the nervous synapses are plastic and malleable is also writ large in this early medical literature striving to define precisely what 'nervous bodies' were and how their more abstract creative and intellectual selves depended on the bodies' states. Joining physical bodies and abstract selves together was the essential 'fire in the soul,' to echo a phrase of Thomas Willis, which some of these early theorists chased.

Yet 'fire in the soul,' whether configured as animal spirits, vital spirits, ether, phlogiston, electricity, psychogenetic protein, DNA, or Crick's intuitive 'awareness neurons,' was also the sublime stuff of poetry versified by literati of the classical period before the great wave of European Romanticism. Poet-doctors in Britain extending from Samuel Garth and John Armstrong, Richard Blackmore and George Cheyne, as we have seen, down through the century of transformation – the eighteenth – to Erasmus Darwin and the canonical Romantic poets, constructed some part of their identities based on their own nervous bodies. In a naive sort of way they intuited deep-layer connections between nervous anatomy and the unfolding of the creative act. Diverse Enlightenment doctor-authors ranging from Mark Akenside to Tobias Smollett, to name but two, versified the hollow tubes and substances running through them, while searching, often haphazardly and vainly, for the types of nervous plasticity the late nineteenth-century physiologists in the world of Ramon y Cajal would discover. By the 1920s, philosopher-critic Gaston Bachelard contemplated the mythic dimensions of this quest for sublime Promethean fire in the brain.[3] This non-material fluid did not merely enable nervous plasticity of the 'synaptic-self'[4] posited by our most recent neuroscientists. It also enhanced the physiological malleability giving rise to nuanced emotional states and complex moods that only adroit wordsmiths – consummate poets and prose writers – can hope to verbalize.

Exceptionally 'nervous readers' among us may think the construction of historical contexts a subordinate activity in comparison to the need for alleviation of their conditions or just to living as well as we possibly can. What, after all, can be more pressing than remedy if you are emotionally shattered, chronically exhausted or nervously compromised, let alone steeped in severe depression? Yet among historians whose primary task often entails description of the large picture – the broad canvas – extending backwards and forwards in time, and the provision of just contexts, the act of dealing with the past – even *science's* pasts – is perilous: itself the cause of terrific nerves and even consternation.

It is now thirty years since I wrote many of these essays, some longer ago than that. Yet in 2004 I remain as driven by the role of nerves in the evolution of civilization and formation of nervous selves as I was in the 1960s. Herbert Spencer's old 'truths of neurophysiology,' articulated over a century ago in his popularizations of Darwinian evolution, still ring true today even if his rhetoric appears too crude and positivistic for our relativist postmodern ears.[5] When described by Oliver Sacks and other neurologist-writers, nervous states appear more persuasive. The nervous system has not diminished its stranglehold. I doubt it will.

Throughout Western civilization – from the Greeks to the present time – the nervous system has been a crucial factor in the rise of ever-more complex civilizations and technological cultures.[6] More than any other body part, the brain and its nervous vassals hold the key to understanding how far back complexity (the sense that everything is getting more complicated) as well as complexity's resulting strains and stresses (fatigue, depression, stress, even mental illness) and complexity's energized counterparts – vitality, creativity, emotional joviality – extend. The nervous system has also enabled the reproductive system to work as it does by virtue of loading the genitals with more nerve cells than any other part of the body except the brain. In humans this simple anatomical fact has led to the miracle of sex. Yet the nervous system has usually been downgraded in cultural analysis or altogether overlooked; treated as a specialized topic within the history of science and medicine. Even the imagination, such a crucial category for human endeavor and expression, has been approached as if it existed *apart* from this nervous system: an abstract faculty, like the Christian soul, belonging to a dis-embodied nerveless 'mind' rather than the bloody flesh-and-neural body. Crying out for clearer articulation also are the cultural contexts of nerves in society. Over the centuries these too have been narrowly drawn.

The world I inhabit today is paradoxically far more 'nervous' than it was in the eighteenth century, and even more 'nervous' than when I began these investigations in the 1960s. Doubtlessly there is more nervous fatigue and stress, and perhaps even more depression and mental illness, than ever before in history and with no sign of improvement.[7] If the Age of Prozac still epitomizes the extraordinary degree to which stress and depression have invaded our lives, the future probably holds in store still more pharmaceutical panaceas. It now appears self-evident that for the foreseeable future the drug industry has been enthroned as the arbiter of civilization's tranquility or lack of it. Alternatives to pills – diverse regimes for care of the self – emerge weekly, but so far with little empirical proof of lasting success. Recent research has suggested that in a controlled experiment classical music had stimulated the brain's alpha waves to create a feeling of calm.[8] CD companies rushed in and exploited the discovery. But the experiment was judged to be lacking in rigor and empirical proof; besides, the large pharmaceutical companies have deep pockets to invest in advertising campaigns capable of crushing the competition. Another experiment yielded the similar discovery that Mozart's music had soothed the tender nerves of a fetus in its mother's womb.[9] If its brain waves could be controlled in this simple way, what was the implication for adult populations with *developed* brain waves? A new experiment in the UK uses the reverse strategy to drive rebellious adolescents from supermarket storefronts by blasting classical music at the entry to arouse (or deaden) their nerves and force them to disperse. The moral is clear: why raise the national debt on antidepressants when you can go to sleep listening to Mozart? The CD industry now promotes 'musical meditation tapes' to calm the nerves and lull them to sleep, but no group can compete with our anti-stress pharmaceuticals for the speed and predictability with which they act. So goes the argument anyway. What is self-evident is that we are a very 'nervous' society, and were so even before 9/11.

Yet if 'nervous selves' were also 'constructed subjectivities' that came into fashion by the mid-eighteenth century, they have not dropped out since then and what the Scottish philosophers of Common Sense, especially Thomas Reid and Lord Kames, had then taught about the discoveries of the mind proceeding from anatomical dissections of the body, continues to find voice today in the experiments of neuro-scientists appealing to Joseph LeDoux's 'emotional brain.'[10] Almost every writer of that eighteenth-century Era of Sensibility had to make peace with, or accommodate, the newly nervous components of culture, as I have aimed to show; and once they had entered the mainstream of thought, nerves were not about to be put out to graze. Correlatively, neuroscience today plays a much larger role in society, most of all pharmacologically, than it has in any previous era. But neuroscience also embodies those truths about consciousness and cognition awaiting detection and disclosure. The neurologist's secrets are so sought after and precious in our time that collectively we will go to any length to discover what they are. Oliver Sacks' insightful neurological writings offer a case in point: bestsellers divulging the inner worlds of his patients, they legitimate the point about the public's thirst for neurological knowledge. More anecdotally, it seems I cannot open a daily newspaper or magazine without reading about similar manifestations, whether in unrelenting articles on the deleterious effects of stress in modern life or – more creatively and oppositionally – in some advertisement or other about the *value* of nervous excitation – whether through travel to remote places or, more locally, in a recent appeal for members to join THE NERVOUS CLUB.[11] Some of our most productive neuroscientists are leaving their laboratories in order to communicate to the public that the brain is the last frontier: the ultimate border for cognition and consciousness, emotion and feeling.

So I return to my question, who among us would have a brain transplant? The idea that *we are our brains* gathers intensity apace when we recognize not merely the central role of feeling and memory in daily life, but also the underlying conditions necessary for artistic creation and the new plasticity of the ever-learning brain. We can begin by memorializing poet Emily Dickinson's pronouncement: 'The BRAIN—is wider than the sky—.'[12] It may be so, but the memory of great writers and composers also haunts us for the way in which they were governed, even undone, by their nervous systems: the body of Romantic composer Robert Schumann, for instance, gripped by heightened states of emotion before mental illness wrecked him and caused him to throw himself into the Rhine. Or Virginia Woolf, another suicide whose diaries and memoirs abound with references to her nuanced nervous states and 'nervous exhaustion.' Erroneously she believed the latter to be a new psychological condition holding the key to emotion and feeling (here she was sound and would be confirmed by the best of twentieth-century neuroscience). Woolf's usage is not limited to the body in pain, even 'creative pain' that occurs during moments of composition. She glimpsed the *aesthetic* side of nerves all too well: its beauty and sublimity as when she stumbled upon 'moments of bare and nervous beauty' in Hemingway's works to which she refers in *Granite and Rainbow: An Essay on Criticism*, or when noticing in her diary the 'celebrated sensibility of my [own] nervous system.'[13] These musings are more recent textual embodiments of eighteenth-century sentiments – in the Smarts, Collinses, and Cowpers – who were also preoccupied with their nervous states as conditions for selfhood.

Recently the ethical dimension of human nervous organization has made little progress. Perhaps it has even regressed. Yet our systems of law make no allowance for crises of stress that cause impulsive actions. The onus on each individual, it seems, is to care for the self by ensuring an equilibrium of nervous stress that guarantees appropriate legal behavior in all of our actions. Consequently, the sentences for verdicts of 'temporary insanity' are harsh – even fiercer now – than they were when I started out in the 1960s. Recent conditions such as 'road rage' and other temporary 'derangements' clearly implicate the nervous system in altered states, sometimes under the influence of drugs and substances, but the fact of their 'high temperature' – as one journalist recently put it – is of no interest to lawyers and judges, nor is the underlying cause to policy makers and statesmen. There is little impetus to grapple with the base root of what stresses out citizens and propels them into acts they later regret. Nor has the law yet caught up with contemporary neuroscience. If it is informed by the implications of LeDoux's 'emotional brain' – a big 'if' – it has not yet adjusted its legal code to take account of natural responses to feverish anxiety and murderous fear. The only remedy, monotonously, is pharmacological, to which realm we collectively and inevitably continue to resort as if it were the only solution because its effects are so reliable.

We also seem to think that the appropriate response to the ethical dimension lies along pharmaceutical lines: i.e., if people are stressed or depressed, they should be medicated. Our age has the largest selection of panaceas to treat nervous conditions, not merely our pandemic stress and depression. But these pharmacological remedies, however well tested and reliable, do not address the underlying causes nor does it seem possible to confront the menace itself – so complex has modern life become. Yet in moving in parallel between now and then it is hard not to see how the long eighteenth century, the chronological heartland of the above essays, held similar attitudes, albeit still inchoate and anticipatory of what was to come. We can already perceive the tendency to medicate on a large scale, while labeling the nervous condition as pathological. Today, at the other end of the time warp, there is no evidence that labeling or medicating is on the wane. It may be that we have lost hope that we *can* be cured by any other means than drugs. What would it mean, anyway, to be 'cured' of 'modern nervousness'?

The future of nerves is necessarily connected to its long-range evolution: anatomical, biological, physiological, but also artistic, intellectual, discursive. Even so, the recognition that 'nerves' are changing in precisely the way that LeDoux and his cohorts have recently described should give us pause about future directions. The idea of the 'learning brain' takes on extraordinary significance today as the brain's synapticity becomes ever more impressive.[14] The biological implications soar and the social construction of our nervous evidence expands into the cultural domain, for now we have bioanatomical evidence for the cultural change surmised long ago, but which had hitherto been incapable of proof.

Future cultural approaches to nerves will be diverse: ranging from the anthropological (the nerves in relation to different societies and their religions and values) to the socio-economic (the notion that systems of civic organization and material possession inherently affect nervous conditions of mankind) and the equally important aesthetic (artifacts that intensify or diminish our nervous states). In the end it may be that neurology is born not so differently from poetry and myth: thriving for understanding on symbol and language and on the dialects of analogy and metaphor that have been formulated, in any case, by the brain itself.

What is our current responsibility in the light of this state of affairs vis-à-vis the nerves in history, and what is the future of nerves? Futures are determined by pasts, even if you believe – in common with Francis Fukuyama – in the end of history. As LeDoux has written: '. . . brain systems that generate emotional behaviours are highly conserved through many levels of evolutionary history.'[15] The essays above have pinpointed a transformative moment in that long 'evolutionary history.' If they alert students of the past to the continuing centrality of nerves in history and nerves in literature the effort will have been worthwhile. 'Nervous' concerns, no matter how transformed, continued to endure in a prominent position, albeit in altered states, long after the gradual displacement of the Romantic ethos in the late nineteenth century, and they have continued to assert their centrality in the creative endeavors of Modernism and Postmodernism. If Edelman, LeDoux and their brain colleagues are right about the system's evolutionary longevism, nerves will continue to haunt us as long as creatures breathe and their brains generate languages.

Notes

1. On 28 July 2004, aged 88.
2. See Francis Crick (1994), passim.
3. See Gaston Bachelard (1964).
4. The phrase is Joseph LeDoux's. He uses it as the title of his 2002 book.
5. He uses this phrase repeatedly in his *Principles of Psychology* (London, 1855; rev. edn, 1872).
6. Elizabeth Wilson (1998) has launched an eloquent case, but see also G. Edelmann (1987) for the future of nerves.
7. The evidence is divided on this point, with roughly half of demographers believing there is more.
8. *Times*, 8 March 2004 and widely reported that day in the media.
9. Sheepdrove Trust Newsletter, Bristol England: Winter 2003, n. p.
10. J. LeDoux (1996); see also his revisions in Le Doux (2002).
11. *Cambridge University Alumni Magazine*, no. 39 Easter Term 2003, p. 48.
12. T. H. Johnson (ed.), *The Poems of Emily Dickinson* (Cambridge, MA: Belknap, 1955), poem number 632.
13. A. O. Bell (1920), III, 1925, 2 June 1931.
14. The most enviable synthesis of the history of this development is found in LeDoux (1996).
15. Ibid., p. 17.

Bibliography

Note: Nor have many of the works cited in those essays. Places of publication and the names of publishers for works printed before 1900 are generally omitted for reasons of space, except in cases where they are judged to be useful for location, or where a particular edition is of interest. Journal names are abbreviated except in cases where the abbreviation is not clear from the reference itself. The bibliography is not a map of the terrain of nerves in 2005 despite containing many references to works published since 2000; instead it includes works I used over the course of three decades when the above essays were written and aims to provide the reader with a sense of the mindset of that time. No attempt has been made to list all the primary works of Thomas Willis, who features prominently in this collection; for that see chapters 2 and 5 above. In this sense, the bibliography aims to function as an historical tool as well as reference source for those wishing to pursue further the nerves and the rise of sensibility.

Abrams, M. H. *The Mirror and the Lamp: Romantic Theory and the Critical Tradition.* New York: Norton, 1953.

Abrams, M. H. *Natural Supernaturalism: Tradition and Revolution in Romantic Literature.* New York: Norton, 1971.

Ackerman, Diane. *A Natural History of the Senses.* London: Phoenix, 1996.

Adair, James Makittrick. *Dissertatio... de febre Indiae Occidentalis maligna flava, etc.* Edinburgh, 1766.

Adair, James Makittrick. *Medical Cautions, for the Consideration of Invalids; those especially who resort to Bath.* Bath, 1786.

Adair, James Makittrick. *A Philosophical and Medical Sketch of the natural history of the human body and mind...* Bath, 1787.

Adair, James Makittrick. *Essays on Fashionable Diseases.* London, 1787; 1790.

Albert, J. *Essai sur l'hypochondrie.* Paris, 1813.

Allen, Benjamin. *The Natural History of the Chalybeat and Purging Waters of England....* London, 1699.

Allen, Richard C. *David Hartley on Human Nature.* Albany: State University of New York Press, 1999.

Allen, Richard O. 'If you have Tears: Sentimentalism as Soft Romanticism.' *Genre* VIII (1975): 119–45.

Allman, J. M. *Evolving Brains.* New York: Scientific American Library, 1999.

Alpina, Prosper. *The Presages of Life and Death in Diseases.* London: G. Stahan, 1746.

Alston, Charles. *Lectures on the Materia Medica.* London: E&C Dilly, 1770.

Altschule, Mark, M. D. 'George Cheyne and his *English Malady*.' *Origins of Concepts in Human Behavior: Social and Cultural Factors.* Washington: Hemisphere, 1977.

Amann, Joseph. *Ueber den Einfluss der weiblichen Geschlechts-krankheiten auf das Nervensystem mit besond. Berücksicht. des Wesens und der Erscheinungen der Hysterie.* Erlangen, 1868.

Anon. 'Descant on Sensibility.' *London Magazine*. May 1776: 33–4.

Anon. 'Histories of Gouty Bilious, and Nervous Cases . . . related by the Patients Themselves to John Scat, M. D.' *Critical Review*. October 1780: 318–19.

Anon. *A letter of congratulation, and advice from the Devil to the Inhabitants of London . . . on their conduct before . . . the late earthquakes. Printed by a fool*. London, 1750.

Anon. *A treatise of diseases of the head, brain & nerves. With directions for their cure . . . To which is subjoin'd a discourse of the nature, cause and cure of melancholy and vapours. By a physician*. London, 1711.

Anon. *All Men Mad: or, England A Great Bedlam*. London, 1704.

Anon. *Anatomical Dialogues; or, a Breviary of Anatomy, by a Gentleman of the Faculty*. London: G. Robinson, 1778.

Anon. *Apollo Mathematicus: or the Art of curing Diseases By the Mathematicks, According to the Principles of Dr. [A.] Pitcairne. To which is subjoined a Discourse of Certainty*. 1695.

Anon. *Necessary Information for Hypochondriacks*. London, 1739.

Anon. *Observations on the Present Epidemic Fear*. 1741.

Anon. 'Of the Hypp.' *Universal Spectator*. 18 November 1732: 1062–3.

Anon. *Riverius Reformatus*. London: R. Wellington, 1713.

Anon. *The Fashionable Tell-Tale; containing . . . anecdotes and bon mots, expressive of the characters of persons and rank*. London: H. Gardner, 1787.

Anon. *The Female Monitor, or The History of Arabella and Lady Gay*. London: W. Richardson, 1781.

Armand-Leroche, J. *Brillat-Savarin et la médecine*. Paris: Le François, 1931.

Armstrong, D. M. *The Mind–Body Problem: an Opinionated Introduction*. Boulder, CO: Westview Press, 1999.

Armstrong, John. *Medical Essays . . . in Two Volumes*. Edinburgh, 1773.

Armstrong, John. *The Art of Preserving Health*. London: T. Cadell, etc., 1796. [originally 1744]

Arnisaeus, F. *Disputatio medica – De melancholia hypochondriaca*. Copenhagen, 1654.

Arthos, John. 'Poetic Diction and Scientific Language.' *Isis*. 32 (1932): 324–38.

Ashton, H. *Madame de La Fayette*. Cambridge: Cambridge University Press, 1922.

Astruc, J. *Traité des maladies des femmes*, 6 vols. Paris: Cavelier, 1761–5.

Astruc, Jean. *A Treatise on the Venereal Disease, in six Books . . . Together with a short Abstract of the lives of the authors who have wrote on those Diseases*. London: W. Innys and J. Richardson, 1737.

Astruc, Jean. *A Treatise of Venereal Diseases*. London: W. Innys and J. Richardson, 1754.

Austen, Jane. *Mansfield Park*. Vol. III of *The Novels of Jane Austen*, 3rd edn. Oxford, 1952.

Avicenna. *A Treatise on the Canon of Medicine of Avicenna*. Trans. O. Cameron Gruner. New York: AMS Press, 1973.

Babcock, R. W. 'Benevolence, Sensibility and Sentiment in Some Eighteenth-Century Periodicals.' *Modern Language Notes* 62 (1947): 394–7.

Babinski, J. and Froment, J. *Hysteria or Pithiatism and Reflex Nervous Disorders in the Neurology of War. Military Medical Manuals*. London, 1918.

Bachelard, Gaston. *The Psychoanalysis of Fire*. Translated by Alan C. M. Ross. London: Routledge & Kegan Paul, 1964.

Baglivi, George. *The Practice of Physick, reduc'd to the ancient Way of Observations, containing a Just Parallel between the Wisdom of the Ancients and the Hypothesis's of Modern Physicians.* London, 1704.

Bahri, Deepika. '"Disembodying the Corpus": Postcolonial Pathology in Tsitsi Dangaremba's *Nervous Conditions.*' *Postmodern Culture* 5(1) (1994): 1–9.

Baillie, John. *An Essay on the Sublime.* London, 1747.

Bain, Alexander. *The Senses and the Intellect.* London: J. W. Parker, 1855.

Balderston, K. C. (ed.). *Thraliana: the Diary of Mrs. Hester Lynch Thrale (later Mrs Piozzi) 1776–1809,* 2 vols. Oxford: Clarendon Press, 1942; rev. edn, 1951.

Barbauld, A. R. (ed.). *The Correspondence of Samuel Richardson,* 6 vols. London, 1804.

Barine, Arvede. *Poètes et névroses.* Paris: Hachette, 1898.

Barker, John. *An Essay on the Agreement betwixt ancient and modern Physicians: or a Comparison between the practice of Hippocrates, Galen, Sydenham, and Boerhaave, in acute diseases, etc.* London: G. Hawkins, 1747.

Barker-Benfield, G. J. *The Culture of Sensibility: Sex and Society in Eighteenth-Century Britain.* Chicago: University of Chicago Press, 1992.

Barkow, J. H., L. Cosmides and J. Tooby, eds. *The Adapted Mind: Evolutionary Psychology and the Generation of Culture.* New York: Oxford University Press, 1992.

Barry, Kevin. *Language, Music and the Sign: a Study in Aesthetics, Poetics and Poetic Practice from Collins to Coleridge.* Cambridge: Cambridge University Press, 1987.

Bate, Walter Jackson. *The Achievement of Samuel Johnson.* New York: Oxford University Press, 1961.

Bath, R., Bradley Thomas, and Nalhden, A. A. (eds). *The Medical and Physical Journal.* London: R. Phillips, 1802.

Battie, William. *A Treatise on Madness.* London: Dawsons, 1962.

Bear, M. F. and Malenka, R. C. 'Synaptic Plasticity: LTP and LTD.' *Current Opinions in Neurobiology* 4 (1994): 389–99.

Beard, George. *A Practical Treatise on Nervous Exhaustion (Neurasthenia), its Symptoms, Nature, Sequences and Treatment.* New York: W. Wood, 1880.

Beard, George. *American Nervousness.* New York, 1881.

Beard, George. *Sexual Neurasthenia (Nervous Exhaustion): its Hygiene, Causes, Symptoms and Treatment with a Chapter on Diet for the Nervous Edited, with Notes and Additions, by A. D. Rockwell.* New York: E. B. Treat, 1898.

Beauchene, Edme P. C. de. *De l'influence des affections de l'âme dans les maladies nerveuses de femmes, avec le traitement qui convient à ces malades . . . Nouvelle edition revue, et augmentée du traitement des maux de nerfs des femmes enceintes.* Amsterdam and Paris: Mequignon l'aine, 1783.

Becker, Bernard. *Scientific London.* New York, 1874.

Beckford, William. *Modern Novel Writing, or the Elegant Enthusiast . . . and Interesting Emotions of Arabella Bloomville, by Lady Margaret Marlow . . . A Rhapsodical Romance.* London, 1796.

Beddoes, Thomas. *Hygëia: or Essays Moral and Medical on the Causes affecting the Personal State of our Middling and Affluent Classes,* 3 vols. Bristol: Phillips, 1802.

Beiser, Frederic. *Enlightenment, Revolution, and Romanticism.* Cambridge, MA: Harvard University Press, 1992.

Bekker, Balthazar. *The World Bewitch'd; or an Examination of the Common opinions concerning Spirits: their Nature, Power, Administration, and Operations. As also, the effects Men are able to produce by their Communication. Vol. 1. Translated from a French Copy, approved and subscribed by the Author's own Hand.* London, 1695.

Bell, A. O. (ed.). *The Diaries of Virginia Woolf*, 4 vols. London: Hogarth Press, 1920.

Bell, Charles. *The Anatomy of the Brain, Explained in a Series of Engravings.* London: Longman, 1802.

Bell, Charles. *The Nervous System of the Human Body embracing the Papers delivered to the Royal Society on the subject of nerves.* London, 1824, 1830.

Bell, Sir Charles. *The Anatomy and Philosophy of Expression as Connected with the Fine Arts.* London: J. Murray, 1844.

Bellers, John. *An Essay on Physick.* London, 1714.

Belloste, Augustin. *Le Chirurgien d'hospital.... Paris, 1714.

Below, J. C. *Dissertatio casum matronae hypochondriacae exhibens.* Erfurt, 1685.

Ben, J. *De suffocatione hypochondriaca.* Leiden, 1683.

Benedetti, Alessandro. *Anatomice, siue de hystoria corporis humani, libri quinque.* Rome, [1528?].

Benjamin, Marina. ed. *Science and Sensibility: Gender and Scientific Enquiry, 1780–1945.* Oxford: Basil Blackwell, 1991.

Bennett, Cyril. *The Modern Malady or, Sufferers from 'Nerves'.* London, 1890.

Bergson, Henri. *Creative Evolution.* London: Routledge and Kegan Paul, 1911.

Bernal, J. D. *The World, the Flesh and the Devil: an Inquiry into the Future of the Three Enemies of the Rational Soul.* London: Cape, 1970.

Bernbaum, Ernest. *The Drama of Sensibility....* Boston: Harvard Studies in English, 1915.

Bethum, John. *A Short View of the Human Faculties and Passions....* London, 1770.

Bett, Walter R. *The History and Conquest of Common Diseases.* Norman, OK: Oklahoma University Press, 1954.

Bewick, Thomas. *Memoirs of the Life of the Rev. Dr. Trusler.* Bath: John Browne, 1806.

Bianchini, Gio. Fortunato. *Saggio d'Esperienze intorno la Medicina Elettrica fatte in Venezia da alcuni Amatori di Fisica.* Venice: Presso Pasquali, 1749.

Binswanger, Ludwig. *Melancholie und Manie.* Pfullingen, 1960.

Birotheau, A. L. *Sur l'hypochondrie.* Paris, 1830.

Blackmore, Richard. *A Critical Dissertation upon the Spleen.* London, 1725.

Blackmore, Richard. *A Treatise of the Spleen and vapours: or, hypochondriacal and hysterical affections.* London: J. Pemberton, 1725.

Blackmore, Richard. *Discourses on the Gout... and the King's Evil.* London: J. Pemberton, 1726.

Blackmore, Richard. *The Poetical Works of Richard Blackmore... containing Creation: a philosophical poem... to which is prefixed the life of the author.* London: C. Cooke, 1797.

Blake, William. *The Four Zoas. A Photographic Facsimile....* London: n. p., 1987.

Blum, E. *De dolore hypochondriaco, vulgo sed falso putato splenetico.* Leipzig, 1671.

Blumenbach, Johann Friedrich. *Elements of Physiology.* Trans. Charles Caldwell. Philadelphia: Thomas Dobson, 1795.

Bock, Martin. *Crossing the Shadow-Line: the Literature of Estrangement.* Columbus: Ohio State University Press, 1989.

Boerhaave, Herman. *Boerhaave's Aphorisms: Concerning the Knowledge and Cure of Disease.* London: B. Course, 1715.

Bondt, Jakob de. *An Account of the Diseases, Natural History and Medicines of the East Indies* ... London: T. Motman, 1769.

Bonnet, Charles. *Essai analytique sur les facultés de l'âme*. Copenhagen, 1760.

Boring, E. G. *A History of Experimental Psychology*. New York: Appleton-Century-Crofts, 1950.

Boscovich, R. J. *A Theory of Natural Philosophy*. Venice, 1763. [*Theoria philosophiae naturalis*]

Bostetter, E. E. 'Coleridge's Manuscript Essay: "On the Passions".' *JHI* 31 (1970): 99–108.

Boswell, James. *Boswell's Column: Being his Seventy Contributions to the London Magazine under the Pseudonym of the Hypochondriak* .... London: Kimber, 1951.

Bourgeois, Susan. *Nervous Juyces and the Feeling Heart: the Growth of Sensibility in the Novels of Tobias Smollett*. New York: Lang, 1986.

Boursier, Laurent François. *Mémoire théologique sur ce qu'on appelle les secours violens dans les convulsions*. Paris, 1788.

Bouvier, Auguste. *J. G. Zimmermann: un représentant suisse du cosmopolitisme littéraire au xviiie siècle*. Geneva: Georg & Co., 1925.

Brand, P. *De malo hypochondriaco*. Copenhagen, 1676.

Bredvold, L. I. *The Natural History of Sensibility*. Detroit: Wayne State University Press, 1962.

Brendel, J. G. *Opusculorum mathematici et medici argumenti pars I. (–III.)*, 3 vols. Göttingen, 1769–75.

Bressy, J. *Recherches sur les vapeurs*. London, 1719.

Brewster, Sir David. 'Reflexions on the Decline of Science in England, and on some of its Causes by Charles Babbage.' *Quarterly Review* XLIII (1830): 320–41.

Bridges, Thomas. *The Adventures of a Bank Note*. 1770.

Brissenden, Robert. *Virtue in Distress*. London: Macmillan, 1974.

Brocklesby, William. *Ancient and Modern Music*. 1749.

Brodie, B. C. *Lectures Illustrative of Certain Local Nervous Affections*. London: Longman, 1837.

[Brodum]. *A Guide to Old Age, or A Cure for the Indescretions of Youth*, 2 vols. London, 1790.

Brooke, Frances. *The History of Lady Julia Mandeville*. 1763.

Brooke, Frances. *Emily Montague*. 1769.

Brophy, Elizabeth Bergan. *Samuel Richardson: the Triumph of Craft*. Knoxville: University of Tennessee Press, 1974.

Brown Clark, Stephanie. 'The Birth of Psychiatry in the Age of Romanticism: the Problem of the Psyche in English Medicine and Literature 1790–1840,' PhD thesis, University of Leiden, 1997.

Brown, Ralph. *Modern Neurasthenia*. London, 1894.

Brown, Theodore M. 'From Mechanism to Vitalism in Eighteenth-Century English Physiology.' *Journal of the History of Biology* 7 (1974): 179–216.

Brown, Thomas the elder. *Bath; a Satirical Novel*. London: Sherwood, Neely and Jones, 1818.

Brown, Thomas. *A Treatise on the Philosophy of the Human Mind*, 2 vols. Cambridge, MA: Hilliard and Brown, 1827.

Brown, W. H. *The Power of Sympathy*. Edinburgh, 1789 [rep. 1996]

Browne, Richard. *Medicina Musica*. London: J. Cooke, 1729.

Brunschwig, Henri. *Enlightenment and Romanticism in Eighteenth-Century Prussia.* Trans. Frank Jellinek. Chicago: University of Chicago Press, 1977.

Buchan, William. Dr. *Buchan's Family Medical Works: Containing the Domestic Medicine, Enlarged: and the Advice to Mothers, on the Subject of Their Own Health;* .... London, 1747; Charleston, SC: John Hoff, 1807.

Bucher, Heinrich Walther. "Tissot und sein Traité des Nerfs." *Züricher Medizingeschictlicher Abhandlungen.* Ed. E. H. Ackerknecht. Zurich, 1958.

Budd, Malcolm. *Music and the Emotions: the Philosophical Theories.* London: Routledge & Kegan Paul, 1992.

Buren, D. van. *De affectione hypochondriaca.* Leiden, 1711.

Burghart, C. G. *De malo sic dicto hypochondriaco.* Wittenberg, 1703.

Burette, Pierre-Jean. *Mémoire pour servir à l'histoire de la lutte des anciens.* Paris, 1746.

Burette, Pierre-Jean. *Paragone dell'antica colla moderna musica.* Venice, 1748.

Burke, Peter and Roy Porter (eds). *Language, Self and Society.* Cambridge: Polity Press, 1991.

Burn, I. H. *The Automatic Nervous System.* Oxford: Blackwells, 1963.

Burney, Frances. *Evelina.* London: Oxford English Novels, 1968.

Burwick, Frederick et al. (eds). *The Crisis in Modernism: Bergson and the Vitalist Controversy.* Cambridge: Cambridge University Press, 1992.

Butler, Joseph. *The Analogy of Religion Natural and Revealed ... to the Constitution and Recourse of Nature. To which are added ... Dissertations I of Personal Identity. II of the Nature of Virtue,* 5th edn, London, 1736.

Byman, Seymour. 'Child Raising and Melancholia in Tudor England.' *J. Psychohistory* 6 (1978): 67–92.

Bynum, William. 'The Nervous Patient in 18th and 19th Century Britain: the Psychiatric Origins of Neurology.' *Anatomy of Madness.* 3 vols, ed. Roy Porter, W. Bynum and M. Shepherd. London: Tavistock, 1985. 1: 89–102.

Cabanis, Georges. *Revolutions of Medical Science.* 1806.

Cabanis, Pierre-Jean-Georges. *On the Relations Between the Physical and Moral Aspects of Man.* Trans. Duggan Saidi, Margaret, 2 vols. Baltimore: John Hopkins University Press, 1981.

Cadogan, William. *An Essay upon Nursing and the Management of Children.* London: J. Roberts, 1748.

Calmeil, L. F. *De la Folie,* 2 vols. Paris, 1845.

Cameron, Kenneth Neill. *Shelley: the Golden Years.* Cambridge, MA: Harvard University Press, 1974.

Camporesi, Piero. *The Incorruptible Flesh: Bodily Mutation and Mortification in Religion and Folklore.* Cambridge: Cambridge University Press, 1988.

Camporesi, Piero. *The Anatomy of the Senses: Natural Symbols in Medieval and Early Modern Italy.* Cambridge: Polity Press, 1994.

Camus, Le. *Medicina de l'espirit.* Paris, 1769.

Canaveri, Francesco. *Neuronomia.* Taurini, Italy, 1836.

Canguilhem, Georges. 'La Constitution de la Physiologie Comme Science.' *Études d'Histoire et de Philosophie des Sciences.* Paris: Vrin, 1968.

Canguilhem, Georges. *The Normal and The Pathological.* Trans. C. R. Fawcett. New York: Zone, 1989.

Caramagno, Thomas. *The Flight of Mind: Virginia Woolf's Art and Manic-Depression.* Berkeley and Los Angeles: University of California Press, 1991.

Carlino, Andrea. *Books of the Body: Anatomical Ritual and Renaissance Learning.* Chicago: University of Chicago Press, 1999.

Carlson, E. T. and Dain, N. 'The Meaning of Moral Insanity.' *Bull. Hist. Psychiatry* 43 (1969): 101–15.

Carlson, Eric T. and Meribeth M. Simpson. 'Models of the Nervous System in Eighteenth-Century Psychiatry.' *Bulletin of the History of Medicine* 43 (1969): 101–15.

Carlyle, Thomas. *Critical and Miscellaneous Essays*, 4 vols. Boston, 1860.

Carpenter, William B. *Principles of Human Physiology.* London: J. Churchill, 1842.

Carter, Elizabeth. *A Series of Letters between Mrs. Elizabeth Carter and Miss Catherine Talbot from the year 1741 to 1770*, 4 vols. London: J. Rivington, 1809.

Carter, Rita. *Mapping the Mind.* Berkeley, CA: University of California Press, 1999.

Cartheuser, Johann Friedrich. *Dissertatio de typhomania.* Frankfort on Oder, 1750.

Cartwright, Frederick F. *Disease and History.* London: Hart-Davis, 1972.

Cartwright, Frederick F. *A Social History of Medicine.* London: Longman, 1977.

Carusi, Paola. *Les cinq sens entre philosophie et médecine (Islam Xe–XIIe siècles).* *Micrologus* 10 (2002) [special issue on the religious dimensions of sense and sensation in Islamic cults]

Castells, Manuel. *The Rise of the Network Society.* Oxford: Blackwell, 1996.

Chabert, M. *De hypochondria.* Paris, 1805.

Changeux, Jean-Pierre. *Neuronal Man.* Princeton: Princeton University Press, 1997.

Changeux, Jean-Pierre. *The Physiology of Truth: Neuroscience and Human Knowledge.* Cambridge, MA: Harvard University Press, 2004.

Chapman, Gerald. *Literary Criticism in England 1660–1800.* New York: Knopf, 1966.

Cheyne, George. *Remarks on Two Late Pamphlets Written by Dr. Oliphant Against Dr. Pitcairn's Dissertations . . .* Edinburgh, 1702.

Cheyne, George. *Observations concerning the nature and due method of treating the gout; for the use of my worthy friend, Richard Tennison, Esq.: together with an account of the nature and qualities of the Bath Waters.* London, 1720.

Cheyne, George. *The English Malady: Or a Treatise of Nervous Diseases of All Kinds, as Spleen, Vapours, Lowness of Spirits, Hypochondriacal, and Hysterical Distempers, &c.* London: 1733. [ed. Roy Porter 1989, Tavistock/Routledge]

Cheyne, George. *Essay on Regimen . . . the principles and theory of philosophical medicine*, 2nd edn. London, 1740.

Cheyne, George. *Own Account of Himself and of his Writings . . . To which are added, . . . An Account of Dr. Pitcairn . . .* London, 1743.

Cheyne, George. *The Letters of Dr. George Cheyne to Samuel Richardon (1733–1743).* Ed. C. E. Mullett. Columbia, Missouri: University of Missouri Studies, vol. XVIII, no. 1. 1943.

Cheyne, George. *The Letters of Dr. George Cheyne to the Countess of Huntington.* Ed. C. E. Mullett. San Marino, CA: Huntington Library, 1940.

Churchland, Patricia S. *Neurophilosophy: Toward a Unified Science of the Mind/ Brain.* Cambridge, MA: MIT Press, 1986.

Churchland, Patricia S. *The Engine of Reason, the Seat of the Soul: a Philosophical Journey into the Brain.* Cambridge, MA: MIT Press, 1995.

Cicero. *Offices and Select Letters. Everyman's Library.* London: J. M. Dent, 1953.

Clark, Sir James. *Treatise on Pulmonary Consumption*. London, 1835.

Clarke, Basil. *Mental Disorder in Earlier Britain*. Cardiff: University of Wales Press, 1975.

Clarke, Edwin. 'The Doctrine of the Hollow Nerve in the Seventeenth and Eighteenth Centuries.' In Lloyd G. Stevenson and Robert P. Multhauf (eds), *Medicine, Science, and Culture: Historical Essays in Honor of Owsei Temkin*, Baltimore: John Hopkins University Press, 1968.

Clarke, Edwin and Jacyna, L. S. *Nineteenth-Century Origins of Neuroscientific Concepts*. Berkeley: University of California Press, 1987.

Clarke, William, MD. *A Medical Dissertation Concerning the Effects of the Passions on Human Bodies*. 1753.

Clifford, James L., ed. *Eighteenth-Century English Literature: Modern Essays in Criticism*. New York: Oxford University Press, 1959.

Clitorides, Philogynes. *The natural history of the frutex vulvaria, or flowering shrub: as it is collected from the best botanists both ancient and modern*. London: W. James, 1732.

Clower, William T. 'The Transition from Animal Spirits to Animal Electricity: a Neuroscience Paradigm Shift.' *Journal of the History of the Neurosciences* 7(3) (December 1998): 201–18.

Coates, Kimberley Engdahl. 'Exposing the "Nerves of Language."' *Literature and Medicine* 21(2) (2002): 242–63.

Cobbold, Richard. *Geoffrey Gambado: or, A simple remedy for hypochondriacism . . . By a humorist physician*. London: Dean, 1865.

Cobe, George and Andrew (eds). *On the Functions of the Cerebellum by Drs. Gall, Vimont, and Broussais . . . Also Answers to the Objections Urged Against Phrenology by Drs. Roget, Rudolphi, Prichard, and Tiedmann*, 8 vols. Edinburgh: Maclachlan & Stewart, 1838.

Cockburn, Catherine. *Fatal Friendship: a Tragedy*. London, 1698.

Coe, Thomas. *A Treatise on Biliary Concretions*. London: D. Wilson, 1757.

Cohen, I. Bernard. *The Newtonian Revolution*. Cambridge: Cambridge University Press, 1980.

Colbatch, Sir John. *A Dissertation Concerning Misletoe*, 6th edn. London: D. Browne, 1730.

Collins, Joseph. *The Way with Nerves; Letters to a Neurologist on Various Modern Nervous Ailments, Real and Fancied, with Replies thereto Telling of Their Nature and Treatment*. New York and London: G. P. Putnam's Sons, 1911.

Collins, Robert. *A Short Sketch of the Life and Writings of the Late Joseph Clarke*. London: Longman, 1849.

Colohri, M. A. *De passione hypochondriaca*. Frankfurt, 1751.

Colombo, Realdo. *De Re Anatomica*. Venice, 1559.

Combe, Andrew. *The Principles of Physiology Applied to the Preservation of Health, and to the Improvement of Physical and Mental Education*. New York, 1839.

Condillac, Etienne Bonnot de. *Traité des sensations*. Paris, 1754.

Conry, Yvette. 'Thomas Willis ou le premier discours rationaliste en pathologie mentale.' *Revue d'histoire des sciences* 31 (1978): 193–231.

Cook, Harold J. 'Boerhaave and the Flight from reason in Medicine.' *Bulletin of the History of Medicine* 73 (2000): 221–40.

Cook, Matt. *London and the Culture of Homosexuality, 1885–1914*. Cambridge: Cambridge University Press, 2003.

Cooke, John. *Treatise on Nervous Diseases in Two Volumes.* London, 1823; Boston, 1824.

Copleston, F. C. *A History of Philosophy: Volume IV, Descartes to Leibniz.* New York: Garden Books, 1963.

Copley, J. *Shift of Meaning.* London: Oxford University Press, 1961.

Corbet, D. *De hypochondriaci.* Edinburgh, 1821.

Corner, George W. *Anatomical Texts of the Earlier Middle Ages.* Washington, DC: Carnegie Institute of Washington, 1927.

Cornoldi, Cesare. *Stretching the Imagination: Representation and Transformation in Mental Imagery.* New York: Oxford University Press, 1996.

Corsi, Pietro. *The Enchanted Loom: Chapters in the History of Neuroscience.* New York: Oxford University Press, 1991.

Corsi, Pietro (ed.). *The Mill of Thought: From the Art of Memory to the Neurosciences.* Milan: Electa, 1989.

Cotta, G. S. *Dissertatio aegrum chylificatione laesa hypochondriaca laborantem exhibens.* Jena, 1689.

Cotugno, Domenico. *De ischiade nervosa commentarius. Ex typographia Sancti Thomae Aquinatis.* Naples, 1775.

Cowling, J. *De hypochondriasi.* Edinburgh, 1768.

Cowper, William. *The anatomy of humane bodies, with figures drawn after the life . . . by some of the best masters in Europe . . .* Oxford: Walford, 1698.

Cowper, William. *The anatomy of the brain. Containing its mechanism and physiology; together with some new discoveries and corrections of ancient and modern authors upon that subject to which is annex'd a particular account of animal functions and muscular motion . . .* London: Sam. Smith and Walford, 1695. [Humphrey Ridley]

Crenner, Christopher W. *Nervousness and Race,* Minnesota Series in the History of Medicine. Minneapolis: University of Minnesota Press, 2002.

Crick, Francis. *The Astonishing Hypothesis: the Scientific Search for the Soul.* Cambridge: Cambridge University Press, 1994.

Critchley, Macdonald. *Historical Aspects of the Neurosciences: a Festschrift.* New York: Raven Press, 1982.

Crooke, Helkiah. *Microcosmographia: a Description of the Body of Man.* London: W. Jaggard, 1618.

Cullen, William. *Synopsis Nosologicae Methodicae.* Edinburgh, 1780.

Cullen, William. *Institutions of medicine. Part I, Physiology,* 3rd edn. Edinburgh: T. Cadell, 1785.

Culpepper, Nicholas. *Culpepper's last legacy: left and bequeathed to his dearest wife for the publick good. Being the choicest and most profitable of those secrets, which while he lived were lock'd in his breast, . . . Containing sundry admirable experiences in several sciences, more especially in chyrugery and physick: . . . With an addition of two hundred choice receipts, . . . The seventh impression; whereunto is added an extract and perfect treatise of anatomy of the reins and bladder, brain and nerves . . .* London: J. Phillips, H. Rhodes and J. Taylor, 1702.

Cumberland, Richard. *The Fashionable Lover. A Comedy.* London: John Bell, 1793.

Cunningham, Andrew. *The Anatomical Renaissance: the Resurrection of the Anatomical Projects of the Ancients.* Aldershot: Scolar Press, 1997.

Curtis, Lewis Perry, ed. *Letters of Laurence Sterne.* London: Oxford University Press, 1967.

Cusanus, Nicolaus de. *The idiot in four books . . . the third of the minde . . .* London: W. Leake, 1650.

D'Holbach, Baron. *The System of Nature.* Boston: J. P. Mendum, 1889.

da Carpi, Berengario. *A Short Introduction to Anatomy.* Trans. L. R. Lind and ed. Paul G. Roofe. New York: Kraus Reprint Co., 1969.

da Vinci, Leonardo. *Leonardo da Vinci on the Human Body: the Anatomical, Physiological, and Embryological Drawings of Leonardo da Vinci.* Ed. J. B. de C. M. Saunders and Charles D. O'Malley. New York: Crown Publishers, 1982.

Dacome, Lucia. 'Living with the Chain.' *History of Science* 39 (2001): 499–521.

Dallas, Robert Charles. *The History of the Maroons . . . and the Island of Jamaica,* 2 vols. London, 1803.

Damasio, Antonio R. *Descartes' Error: Emotion, Reason, and the Human Brain.* New York: G. P. Putnam, 1994.

Dangarembga, Tsitsi. *Nervous Conditions: Colonizing the Body.* London: Women's Press, 1988.

Daran, Jacques. *Observations chirurgicales sur les maladies de l'urethre.* Paris: Chez Debure l'Aine, 1758.

Darnton, Robert. *Mesmerism and the End of the Enlightenment in France.* Cambridge, MA: Harvard University Press, 1968.

Darwin, Charles Robert. *The Expression of the Emotions in Man and Animals.* London: John Murray, 1873.

Darwin, Erasmus. 'The Mechanism of the Body.' Huntington Library MSS 490: 1798.

Darwin, Erasmus. *The Temple of Nature.* London: Scolar Press, 1973.

Darwin, Erasmus. *The Botanic Garden, Parts One and Two,* 2 vols. New York: Garland, 1979.

Daudin, Henri. *De Linné à Lamarck. Méthodes de la classification et idée de série en botanique et en zoologie (1740–1790).* Paris: Editions des archives contemporaines, 1926.

Daudin, Henri. *Henri Cuvier et Lamarck, les classes zoologiques et l'idée de seri animale, 1790–1830.* Paris, 1926.

Davach del la Riviera, Jean. *Le miroir des urines.* Paris, 1718.

Davan, Kingsmill. *An Essay on the Passions.* London: Vernor and Hood, 1799.

Davie, Donald. *Science and Literature 1700–1740.* London: Sheed and Ward, 1964.

Davis, Lennard J. *Enforcing Normalcy: Disability, Deafness, and the Body.* London: Verso, 1995.

Davis, Owen. *The Nervous Wreck; a Comedy in Three Acts.* New York: S. French, 1926.

Dawson, Warren R. *The Custom of Couvade.* Manchester: Manchester University Press, 1929.

Decker, Hannah S. *Freud in Germany: Revolution and Reaction in Science.* New York: International University Press, 1977.

Deleuze, Gilles. *Francis Bacon: logique de la sensation.* Paris: Editions de la Différence, 1984.

Deloffre, Frédéric. *Marivaux, Pierre Carlet de Chamblain de, 1688–1763.* Genève: Droz, 1955.

Deloffre, Frédéric. *Marivaux et le marivaudage, une préciosité nouvelle,* 2nd edn. Paris: Librairie A. Colin, 1967.

Deloffre, Frédéric. *La Nouvelle en france à l'âge classique.* Paris: Didier, 1968.

Deloffre, Frédéric. *Théâtre Complet.* Paris: Garnier Freres, 1968.

Deloffre, Frédéric. *Journaux et oeuvres diverses.* Paris: Garnier Freres, 1969.

Dendy, Walter Cooper. *The Philosophy of Mystery.* New York: Harper, 1841.

Dendy, Walter Cooper. 'Fantasy from Sympathy with the Brain.' In *Psyche: A Discourse on the Birth and Pilgrimage of Thought.* London, 1845.

Derham, William. *Physico-theology: or, A demonstration of the being and attributes of God, from his works of creation, 16 sermons at mr. Boyle's lectures. With notes.* London, 1712.

Derivaux, J. B. *Essai sur l'hypochondrie.* Strasbourg, 1836.

Desaive, J. P. *Médecins, climat, et épidémies à la fin du XVIIIe siècle.* Paris: Mouton, 1972.

Descartes, René. *Treatise on Man.* In *The Philosophical Writings of René Descartes.* Cambridge: Cambridge University Press, 1985. I: 99–107.

Descot, Pierre Jules. *Dissertation sur les affections locales des nerfs . . . .* Paris, 1882.

Dewhurst, Kenneth. *The Quicksilver Doctor: a Biography of Thomas Dover, Physician and Adventurer.* Bristol: J. Wright and Sons, 1957.

Dewhurst, Kenneth. *An Illustrated History of Brain Function.* Berkeley and Los Angeles: University of California Press, 1972.

Dictionary of Sensibility. [2002] http://www. engl. virginia. edu/enec981/dictionary/ s_biblio. html [online resource]

Diderot, D. *Encyclopédie, ou, Dictionnaire raisonné des sciences, des arts, et des métiers,* 35 vols. Lausanne et Berne, 1782.

Disraeli, Isaac. 'Medical Music,' in *Curiosities of Literature . . . Three Volumes* (London, 1881), I: 269–74.

Dixon, Thomas. *From Passions to Emotions: the Creation of a Secular Psychological Category.* Cambridge: Cambridge University Press, 2003.

Dodds, E. R. *Pagan and Christian in an Age of Anxiety.* Cambridge: Cambridge University Press, 1965.

Dondt, Jakob de. *An account of the diseases, natural history and medicines of the East Indies.* London, 1769.

Doody, Margaret A. *Frances Burney: the Life in the Works.* New Brunswick: Rutgers University Press, 1988.

Dormandy, Thomas. *The White Death: a History of Tuberculosis.* London: The Hambledon Press, 1999.

Dostrovsky, Sigalia. 'Early Vibration Theory: Physics and Music in the Seventeenth Century.' *Acoustical Science.* 4 (1975): 169–218.

Dougherty, F. W. P. 'Nervenmorphologie und-physiologie in den actziger Jahren des 18. Jahrhunderts. Gottinger Beitrag zur Forschung und Theorie der Neurologie in der vorgalvanischen Ara.' *Gehirn – Nerven – Seele. Anatomie und Physiologie im Umfeld Soemmerrings.* Soemmerring-Forschungen III, Stuttgart: Gustav Fischer, 1987, pp. 286–312.

Doughty, Oswald. 'The English Malady.' *Review of English Studies* 2 (1926): 257.

Douglass, Aileen. *Uneasy Sensations: Smollett and the Body.* Chicago: University of Chicago Press, 1995.

Dover, Thomas. *The Ancient Physician's Legacy to his Country. . . .* London: R. Brady, 1733.

Dowling, John E. *Neurons and Networks: an Introduction to Behavioural Neuroscience.* Cambridge, MA and London: Belknap Press of Harvard University Press, 1992.

Downing, Hugh. *Infancy: a Poem, in Three Cantos.* London: G. Kearsley, 1776.

Drake, James. *Anthropologia Nova: or, A new system of Anatomy,* 2 vols. London, 1707.

Draper, J. W. *The Funeral Elegy and the Rise of English Romanticism.* New York: Macmillan, 1929.

Duden, Barbara. *The Woman Beneath the Skin: a Doctor's Patients in Eighteenth-Century Germany.* Trans. Thomas Dunlap. Cambridge, MA, and London: Harvard University Press, 1991.

Duchesneau, F. 'The Scientific Revolution and the Problematics of the Living Being, with Special Reference to the Concepts of William Harvey and J. B. van Helmont.' *Revue philosophique du Louvain,* 94(4) (November 1996): 568–98.

Dupau, Jean Amédée. *L'éréthisme nerveux, ou analyse des affections nerveuses.* Montpellier: J. Martel, Jnr., 1819.

Du Verney, M. *A Treatise of the Organ of Hearing.* London: S. Baker, 1737.

Eagleton, Terry. *The Ideology of the Aesthetic.* Oxford: Basil Blackwell, 1990.

Eccles, John. *The Neurophysiological Basis of Mind: the Principles of Neurophysiology.* Oxford: Clarendon Press, 1953.

Edelman, Gerald. *Neural Darwinism.* New York: Basic Books, 1987.

Edelman, Gerald. *Bright Air, Brilliant Fire: On the Matter of the Mind.* London: Allen Lane, 1992.

Edwardes, Allen. *The Rape of India: a Biography of Robert Clive and a Sexual History of the Conquest of Hindustan.* New York: The Julian Press, 1966.

Edwards, George. *A Natural History of Uncommon Birds....* London, 1743.

Edwards, George. *Essays Upon Natural History.* London: Printed for J. Robson, 1770.

Egilsrud, J. S. *Le 'Dialogue des Morts' dans les littératures française, allemande et anglaise (1644–1789).* Paris: L'Entente linotypiste, 1934.

Ehrenreich, Barbara. *Blood Rites: Origins and History of the Passions of War.* London, 1997.

Elias, Norbert. *The Civilizing Process.* Oxford: Blackwell, 2000. [rev. edn of 1978–82]

Eliot, T. S. *On Poetry and Poets.* London: Faber, 1957.

Ellenberger, H. T. *The Discovery of the Unconscious.* London: Allen Lane, 1970.

Ellis, William Charles. *A treatise on the nature, symptoms, causes and treatment of insanity, with practical observations on lunatic asylums....* London: S. Holdsworth, 1838.

Ellman, Geoffrey. *Rethinking Innateness.* Cambridge, MA: MIT Press, 1996.

Elton, Oliver. *A Survey of English Literature 1730–1780.* London: Edward Arnold, 1928.

Emden, Cecil S. 'Rythmical Features in Dr Johnson's Prose.' *RES* 25 (1949): 38–54.

Entralgo, Láin Pedro. *Doctor and Patient.* London: Weidenfeld and Nicolson, 1969.

Erämetsa, Erik. *A Study of the Word 'Sentimental' and of Other Linguistic Characteristics of Eighteenth-Century Sentimentalism in England.* Helsinki: n.p., 1951.

Ernoul-Provoté, J. B. *Essai sur l'hypochondrie.* Paris, 1816.

Evelyn, John. *The History of Religion.* Ed. R. M. Evanson, 2 vols. London: Henry Colburn, 1850.

Eze, Emmanuel C. *Race and the Enlightenment: a Reader.* Oxford: Blackwell, 1997.

Fairchild, H. N. *The Noble Savage: a Study in Romantic Naturalism.* New York: Columbia University Press, 1928.

Falret, J. P. *De l'hypochondrie et du suicide*. *Considérations sur les causes, sur le siège et la traitement de ces maladies, sur les moyens d'en arrêter les progrès et d'en prévenir le développement*. Paris: Croullebois, 1822.

Farmer, Hugh. *An Essay on the Demoniacs of the New Testament*. London: G. Robinson, 1775.

Fausset, Hugh J'Anson. *The Proving of Psyche*. London: Jonathan Cape, 1929.

Febvre, Lucien. *Combats Pour l'Histoire*, 2nd edn. Paris: A. Colin, 1953.

Febvre, Lucien. 'Comment reconstituer la vie affective d'autrefais? La sensibilite et l'histoire.' *Combats pour L'Histoire*. Ed. Lucien Febvre. Paris: Librairie Armand Calin, 1953.

Fedida, Pierre. 'Les exercices de l'imagination et la commotion sur la masse des nerfs: un erotisme de tete.' *Oeuvres complètes du Marquies de Sade*, 16 vols. Paris: Au cercle du livre precieux, 1967, pp. 613–25.

Feher, Michel. *Fragments for a History of the Human Body*. New York: Zone Books, 1989.

Feist, J. *Morbi hypochondriaci cum hysterico comparatio*. Berlin, 1819.

Fernel, Jean. *Universa Medicina* .... Geneva, 1643 [written c. 1554]

Ferriar, John. *Medical Histories and Reflections*, 3 vols. Warrington, 1795.

Feuchtersleben, Ernst Freiherr von. *The Principles of Medical Psychology*. Trans. H. E. Lloyd and revised by B. G. Babington. London, 1847.

Feuchtersleben, Ernst Freiherr von. *The Dietetics of the Soul*. London, 1852.

Finn, Michael R. *Proust, the Body and Literary Form*. Cambridge: Cambridge University Press, 1999.

Fish, Stanley. 'Being Interdisciplinary is so very hard to do.' In *There's No Such Thing as Free Speech*. Oxford: Oxford University Press, 1994, pp. 231–42.

Fisher, Michael. *The Vehement Passions*. Princeton and Oxford: Princeton University Press, 2002.

Fizes, Dr. A. *Dissertatio medico-physiolgiae secretione fluidi nervorum, ipsius indole, motu, ac usibus*. Montpellier, 1739.

Flemyng, Malcolm. *Neuropathia; sive de morbis hypochondriacis et hystericis, etc.*. York, 1740.

Flemyng, Malcolm. *The nature of the nervous fluid, or, Animal spirits demonstrated*. London: A. Millar, 1751.

Flemyng, Malcolm. *Introduction to Physiology ... lectures upon the ... animal economy*. London, 1759.

Flemyng, Malcolm. *A Discourse on the Nature, Causes and Cure of Corpulency*. London: A. Millar, 1760.

Floyer, Sir John. *Psychrolusiad or ... History of Cold Bathing, both ancient and modern*. London, 1697.

Flynn, Carol. *The Body in Swift and Defoe*. Cambridge: Cambridge University Press, 1990.

Foot, Jesse. *The Life of John Hunter*. London: T. Becket, 1794.

Fordyce, James. *A Discourse on Pain*. London, 1791.

Fothergill, Samuel. *An Account of a Painful Affection of the Nerves of the Face, Commonly Called Tic Douloureux*. London, 1804.

Foucault, Michel. *The History of Sexuality, Volume One: An Introduction*. London: Penguin, 1979.

Fracassini, A. *Naturae morbi hypochondriaci ejusque curationis mechanica investigatio*. Verona, 1754.

Frank, Robert G. 'Thomas Willis and his Circle: Brain and Mind in Seventeenth-Century Medicine.' In G. S. Rousseau (ed.), *The Languages of Psyche: Mind and Body in Enlightenment Thought*. Berkeley, Los Angeles, and Oxford: University of California Press, 1990, pp. 107–47.

Franklin, Benjamin. *Anniversary Oration Delivered Before The American Philosophical Society*. 1750.

Freed, Lewis. *T. S. Eliot: Aesthetics and History*. La Salle, IL: Open Court, 1962.

French, R. K. *Robert Whytt, the Soul, and Medicine*. London: Wellcome Institute for the History of Medicine, 1969.

French, Roger K. *Dissection and Vivisection in the European Renaissance*. Aldershot: Ashgate, 1999.

Freud, S. 'Mourning and Melancholia.' *Collected Papers*. London: Hogarth Press, 1951.

Freud, S. *The Origins of Psychoanalysis*. New York: Basic Books, 1954.

Frith, Christopher D. and Wolpert, D. M. (eds). *The Neuroscience of Social Interactions: Decoding, Influencing, and Imitating the Actions of Others*. Oxford: Oxford University Press, 2004.

Frye, Northrop. 'Towards Defining an Age of Sensibility.' *English Literary History* 23(2) (June 1956): 144–52.

Fuller, Thomas. *Medicina Gymnastica: Or a Treatise Concerning the Power of Exercise, with Respect to the Animal Oeconomy; and the Great Necessity of it in the Cure of Several Distempers*. London, 1705.

Fulton, John F. *Physiology of the Nervous System*. Cambridge: Cambridge University Press, 1930.

Fulton, John F. *The Great Medical Bibliographers: a Study in Humanism*. Philadelphia: University of Pennsylvania Press, 1951.

Fyfe, Andrew. *A compendious system of anatomy. In six parts. Part II. Osteology. II. Of the muscles, etc. III. Of the abdomen. Part IV. Of the thorax. V. of the brain and nerves. IV of the senses*. Philadelphia: Thomas Dobson, [1790].

Fyfe, Andrew. *A system of anatomy and physiology, with the comparative anatomy of animals*. 3 vols. Edinburgh: William Creech, 1795.

Gadebusch, S. V. *De affectione hypochondriaca*. 1685.

[Galen]. *Galen on the Usefulness of the Parts of the Body*. Trans. Margaret Tallmadge May. Ithaca: Cornell University Press, 1968.

Galt, John. *The Life of Lord Byron*. London: Colburn and Bentley, 1830.

Garboe, P. S. *Experimenta quaedam circa malum hypochondriacum*. Halle, 1762.

García Ballester, Luis. *Galen and Galenism: Theory and Medical Practice from Antiquity to the European Renaissance*. Aldershot: Ashgate, 2002.

Gartzwyler, T. W. *De bile atra, ejusque effectibus*. Leiden, 1742.

Gay, John. *Fables*. London: Tonson, 1727.

Gay, Peter (ed.). *Eighteenth Century Studies: Presented to Arthur M. Wilson*. Hanover: University Press of New England, 1972.

Gazzaniga, M. S. *The Cognitive Neurosciences*. Cambridge, MA: MIT Press, 1995.

Gazzaniga, M. S. *Conversations in the Cognitive Neurosciences*. Cambridge, MA: MIT Press, 1996.

Gazzaniga, M. S. *The Mind's Past*. Berkeley and Los Angeles: University of California Press, 1998.

Geiger, M. *Microcosmus hypochondriacus, siue de melancholia hypochondriaca tractatus*. Munich, 1652.

Geison, G. *Michael Foster and the Cambridge School of Physiology: the Scientific Enterprise of Late Victorian Society.* Princeton: Princeton University Press, 1978.

Genser, J. A. *Dissertatio pathologiam mali hypochondriaci inquirens.* Wittenberg, 1797.

Georget, E. J. *De la Physiologie du Sytème Nevrose.* Paris: H. Baillière, 1821.

Georgii, A. *The Movement Cure.* London: H. Baillière, 1852.

Giacchi, Oscar. *L'isterismo e l'iponchondria avvero il malo nervosa ... Giudizii fisio-clinici-sociali.* Milan, 1875.

Gibbs, James. *Observations of various eminent cures of ... distempers call'd the King's Evil.* London, 1712.

Gibbons, A. C. and Heller, G. 'Music Therapy in Haendel's England: Browne's *Medicina Musica* (1729).' *College Music Symposium* 25 (1985): 59–72.

Gijswijt-Hofstra, Marijke and R. Porter (eds). *Cultures of Neurasthenia from Beard to the First World War.* Clio Medica 63. Amsterdam: Rodopi, 2001.

Gilchrist, Ebenezer. *The Use of Sea Voyages in Medicine.* London: Printed for A. Millar, 1757.

Gillespie, Charles C. *The Edge of Objectivity: an Essay in the History of Scientific Ideas.* Princeton: Princeton University Press, 1960.

Gillespie, R. D. 'Hypochondria: Its Definition, Nosology and Psychopathology.' *Guy's Hospital Rep.* 78 (1928): 408–60.

Giraldus, M. 'De singulari sensibilitate hypochondriacorum ejusque causis.' Göttingen, 1749.

Girard, H. *Considérations physiologiques er pathologiques sur les affections nerveuses, dites hysteriques.* 1841.

Glaser, Gilbert H. 'Epilepsy, Hysteria and Possession.' *Journal of Nervous Mental Disorders* 166: 4 (April 1978): 268–74.

Glynn, Ian. 'Two Millennia of Animal Spirits.' *Nature* 402 (6760) 25 November 1999: 353.

Goldstein, Jan. 'The Wandering Jew and the Problem of Psychiatric Anti-Semitism in Fin-de-Siècle France.' *Journal of Contemporary History* 20 (1985): 521–52.

Goldstein, Jan. *Console and Classify: the French Psychiatric Profession in the Nineteenth Century.* Cambridge: Cambridge University Press, 1987.

Goldstein, Laurence. *The Male Body: Features, Destinies, Exposures.* Ann Arbor: University of Michigan Press, 1994.

Golinski, Jan. *Science as Public Culture.* Cambridge: Cambridge University Press, 1992.

Gordon, John. *Physiology and the Literary Imagination.* Gainesville: University Press of Florida, 2002.

Goreau, Angeline. *Reconstructing Aphra: a Social Biography of Aphra Behn.* New York: The Dial Press, 1980.

Gosse, Edmund. *A Short History of Modern English Literature.* London: Macmillan, 1897.

Gosse, Edmund. *A History of Eighteenth Century Literature.* London: Macmillan, 1907.

Gosse, Philip. *Dr. Viper: The Querulous Life of Philip Thicknesse.* London: Cassell, 1952.

Gottschalk, Louis R. *Jean Paul Marat: a Study in Radicalism.* New York: Allen & Unwin, 1927.

Goubert, Jean Pierre. 'Malades et médecins en Bretagne.' University of Haute-Bretagne, 1974.

Gouk, Penelope. *Musical Healing in Cultural Contexts*. Aldershot: Ashgate, 2000.

Goulard, Thomas. *A Treatise on . . . Lead*. London: Printed for P. Elmsly, 1773.

Gould, Stephen Jay. 'The Analogistic Tradition from Anaximander to Bonnet.' *Ontogeny and Phylogeny*. Harvard: Harvard University Press, 1977, pp. 13–32.

Gould, Stephen Jay. *The Hedgehog and the Fox*. London: Jonathan Cape, 2003.

Graham, John. 'Lavater's Phsiognomy in England.' *JHI* XXII (1961): 561–72.

Grande, Francisco and Maurice B. Visscher (eds). *Claude Bernard and Experimental Medicine: Collected Papers from a Symposium*. Cambridge, MA: Schenkman Publishing Co. Inc., 1967.

Grayson, John. *Nerves, Brain and Man*. London: Phoenix House, 1961.

Greene, Robert. *The Principles of the Philosophy of the Expansive & Contractive Forces*. Cambridge, 1727.

Gregory, John. *A Comparative View of the State and Faculties of Man with Those of the Animal World*, 4th edn. London: J. Dodsley, 1777.

Gregory, John. *The Economy of Nature Explained and Illustrated on the Principles of Modern Philosophy*. London, 1796.

Griffith, Elizabeth. *Lady Juliana Harley II*. London, 1776.

Griffith, Richard. *Something New by Automathes*. Edinburgh: E. C. Dilly, 1772.

Griffiths, Paul E. *What Emotions Really Are*. Chicago: Chicago University Press, 1997.

Groddeck, George. *The Book of the It*. Trans. M. E. Collins. London: Vision Press, 1950.

Groddeck, George. *World of Man*. New York: Vision Press, 1951.

Groneman, Carol. 'Nymphomania: The Historical Construction of Female Sexuality.' *Signs* 19 (1994): 337–67.

Gross, C. G. *De morbis imaginariis hypochondriacorum*. Göttingen, 1755.

Grosskurth, Phyliss (ed.). *The Memoirs of John Addington Symonds*. New York: Random House, 1984.

Grossman, Carl M. *The Wild Analyst: the Life and Work of George Groddeck*. New York: Braziller, 1965.

Grosvenor, Benjamin. *Health, An Essay on its nature, value, uncertainty, preservation and best improvement*. London: A. Wembley, 1716.

Guer, Jean Antoine. *Histoire critique de l'âme des bêtes*. Amsterdam: F. Changuion, 1749.

Guidott, Thomas. *An Apology for the Bath*. London, 1724.

Guidott, Thomas. *The Lives and Characters of the Physicians of Bath*. London, 1724.

Gully, James. *An Exposition of the Symptoms, Essential Nature and Treatment of Neuropathy, or Nervousness*. London, 1837.

Guthkelch, A. C. et al. *A Tale of a Tub . . . and the Mechanical Operation of the Spirit . . .* Oxford: Clarendon Press, 1958.

Guyenot, Emile. *Les Sciences de la vie au xvii et xviii siècles. l'idée d'évolution*. Paris, 1942.

Haffner, A. *De hypochondriasi ut morbo coenaesthesis*. 1808.

Haghen, C. van der. *De melancholia hypochondriaca*. Leiden, 1715.

Hagstrum, Jean H. *Sex and Sensibility: Ideal and Erotic Love from Milton to Mozart*. Chicago: University of Chicago Press, 1980.

366   Bibliography

Haigh, Elizabeth. 'Vitalism, the Soul and Sensibility.' *JHM* 31 (1976): 30–41.

Hale, Thomas. *An account of several new inventions and improvements now necessary for England.* London, 1691.

Hall, M. *On the mimoses: or, A descriptive, diagnostic and practical essay on the affections usually denominated dyspeptic, hypochondriac, bilious, nervous, chlorotic, hysteric, spasmodic, etc.* London: Longmen [& others], 1818.

Hall, Marshall. *Essays on Disorders of the Digestive Organs and General Health.* Keene, NH, 1823.

Haller, Albrecht von. *On the Sensible and Irritable Parts of Animals.* Trans. Tissot. London, 1755.

Haller, Albrecht von. *First Lines of Physiology.* London, 1767.

Haller, Albrecht von. *Epistolarum ab eruditis viris ad Alb. Hallerum scriptarum pars 1,* 6 vols. Bern, 1773–5.

Halsband, Robert, ed. *The Complete Letters of Lady Mary Wortley Montagu,* 3 vols. Oxford: Clarendon Press, 1965–7.

Hamilton, E. 'Mr. Lewes's Doctrine of Sensibility.' *Mind* 4 (1879): 256–61.

Hanen, Marsha P. et al. (eds). *Science, Pseudo-Science & Society: Essays.* Waterloo, Ontario, Canada: Wilfrid Laurier University Press, 1980.

Hanslick, Eduard. *Vom Musikalisch-Schönen. Translated in 1891 by Gustav Cohen as: The Beautiful in Music.* Indianapolis: Bobbs-Merrill Co., 1957.

Hardy, Barbara. 'Dickens and the Emotions.' *Nineteenth-Century Fiction* 24 (1970): 449–66.

Harley, John. *The Old Vegetable Neurotics: Hemlock, Opium, Belladonna . . . and their Physiological Action.* London: Macmillan, 1869.

Harris, Frances. *Transformations of Love: the Friendship of John Evelyn and Margaret Godolphin.* Oxford: Oxford University Press, 2002.

Harris, R. W. *Romanticism and the Social Order 1780–1830.* London: Blandford, 1969.

Hartley, David, M. D. *Observations on Man, his frame . . . pt. 1: Observations on the frame of the human body and mind, and on their mutual connexions and influences.* London, 1749.

Hartley, Lucy. *Physiognomy and the Meaning of Expression.* Cambridge: Cambridge University Press, 2001.

Harvey, Elizabeth D. (ed.). *Sensible Flesh: On Touch in Early Modern Culture.* Philadelphia: University of Pennsylvania Press, 2002.

Harvey, William. *Lectures on the Whole of Anatomy: an Annotated Translation of Prelectiones anatomiae universalis.* Ed. and trans. C. D. O'Malley et al. Berkeley: University of California Press, 1961.

Haslam, Fiona. *From Hogarth to Rowlandson: Medicine and Art in Eighteenth-Century Britain.* Liverpool: Liverpool University Press, 1996.

Haslam, John. *Illustrations of Madness.* London: Routledge, 1810.

Haslam, John. *Observations on Madness & Melancholy.* 1809.

Haussonville, Le Comte d'. *Madame de La Fayette.* Paris, 1891.

Hawes, Clement. *Mania and Literary Style.* Cambridge: Cambridge University Press, 1996.

Haweis, H. R. (Hugh Reginald). *Music and Morals.* London: Daldy, Isbister & Co., 1876.

Hay, William. *Deformity: an Essay,* 3rd edn. London: Dodsley, 1755.

Hebb, Donald. *The Organization of Behaviour: a Neuropsychological Theory.* New York: Wiley, 1949.

Hebb, Donald. *The Conceptual Nervous System*. Oxford: Pergamon Press, 1982.

Heberden, William. 'Miscellaneous items from the William Heberden Collection, presented to the Countway Library, Harvard University, by G. I. W. Cottam, including letters, and observations on infection, nervous disorders...' [no date; manuscripts in the Royal College of Physicians, London]

Helvetius, C. A. *Essays on the Mind*. London: Printed for James Cundee, 1810.

Hemlow, Joyce. *The History of Fanny Burney*. Oxford: Clarendon, 1958.

Henaff, Marcel. *Sade: l'invention du corps libertin*. Paris: Presses Universitaires de France, 1978.

Henderson, Fergus. 'Novalis, Ritter and "Experiment": A Tradition of Active Empiricism.' In E. Schaffer (ed.), *The Third Culture*. Berlin: De Gruyter, 1997, pp. 153–69.

Herfelt, H. G. *De affectione hypochondriaca*. Duisberg, 1678.

Hesse, Mary B. *Models and Analogies in Science*. Notre Dame: University of Indiana Press, 1966.

Heyman, J. G. *De praecipuo literatorum morbo affectu hypochondriaco*. Leiden, 1732.

Hick, John. *Biology and the Soul: Arthur Stanley Eddington Memorial Lecture Number 25*. Cambridge: Cambridge University Press, 1972.

Highmore, Nathaniel. *De hysterica et hypochondriaca passione: Responsio epistolaris ad Doctorem Willis*. London, 1670.

Hill, John Spencer (ed.). *Imagination in Coleridge*. London: Macmillan, 1978.

Hill, John. *Observations on the Greek and Roman Classics*. London, 1753.

Hill, John. *The Actor*. London, 1750.

Hill, John. *The Construction of the Nerves and Causes of Nervous Disorders*. London: R. Baldwin, 1758.

Hill, John. *The Fabrick of the Eye*. London: Printed for J. Waugh...and M. Cooper, 1758.

Hill, John. *The Virtues of Wild Valerian in Nervous Disorders*. London: R. Baldwin, 1758.

Hillis, Fredrick W. and Harold Bloom (eds). *From Sensibility to Romanticism*. Oxford: Oxford University Press, 1965.

Hillman, David and Carla Mazzio (ed.). *The Body in Parts: Fantasies of Corporeality in Early Modern Europe*. London: Routledge, 1997.

Hintzsche, E. 'Neue Funde zum Thema: "L'homme Machine" und Albrecht Haller.' *Gesnerus* 25 (1968): 135–66.

Hippocrates. *Places in Man*. Ed. and trans. Elizabeth M. Clark Oxford: Clarendon Press, 1998.

Hippocrates. *The Prognostics and Prorrhetics*. Trans. John Moffat. London: T. Bensley, 1788.

Hoffman, Paul. *La Femme dans la pensée des lumières*. Paris: Ophris, 1977.

Holbach, P. H. *The System of Nature*. Boston: J. P. Mendum, 1889.

Hollander, Bernard. *Nervous Disorders of Women: the Modern Psychological Conception of their Causes, Effects, and Rational Treatment*. London: Kegan Paul, 1916.

Holmes, Richard. *Shelley: the Pursuit*. London: Weidenfeld and Nicolson, 1974.

Home, Henry [Lord Kames]. *Loose Hints upon Education...Concerning the Culture of the Heart*. Edinburgh, 1781.

Home, Henry [Lord Kames]. *Sketches of the History of Man, Enlarged by Additions and Corrections of the Author*. Edinburgh, 1788.

Howe, Eric Graham. *Invisible Anatomy: a Study of Nerves, Hysteria and Sex*. London: Faber, 1944.

Hughes, J. Trevor. *Thomas Willis 1621–1675: His Life and Work*. London: Royal Society of Medicine, 1991.

Hume, David. *Four Dissertations*. London, 1757.

Hume, David. *Of the Delicacy of Taste and Passion*. London, 1757.

Humphrey, N. *A History of the Mind*. New York: Simon and Schuster, 1992.

Humphrey, N. *How to Solve the Mind–Body Problem*. Thorverton, UK: Imprint Academic, 2000.

Hunt, Leigh. *The Indicator*. London, 1834.

Hunter, John. *Of the heat, &c. of animals and vegetables. By Mr. John Hunter, F. R. S. Read at the Royal Society, June 19, and Nov. 13, 1777*. London, 1778.

Hurley, Kelly. *The Gothic Body*. Cambridge: Cambridge University Press, 1996.

Hurst, M. *Maria Edgeworth and the Public Scene: Intellect, Fine Feeling and Landlordism in the Age of Reform*. Miami: University of Miami Press, 1969.

Hutcheson, Frances. *An Essay on the Nature and Conduct of the Passions and Affections*. London: J. Knapton, 1728.

Huxham, John. *An Essay on Fevers*. London, 1750.

Illie, Paul. 'The Epistemology of Aether.' *Age of Minerva* 2 (1994): 284–5.

Iltis, Cardyn. 'D'Alembert and the Vis Viva Controversy.' *Studs. Hist. Phil. Science* 1 (1970): 135–44.

Inman, Thomas. *On myalgia: its nature, causes, and treatment, being a treatise on painful and other affections of the muscular system, which have been... mistaken for hysterical, inflammatory, hepatic, uterine, nervous, spinal, or other diseases*. London, 1860.

Ireland, William W. 'The Insanity of King Louis II of Bavaria.' *Through the Ivory Gate: Studies in Psychology and History*. Edinburgh, 1889.

Isenflamm, J. F. *Versuch einiger praktischen anmerkungen über die Nerven zur Erläuterung verschiedener Krankheiten derselben, veruchmlich hypochondrisch und hysterischer*. Autälle, 1774.

Isenflamm, J. F. *Uber die Nerven*. Erlangen, 1774.

Ishizuka, Hisao. 'William Blake and Eighteenth-Century Medicine.' PhD thesis, University of Essex, 1999.

Isler, Hansruedi. *Thomas Willis: 1621–1675 – Doctor and Scientist*. New York: Hafner, 1968.

Israel, J. M. *De hypochondriaco malo monita quaedam*. Hamburg, 1798.

Jack, Ian. 'Phrenology, Physiognomy, and Characterization in the Novels of Charlotte Bronte.' *Bronte Society Transactions* 15 (1966): 377–91.

Jackson, Stanley. *Melancholia and Depression*. New Haven: Yale University Press, 1986.

Jackson, Stanley. 'Two Sufferers' perspectives on melancholia'. In E. R. Wallace and L. C. Pressley (eds), *Essays in the History of Psychiatry*. New York: Harcourt Brace 1980, pp. 58–71.

Jackson, Stanley. 'A History of Melancholia and Depression.' *Depression and Stress* 1(1) (1995): 3–42.

Jackson, Wallace. *Immediacy: the Development of a Critical Concept from Addison to Coleridge*. Amsterdam: Rodopi, 1973.

James, Ralph. *The Fashionable Lady: or, Harlequin's Opera*. London: J. Watts, 1730.

James, Robert. *A Treatise on Canine Madness.* London: Printed for J. Newbery, 1760.

James, William. *Principles of Psychology.* New York: Henry Holt, 1890.

Jarrett, Derek. *England in the Age of Hogarth.* London: Hart-Davis, 1974.

Jauffret, Louis François. *The wonders of the human body, a familiar introduction to anatomy and physiology.* London, 1810.

Jay, Martin. *Cultural Semantics: Keywords of Our Time.* London: Athlone, 1998.

Jaynes, Julian. *The Origin of Consciousness in the Breakdown of the Bicameral Mind.* Boston: Houghton Mifflin, 1976.

Jenkins, Richard. *The Victorians and Ancient Greece.* Cambridge, MA: Harvard University Press, 1980.

Jessenwanger, J. N. *Dissertatio sistens morbum hypochondriacum et hystericum.* Halle, 1778.

Johnson, Alexander. *A Short Account of a Society at Amsterdam . . . For the Recovery of Drowned Persons.* London: John Nourse, 1773.

Johnson, James. 1827. *An Essay on Morbid Sensibility of the Stomach and Bowels, as the . . . Cause . . . of Indigestion.* London, 1827.

Johnson, Samuel. *A Dictionary of the English Language. . . .* London, 1827.

Johnson, Samuel. 'Essays from the *Rambler, Adventurer,* and *Idler.'* Ed. Walter Jackson Bate. New Haven: Yale University Press, 1968.

Jorden, Edward. *A Brief Discourse of a Disease Called the Suffocation of the Mother.* London: John Windet, 1603.

Jurin, James. *The Correspondence of James Jurin, 1684–1750.* Ed. A. Rusnock. Amsterdam, NL and Atlanta, GA: Rodopi, 1996.

Kahn, L. J. *Nervous Exhaustion: its Cause and Cure Comprising a Series of Lectures on Debility and Disease . . . with Practical Information on Marriage, its Obligations and Impediments.* New York, 1870.

Kallich, M. *The Association of Ideas and Critical Theory in Eighteenth Century England.* The Hague: Mouton, 1970.

Kalm, Peter. *Kalm's Account of his Visit to England on his Way to America in 1748.* London, 1892.

Kenney, William. 'Addison, Johnson, and the Energetic Style.' *Studia Neophilogica* 33 (1961): 103–14.

King, Lester S. 'Medicine in 1695: Friedrich Hoffmann's "Fundamenta Medicinae."' *Bull. Hist. Med.* 43 (1969): 17–29.

Kinneir, David. *An Essay on the Nerves; and the Doctrine of the Animal Spirits Rationally Considered.* London, 1737. [second edition 1739]

Kinneir, David. *A New Essay on the Nerves and the Doctrine of Animal Spirits.* London, 1738.

Kirkland, Thomas, MD. *A Treatise on Child-Bed Fevers . . . To Which are prefixed two dissertations, the one on the Brain and Nerves; the Other on the Sympathy of the Nerves, and on Different Kinds of Irritability.* London: Baldwin & Dawson, 1774.

Kleinman, Arthur. *Patients and Healers in the Context of Culture: an Exploration of the Borderland between Anthropology, Medicine, and Psychiatry.* Berkeley and London: University of California Press, 1980.

Kleinman, Arthur. *Social Origins of Distress and Disease: Depression, Neurasthenia, and Pain in Modern China.* Ann Arbor: Universiy of Michigan Press, 1998.

Klibansky, Raymond. *Saturn and Melancholy.* London: Nelson, 1964.

Klotz, Oskar. 'Albrecht von Haller (1708–1777).' *Annals of Medical History* 8 (1936): 10–26.

Knapp, Lewis M. (ed.). *The Letters of Tobias Smollett.* Oxford: Clarendon Press, 1970.

Knight, David M. *Ordering the World: a History of Classifying the World.* London: Burnett, 1981.

Knight, Henrietta [Lady Luxborough]. *Letters of Lady Luxborough . . . to the poet William Shenstone.* London, 1775.

Knox, Vicesimus. *Winter Evenings. Or, Lucubrations on Life and Letters.* London, 1790.

Koch, D. A. *De infarctibus vasorum in infimo ventre ceu caussa plurium pathematum chronicorum, speciatim eorum, quae sub mali hypochondriaci nomine veniunt.* Strasburg, 1752.

Kohen, L. I. *De morbo hypochondriaco.* Göttingen, 1729.

Kolle, Kurt. *Grosse Nervenartze,* 2 vols. Stuttgart: Thieme, 1956–63.

Kuhn, Thomas S. *The Structure of Scientific Revolutions.* Chicago: University of Chicago Press, 1962; rev. edn, 1970.

Kuhn, Thomas S. *The Essential Tension: Selected Studies in Scientific Tradition and Change.* Chicago: University of Chicago Press, 1977.

Ladee, G. A. *Hypochondriacal Syndromes.* Amsterdam: Elsevier, 1966.

Lamb, Charles. *Selected Prose.* London: Penguin Press, 1985.

Lamb, William. 'On the Melancholy of Tailors.' *Works of C. Lambe.* London, 1983.

La Mettrie, Julien Offray de. *L'école de la volupté.* n. p., 1747.

La Mettrie, Julien Offray de. *L'Homme Machine: a Study in the Origins of an Idea: Critical Edition with an Introductory Monograph and Notes by Aram Vartanian.* Princeton: Princeton University Press, 1960.

La Mettrie, Julien Offray de. *Man a machine . . . Translated from the French of Mons. de La Mettrie,* 2nd edn. London: G. Smith, 1750.

La Mettrie, Julien Offray de. *Le traité de l'âme de La Mettrie with commentary by Theodorus Hendrikus Maria Verbeek.* Utrecht: OMI/Grafisch Bedrijf, 1988.

Langhorne, John. *Letters on Religious Retirement, Melancholy, and Enthusiasm.* London, 1762.

Langstaff, George. *The Village doctor; or, The art of curing diseases rendered familiar and easy: with select receipts, from the practice of the most eminent physicians and surgeons . . . 2nd edn.* London: Knight and Lacey, 1825.

Laqueur, Thomas. *Making Sex: Body and Gender from the Greeks to Freud.* Cambridge, MA: Harvard University Press, 1990.

Lavater, Johann Caspar. *Essays on physiognomy for the promotion of the knowledge and the love of mankind/written in the German language by J. C. Lavater; abridged from Mr. Holcrofts translations . . . London, 1750.*

Lawrence, Christopher. 'Ornate physicians and learned artisans: Edinburgh medical men, 1726–1776.' *William Hunter and the Eighteenth-Century Medical World.* Eds. W. F. Bynum and R. Porter. New York: Cambridge University Press, 1985, pp. 153–75.

Lawrence, Christopher. 'The Nervous System and Society in the Scottish Enlightenment.' *Natural Order: Historical Studies of Scientific Enlightenment.* Ed. Barry Barnes and Steven Shapin. Beverly Hills: Sage Publications, 1979, pp. 19–40.

Laycock, Thomas. *A Treatise on the Nervous Diseases of Women; Comprising an Inquiry into the Nature, Causes, and Treatment of Spinal and Hysterical Disorders.* London: Longman, 1840.

Le Pois, Charles (Carolus Piso). *Selectiorum observationum et consiliorum de prae-tervisis hactenus morbis affectibusque praeter naturam, ab aqua, seu serosa colluvie et diluvie ortis, liber singularis,* new edn. Leiden, 1714.

Leavis, F. R. 'Johnson as Critic.' *Samuel Johnson: a Collection of Critical Essays.* Ed. D. J. Greene. Englewood Cliffs: Prentice-Hall, 1965.

Leavis, F. R. *Nor Shall my Sword: Discourses on Pluralism, Compassion and Social Hope.* London: Chatto & Windus, 1972.

LeDoux, J. E. *The Emotional Brain: the Mysterious Underpinnings of Emotional Life.* New York: Simon & Schuster, 1996.

LeDoux, J. E. 'Emotion Circuits in the Brain.' *Annual Review of the Neurosciences* 23 (2000): 155–84.

LeDoux, J. E. *Synaptic Self: How Our Brains Become Who We Are.* New York: Viking, 2002.

Lee, Harriet. *The Errors of Innocence.* London, 1786.

Lee, Sophia. *The Recess.* London: T. Cadell, 1783.

Legée, Georgette. 'Influence du vitalisme montpelliérain sur la neurophysiologie de Pierre Flourens.' *Hist. et Nat.* 21 (1982): 13–48.

Leland, Charles G. *The Alternate Sex, or, the Female Intellect in Man and the Masculine in Woman.* London: Hutchimson, 1904.

Lepenies, Wolf. *Melancholy and Society.* Trans. Jeremy Gaines and Doris Jones. Cambridge, MA: Harvard University Press, 1992.

Lewis, I. *Nervousness: its Causes and Remedies, Considered Religiously, Philosophically, and Medically.* London, 1864.

Lidderdale, Jane and Mary Nicolson. 'Mr. Joyce's Dreadful Eye Attack.' *James Joyce Quarterly* 7 (1970): 186–90.

Lilar, Suzanne. *Aspects of Love in Western Society.* London: Thames and Hudson, 1965.

Lind, L. R., ed. and trans. *Studies in Pre-Vesalian Anatomy.* Philadelphia: American Philosophical Society, 1975.

Lindberg, David C. *Theories of Vision from al-Kindi to Kepler.* Chicago: University of Chicago Press, 1976.

Lindeboom, G. A. 'Pitcairne's Leyden Interlude Described from the Documents.' *Annals of Science* 19 (1963): 273–84.

Lindeboom, G. A. 'Boerhaave's Concept of the Basic Structure of the Body.' *Clio Medica* 5 (1970): 203–8.

Lobb, Theophilus. *Medical Principles and Cautions.* London: J. Buckland, 1753.

Logan, Peter. *Nerves and Narrative: a Cultural History of Hysteria in Nineteenth-Century Prose.* Berkeley and Los Angeles: University of California Press, 1997.

Lorry, Anne-Charles. *Sur les mouvemens du cerveau et de la dure-mere: premier mémoire, sur le mouvement des parties contenues dans le crâne, considérées dans leur état naturel.* Paris, 1760.

Lorry, Anne-Charles. *De Melancholia et Morbis Melancholicis.* Paris: P. Guillelmum Cavelier, 1765.

Love, R. 'Some Sources of Hermann's Boerhaave's Doctrine of Fire.' *Ambix* 19(1972): 157–74.

Love, R. 'Hermann Boerhaave and the Element-Instrument-Concept of Fire.' *Annals of Science* 31 (1974): 547–59.

Lowenberg, Richard. 'The Significance of the Obniaus: an Eighteenth-Century Controversy on Pyschosomatic Principles.' *Bull. Hist. Med.* X (1941): 666–79.

Lowenthal, Gustavus Julius. *Dissertatio inauguralis medico-practica de hysteria.* Moscow, 1825.

Ludwig, Christian Friedrich. *Scriptores neurologici minores selecti sive Opera minora ad anatomiam physiologiam et pathologiam nervorum spectantia,* 4 vols. Leipzig, 1791–5.

Lumsden, C. E. et al. *Pathology of Tumours of the Nervous System . . . with a Chapter on the Study by Tissue Culture of Tumours of the Nervous System.* London: Edward Arnold, 1971.

Lutz, Tom. *American Nervousness, 1903: an Anecdotal History.* Ithaca: Cornell University Press, 1991.

Lyles, Albert M. *Methodism Mocked: the Satiric Reaction to Methodism in the Eighteenth Century.* London: Epworth, 1960.

Mackenzie, Henry. *The Man of Feeling.* London, 1771.

Mackenzie, Henry. *Julia de Roubigné, a tale. In a series of letters. Published by the author of The man of feeling, and The man of the world . . . .* London, 1777.

Mackenzie, Henry. *The Works of Henry MacKenzie, Esq. in Eight Volumes,* vol. 7 of 8. Edinburgh: James Ballantyne and Co., 1808.

McKenzie, John Grant. *Nervous Disorders and Religion: a Study of Souls in the Making.* London: Allen and Unwin, 1951.

Machamer, Peter K. et al. (eds). *Theory and Method in the Neurosciences.* Pittsburgh: University of Pittsburgh, 2001.

Madden, Richard Robert. 'Nervous States – Inspired Religious Vision.' In *Phantasmata or Illusions and Fanaticisms of Protean Forms,* 2 vols. London: T. C. Newby, 1857. 2: 517–19.

Magendie, François. *Anatomie des systems nerveux des animaux.* Paris, 1825.

Maimonides, Moses. *The Medical Aphorisms of Moses Maimonides,* 2 vols. Ed. and trans. F. Rosner and S. Munter. New York: Yeshiva University Press, 1970.

Malouf, David. *An Imaginary Life.* London: Picador, 1978.

Malcolm, J. P. *Anecdotes of the Manners and Customs of London during the Eighteenth Century.* London, 1808.

Mandeville, Bernard. *A Treatise of the Hypochondriack and Hysterick Diseases.* London, 1730.

Mandler, George. *Consciousness Recovered.* Amsterdam: John Benjamins, 2002.

Manningham, Sir Richard. *The Symptoms, Nature, Causes and Cure of the Febricula, or little fever: commonly called the nervous or hysteric fever . . . the fever on the spirits; vapours, hypo, or spleen.* London: 1746.

Martensen, Robert. '"Habit of Reason": Anatomy and Anglicanism in Restoration England.' *Bulletin of the History of Medicine* 66(4) (1992): 511–35.

Martin, B. (Freke). *A Supplement . . . on a Rhapsody of Adventures of a Modern Knight-errant in Philosophy.* London, 1746.

Marx, Karl Friedrich H. *On the Decrease of Disease Effected by the Progress of Civilization.* Trans. R. Willis, M. D. London, 1844.

Marx, Karl Friedrich H. *The Moral Aspects of Medical Life.* Trans. James Macliness. London, 1846.

Mason, Simon. *Practical Observations in Physic . . . wherein is exhibited the ætiology, or the rise and nature of the most prevalent distempers, with a plain, rational and concise method of treating them . . .* Birmingham: T. Warren, 1757.

Matthews, R. C. O. *Animal Spirits: the John Maynard Keynes British Academy Lecture in Economics.* London: The British Museum, 1986.

Mayer, H. C. *De splenetico malo.* Göttingen, 1719.

Mazzolini, Renato G. 'Schemes and Models of the Thinking Machine (1662–1762).' In Pietro Corsi (ed.), *Enchanted Loom: Chapters in the History of Neuroscience.* New York: Oxford University Press, 1991, pp. 68–83.

McElroy, D. D. *Scotland's Age of Improvement: a Survey of Eighteenth-Century Literary Clubs and Societies.* Washington: Washington State University Press, 1969.

McGuire, J. E. 'Atoms and the "Analogy of Nature": Newton's Third Rule of Philosophizing.' *Studies in the History and Philosophy of Science* 1 (1970): 3–58.

McIver, George. *Neuroomia: a New Continent.* London, 1894.

McKenzie, A. T. *Certain, Lively Episodes: the Articulation of Passion in Eighteenth-Century Prose.* Athens, GA and London: University of Georgia Press, 1990.

McKenzie, G. *Critical Responsiveness: a Study of the Psychological Current in Later Eighteenth-Century Criticism.* Berkeley and Los Angeles: University of California Press, 1949.

McKeown, Thomas. *The Role of Medicine: Dream, Mirage, or Nemesis?* Oxford: Blackwell, 1976; 2nd edn, 1979.

McKillop, A. D. *English Literature from Dryden to Burns.* New York: Appleton-Century-Crofts, 1948.

McLuhan, Marshall. *Through the Vanishing Point: Space in Poetry and Painting.* New York: Harper and Row, 1968.

McMaster, Juliet. 'Uncrystalized Flesh and Blood: The Body in *Tristram Shandy.*' *Eighteenth-Century Fiction* 2(1990): 197–214.

Mead, Richard. *A Treatise concerning the influence of the sun and moon upon human bodies, and the diseases thereby produced.* London: J. Brindley, 1748.

Mead, Richard. *The Medical Works of Dr. Richard Mead.* Edinburgh: A. Donaldson and J. Reid, 1765.

Medawar, Sir Peter. 'Vitalism.' In Allan Bullock (ed.), *The Harper Dictionary of Modern Thought.* New York: Harper & Row, 1988.

Mede, Joseph. *The Works of... Joseph Mede. Corrected and enlarged according to the author's own manuscripts [by J. Worthington].* London, 1672.

Mee, Jon. *Dangerous Enthusiasm: William Blake and the Culture of Radicalism in the 1790s.* Oxford: Clarendon Press, 1992.

Meersch, J. van der. *De hypochondria.* Leiden, 1817.

Mehnert, Henning. *Melancholie und Inspiration: Beitrage zur neueren Literaturgeschichte,* vol. 35. Heidelberg: Winter, 1978.

Meighan, Sir Christopher, MD *A Treatise of the Nature and Powers of Bareges's Baths and Waters, wherein their superiour Effects for the Cure of Gun-Shot Wounds, with all their complications of inveterate Ulcers, Caries's [sic] of the Bones, Fistula's, Contractions of the Nerves and Tendons... are clearly demonstrated... with an Enquiry into the Cause of the Heat of Thermal Waters in General.* London, 1742.

Meineke, A. C. *De vera morbi hypochondriaci sede indole, ac curatione.* 1719.

Mewis, Eberhardus Wilhelmus Ludovicus. *Dissertatio inauguralis medica de differentia passionis hystericae a morbis convulsivis reliquis.* Duisburg, 1780.

Meyers, J. A. *Tuberculosis, a Half Century of Study and Conquest.* St Louis: Warren H. Green, 1970.

Micale, Mark S. (ed.). *The Mind of Modernism: Medicine, Psychology, and the Cultural Arts in Europe and America, 1880–1940.* Stanford, CA: Stanford University Press, 2004.

Midriff, John. *Observations on the spleen and vapours, containing cases of persons afflicted since the fall of South sea stocks.* London, 1721.

Miller, Karl. 'Andante Capriccioso.' *Authors*. Oxford: Clarendon Press, 1989.

Moir, John. *Anatomical Education in a Scottish University, 1620: an Annotated Translation of the Lecture Notes of John Moir*. Ed. and trans. R. K. French. Aberdeen: Equipress, 1975.

Moises, Hugh. *Treatise on the Blood; or, General Arrangement of Many Important Facts Relative to the Vital Fluid, With Some Observations on the Theory of Animal Heat . . . from the Inductions of Modern Chemistry*. London, 1794.

Moll, Albert, ed. *Berühmte Homosexuelle: Grenzefragen des Nerven- und Seelenlebens*. Wiesbaden: Bergmann, 1910.

Monboddo, James Burnett, Lord. *Of The Origin and Progress of Language*. Edinburgh, 1773–92.

Mongin-Montrol, C. *Sur l'hypochondrie*. Paris, 1823.

Monro, Alexander. *Experiments on the nervous system, with opium and metalline substances; made chiefly with the view of determining the nature and effects of animal electricity*. Edinburgh: Adam Neill, 1793.

Monro, Alexander. *A system of anatomy and physiology, with the comparative anatomy of animals*. Edinburgh: William Creech, 1795.

Montegut, Émile. 'Les Confidences d'un hypochondriaque.' *Types littéraires et fantaisies esthetiques*. Paris, 1882.

Moore, Cecil. *Backgrounds to English Literature 1700–1760*. Minneapolis: University of Minnesota Press, 1953.

Moore, John. *A View of Society and Manners in Italy: With Anecdotes Relating to Some Eminent Characters*. Boston: Belknap, 1792.

Moore, John. *Medical Sketches in two parts*. Providence: Carter and Wilkinson, 1794.

Moorman, Lewis Jefferson. *Tuberculosis and Genius*. Chicago: Chicago Press, 1940.

Moravia, S. 'From Homme Machine to Homme Sensible: Changing Eighteenth-Century Models of Man's Image.' *Journal of the History of Ideas* 39 (1978): 45–60.

More, Hannah. *Sacred Dramas . . .* London, 1782.

Morgagni, Giovanni Battista. *The Seats and Causes of Diseases investigated by anatomy . . .* London: A. Millar, 1769.

Morgan, Thomas. *The Mechanical Practice of Physick . . . and the Bellinian hypothesis of animal secretion and muscular motion . . . .* London: 1735.

Moriceau, François. *Traité des maladies des femmes grosses*. Paris: Chez Jean Geoffroy Nion . . ., 1712.

Moore, John. *Medical Sketches. With some original remarks on the nervous system and its interacting effects*. Printed for A. Strahan and T. Cadell, 1786.

Morris, John N. *Versions of the Self: Studies in English Autobiography from John Bunyan to John Stuart Mill*. New York: Basic Books, 1966.

Morton, Richard. *Phthisiologia: or, A treatise of consumptions. Tr. from the original*. London, 1694.

Mossdorff, J. F. *De valetudinariis imaginariis, von Menschen, die aus Einbildung kranck werden*. Halle, 1721.

Mosse, George L. *The Image of Man: The Creation of Modern Masculinity*. New York: Oxford University Press, 1996.

Motherby, George. *A New Medical Dictionary, or general repository of physic*. London: J. Johnson, 1755.

Mullan, John. 'Hypochondria and Hysteria: Sensibility and the Physicians.' *The Eighteenth Century: Theory and Interpretation* 25(2) (1984): 141–74.

Mullan, John. *Sentiment and Sociability: the Language of Feeling in the Eighteenth Century*. Oxford: Clarendon Press, 1990.

Müller, J. F. *Dissertatio sistens casum peculiarem de morbo motuum habituali ex imaginatione, sub ructuum schemate enato*. Halle, 1732.

Mullett, Charles F. (ed.) *The Letters of Dr. George Cheyne to the Countess of Huntingdon*. San Marino, CA: The Huntington Library, 1940.

Mullett, Charles F. (ed.). *The Letters of Doctor George Cheyne to Samuel Richardson (1723–1743)*. Columbia: University of Missouri Press, 1943.

Nadelhaft, J. 'The English Malady, Corrupted Humors, and Krook's Death.' *Studies in the Novel* 1 (1968): 230–9.

Nass, Lucien. *La névrose révolutionnaire/Docteurs Cabanès et L. Nass*. Paris: A. Michel, [1924].

Nass, Lucien. *Les névroses de l'histoire*. Paris, 1908.

Neale, John. *Practical Dissertations on Nervous Complaints and Other Diseases incident to the Human Body with a Historical Investigation of Their Causes and Cure in Which Are Interspersed Some Singular Cases*. London: printed for the author, 1788.

Newnham, W. *Essay on the Disorders incident to Literary Men: and on the best means of preserving their health, read before the Royal Society of Literature, Nov. 5, 1834; and dedicated, by permission, to the Lord Bishop of Salisbury*. London: John, 1836.

Nicolson, M. H. *Newton Demands the Muse*. Princeton: Princeton University Press, 1949.

Nicolson, M. H. *Mountain Gloom and Mountain Glory: the Development of the Aesthetics of the Infinite*. Ithaca: Cornell University Press, 1959.

Nollet, Jean Antoine. *Essai sur l'électricité des corps...* Paris: Guerin, 1746.

Nollet, Jean Antoine. *Leçons de physique expérimentale*. Paris: Guerin, 1749–64.

Nollet, Abbé (Jean Antoine). *Lectures in Experimental Philosophy*. Trans. John Colson. London: Printed for J. Wren, 1752.

Norton, Jane E. (ed.). *The Letters of Edward Gibbon*, 3 vols. London and New York: Macmillan, 1956.

Novalis, Friedrich. 'Nervenkunst' in *Philosophical writings; translated and edited by Margaret Stoljar*. Albany, NY: State University of New York Press, 1997.

Nutton, Vivian. *Galen on Prognosis: Edition, Translation, and Commentary*. Berlin: Akademie Verlag, 1979.

Nutton, Vivian (ed.). *Cambridge Conference on Galen, Problems and Prospects (1979)*. London: Wellcome Institute for the History of Medicine, 1981.

Nutton, Vivian (ed.). *On My Own Opinions: Galen; Edition, Translation and Commentary*. Berlin: Akademie Verlag. 1999.

Nutton, Vivian. *Ancient Medicine*. New York: Routledge, 2004.

O'Brien, Gordon W. 'The Genius and the Mortal Instruments: Mind and Body in the Romantic Imagination.' *The Minnesota Review* 6 (1966): 316–52.

O'Halloran, Sylvester. *A New Treatise on the Glaucoma*. Dublin: S. Powell, 1750.

O'Leary, John. 'Teachers Find Mozart Soothes the Savage Beast.' *Times*, 11 September 1997: 23–4.

Ochlitius, S. *De passione hypochondriaca*. Jena, 1666.

O'Malley, Donald C. *The Human Brain and Spinal Cord: a Historical Study Illustrated by Writings from Antiquity to the Twentieth Century 2nd ed., with a New Preface by Edwin Clarke*. San Francisco: Norman, 1996.

O'Malley, Donald C. *Leonardo da Vinci on the Human Body: the Anatomical, Physiological, and Embryological Drawings of Leonardo da Vinci*. New York: H. Schuman, 1952.

Oppenheim, Janet. *'Shattered Nerves': Doctors, Patients, and Depression in Victorian England*. Oxford: Oxford University Press, 1991.

Oppenheimer, Jane M. 'Aristotle as a Biologist,' *Scientia*. Milano: Arti Grafiche, 1971: 9–10.

Osterud, Erik. 'Electricity and Nerves: an Aesthetic Central Theme in Strindberg's "Ved Havbrynet."' *Kritik* 155/6 (2002): 37–46.

Otto, J. J. *De malo hypochondriaco*. 1722.

Paget, Sir James. 'Nervous Mimicry.' In Stephen Paget (ed.), *Selected Essays and Addresses by Sir James Paget*. London: Longmans, 1902, chap. 7, pp. 73–143.

Palante, G. *La Sensibilité Individualiste*. Paris: Bibl. de phil. Contemp., 1909.

Paley, William. *Natural Theology*. London: R. Faulkner, 1802.

Papineau, David. *Thinking about Consciousness*. Oxford: Clarendon Press, 2002.

Paracelsus. *Medicina diastatica or, sympatheicall mumie: abstracted from the works of Theophr. Paracelsus . . .* London, 1653.

Pargeter, William. *Observations on Maniacal Disorders*. Reading, 1792.

Park, Katharine. 'The Criminal and Saintly Body: Autopsy and Dissection in Renaissance Italy.' *Renaissance Quarterly* 47 (1994): 1–33.

Park, Katharine. 'The Life of the Corpse: Division and Dissection in Late Medieval Europe.' *Journal of the History of Medicine and Allied Sciences* 50 (1995): 111–32.

Park, Roberta J. 'Physiologists, Physicians, and Physical Educators,' *Journal of Sports History* 20 (1997): 28–60.

Parker, Patricia. 'Virile Style.' In. L. O. Aranye Fradenburg et al. (eds), *Premodern Sexualities*. New York and London, Routledge, 1996, pp. 201–22.

Parkinson, James. *Observations on Dr. Hugh Smith's Philosophy of Physic*. London, 1780.

Parsons, James. *Philosophical Observations on the Analogy between the Propagation of Animals and that of Vegetables in which are answered some objections against the indivisibility of the soul . . . with an Explanation of the Manner in which each Piece of a divided Polypus becomes another perfect Animal of the same Species*. London: Davis, 1752.

Paster, Gail Kern. 'Nervous Tension.' In. David Hillman and Carla Mazzio (eds), *The Body in Parts*. London: Routledge, 1997. 107–28.

Paulson, Ronald. *Popular and Polite Art in the Age of Hogarth and Fielding*. Notre Dame: University of Notre Dame Press, 1979.

Peacock, Thomas. *Nightmare Abbey*. London: Penguin, 1948.

Peart, Edward. *Physiology; or, an attempt to explain the functions and laws of the nervous system; the contraction of muscular fibres; and the constant and involuntary actions of the heart, the stomach, and organs of respiration, . . . To which are added, observations on the intellectual operations of the brain; . . . with remarks on the effects of poisons; and an explanation of the experiments of Galvani and others on animal electricity*. London: W. Miller, Murray and Highley, 1798.

Pelgrom, A. F. *De morbo hypochondriaco*. Leiden, 1759.

Penfield, William. 'Science, the Arts and the Spirit.' *Transactions of the Royal Society of Canada* 7 (1969): 73–83.

Pera, M. and W. R. Shea. *Persuading Science: the Art of Scientific Rhetoric*. Canton, MA: Science History, 1991.

Perfect, Wm. *Cases of Insanity... Hypochondriacal Affection...* London, 1781.

Perry, Charles. *A Mechanical Account and Explication of the Hysteric Passion... comprehending a general account of all other nervous diseases.* London, 1755.

Pettigrew, Thomas J. *Memoirs of the Life and Writings of J. C. Lettsom,* 3 vols. London: Nicholson and Bentley, 1817.

Philander, Misaurus. *The Honour of the Gout. Or, a rational discourse, demonstrating, that the gout is one of the greatest blessings which can befal [sic] mortal man.* London, 1720.

Philip, Alexander. *A Treatise on Indigestion and its Consequences, Called Nervous and Bilious Complaints.* London: T. and G. Underwood, 1823.

[Phyllosan]. *Phyllosan: a Doctor Telling a Couple to take Phyllosan to Compensate for the Stress of Modern Life.* n.p., 1937.

Pinero, J. M. Lopez. *Historical Origins of the Concept of Neurosis.* Trans. German Berrios. Cambridge: Cambridge University Press, 1983.

Pinto-Correia, Clara. *The Ovary of Eve: Egg and Sperm and Preformation.* Chicago: University of Chicago Press, 1997.

Piquer, Andres. *Discurso sobre la applicacion de la philosophia a los assuntos de religion para la juventud.* Madrid, 1757.

Plasha, Wayne W. 'The Social Construction of Melancholia in the Eighteenth Century: Medical and Religious Approaches to the Life and Work of Samuel Johnson and John Wesley,' DPhil thesis, University of Oxford, 1993.

Pomme, Pierre, M. D. *On Hysteric and Hypochondriac Diseases.* London, 1777.

Pomme. *Traité des Affections Vaporeuses de deux sexes, ou Maladies Nerveuses vulgairement apelés de nerfs.* Paris, 1782.

Popper, Karl R. and J. C. Eccles. *The Self and its Brain.* Berlin and New York: Springer International, 1977.

Port, Robert F and Timothy van Gelder (eds). *Mind as Motion: Explorations in the Dynamics of Cognition.* Cambridge, MA: MIT Press, 1995.

Porta, Giambattista della. *De Humana Physiogonomia.* 1586.

Porter, Roy. *The Making of Geology.* Cambridge: Cambridge University Press, 1977.

Porter, Roy. *Mind Forg'd Manacles: a History of Madness in England from the Restoration to the Regency.* London: Athlone, 1987.

Porter, Roy. 'Addicted to Modernity: Nervousness in the Early Consumer Society.' In J. Melling and J. Barry (eds), *Culture in History.* Exeter: University of Exeter Press, 1992, pp. 180–94.

Porter, Roy. 'Medical Lecturing in Georgian London.' *British Journal of the History of Science* 28 (1995): 91–100.

Porter, Roy. 'Nervousness, Eighteenth and Nineteenth Century Style: From Luxury to Labour.' In M. Gijswijt-Hofstra and R. Porter (eds), *Cultures of Neurasthenia* (2001), pp. 31–50.

Porter, Roy. *Flesh in the Age of Reason.* London: Penguin, 2003.

Pott, Percivall. *Observations... on injuries to... the head...* London: Hawes, Clarke and Collins, 1768.

Pratt, Samuel Jackson. *Emma Corbett: or, The miseries of civil war, by the author of Liberal opinions.* Dublin, 1780.

Pratt, Samuel Jackson. *The Paternal Present.* London, 1802.

Praz, Mario. *The Romantic Agony.* London: Oxford University Press, 1933.

Prichard, James Cowles. *A Treatise on Diseases of the Nervous System.* London: Underwood, 1822.

Procháska, George. *Dissertation on the Functions of the Nervous System.* London, 1784.

Pulteney, Richard. *A General View of the Writings of Linnaeus.* London, 1805.

Purcell, John. *A Treatise of Vapours, or, Hysterick Fits,* 2nd rev. edn. London: J. Johnson, 1707.

Putscher, Marielene. *Pneuma, Spiritus, Geist.* Wiesbaden: Steiner, 1974.

Quilley, Claude. *Callipaediae: or, An art how to have handsome children.* London: John Morphew, 1710.

Racamier, P. *Le psychanalyste sans divan: la psychanalyse et les institutions de sains psychiatriques.* Paris: Payat, 1970.

Rahman, F. *Avicenna's Psychology: an English Translation of Kitab al-Najat, Book II, Chapter VI with Historico-Philosophical Notes.* London: Oxford University Press, 1952.

Ralli, Augustus. *Critiques.* London: Longmans, 1927.

Ralph, James. *The fashionable Lady; or Harlequin's Opera.* London: [J. Watts], 1730.

Ralston, Gulliver. 'Richard Wagner's Pills and Potions'. Talk delivered at Oxford University, 4 May 2004.

Ramon y Cajal, Santiago. *Degeneration & Regeneration of the Nervous System translated and edited by Raoul M. May.* London: Oxford University Press, 1928.

Ramon y Cajal, Santiago. *Ramon y Cajal's Contribution to the Neurosciences Proceedings of the Symposium 'Horizons in Neuroscience', Honoring the 100th Aniversary of Santiago Ramon y Cajal's Research Career, 25–27 March, 1982, edited by Santiago Grisolla . . . et al.,* Amsterdam and New York: Elsevier, 1982.

Rather, L. J. 'G. E. Stahl's Psychological Physiology.' *Bulletin of the History of Medicine* 35 (1961): 37–49.

Rather, L. J. *Mind and Body in Eighteenth-Century Medicine: a Study Based on Jerome Gaub's De regimine mentis.* London: Wellcome Historical Medical Library, 1965.

Rather, L. J. 'Old and New Views of the Emotions and Bodily Changes,' *Clio Medica* I (1965): 1–25.

Rather, L. J. 'Some Relations between Eighteenth-Century Fiber Theory and Nineteenth-Century Cell Theory.' *Clio Medica* 4 (1969): 191–202.

Raulin, J. *Traité des affections vaporeuses du sexe.* Paris, 1758.

Read, Alexander. *A treatise of all the muscles of the whole body.* London, 1650.

Read, Herbert. *The Sense of Glory: Essays in Criticism.* Cambridge: Cambridge University Press, 1930.

Reichel, E. H. *De hypochondria et hysteria.* Jena, 1803.

Reid, J. *Essays on Hypochondriacal and Other Nervous Affections.* Philadelphia: M. Carey & Son, 1817.

Rey, Roselyne. *The History of Pain.* Trans. Louise Elliot Wallace et al. Cambridge, MA: Harvard University Press, 1993.

Reynolds, W. V. 'Johnson's Opinions on Prose Style.' *Review of English Studies* 9 (1933): 433–46.

Reynolds, W. V. 'The Reception of Johnson's Prose.' *Review of English Studies* 11 (1935): 145–62.

Richards, Anna. *The Wasting Heroine in German Fiction by Women 1770–1914.* Oxford: Clarendon Press, 2004.

Richardson, Alan. *British Romanticism and the Science of the Mind.* Cambridge: Cambridge University Press, 2001.

Richter, Simon. 'On the Threshold: G. S. Rousseau and the Discourses of Then and Now.' *The Eighteenth Century: Theory and Interpretation* 34(1) (1993): 85–95.

Ridgway, Ronald S. *Voltaire and Sensibility*. Montreal: McGill-Queen's University Press, 1973.

Riese, Walther. *Principles of Neurology in the Light of History and their Present Use*. New York: Coolidge Foundation Publishers, 1950.

Riese, Walther. *A History of Neurology*. New York: MD Publications, 1959.

Riskin, Jessica. *Science in the Age of Sensibility: the Sentimental Empiricists of the French Enlightenment*. Chicago: University of Chicago Press, 2002.

Risse, Guenter B. *Hospital Life in Enlightenment Scotland: Care and Teaching at the Royal Infirmary of Edinburgh*. Cambridge: Cambridge University Press, 1986.

Roberts, K. B. and Tomlinson, J. D. W. *The Fabric of the Body: European Traditions of Anatomical Illustration*. Oxford: Oxford University Press, 1992.

Robertson, William. *De hysteria*. Edinburgh, 1790.

Robinson, Brian. *A Treatise of the Animal Oeconomy*. Dublin: S. Powell, 1734.

Robinson, Mary. *Walshingham: or the Pupil of Nature, A Story*. London, 1797.

Robinson, Nicholas. *A New Method of Treating Consumptions ... Wherein all the Decays Incident to Human Bodies, Are Mechanically Accounted for ...* London: A. Bettesworth, 1727.

Robinson, Nicholas. *A new system of the spleen, vapours, and hypochondriack melancholy; wherein all the decays of the nerves, and lowness of the spirits are mechanically accounted for. To which is subjoined, a discourse upon the nature, cause, and cure of melancholy, madness, and lunacy. With a particular dissertation on the origine of the passions; the structure of the nerves ...* London, 1729.

Roelcke, Volker. 'Electrified Nerves, Degenerated Bodies: Medical Discourses on Neurasthenia in Germany, circa 1880–1914.' In Marijke Gijswijt-Hofstra and R. Porter (eds), *Cultures of Neurasthenia: From Beard to the First World War*. Amsterdam: Rodopi, 2001, pp. 177–98.

Roger, Jacques. *The Life Sciences in Eighteenth-Century French Thought; edited by Keith R. Benson; translated by Robert Ellrich*. Palo Alto: Stanford University Press, 1997.

Rogers, Pat. 'Shaftesbury and the Aesthetics of Rhapsody.' *British Journal of Aesthetics* 12 (1972): 244–57.

Rogers, Timothy. *A Discourse Concerning Trouble of Mind, and the Disease of Melancholy. To Which Are Annexed Letters from Several Divines, Relating to the Same Subject*. London: 1691; 2nd corr. edn, 1706.

Romberg, Moritz Heinrich. *A Manual of the Nervous Diseases of Man*. Trans. E. H. Sieveking. London: Sydenham Society, 1853.

Rosario II, Vernon A. 'Review of Sander Gilman, Helen King, Roy Porter, G. S. Rousseau, Elaine Showalter. *Hysteria Beyond Freud*.' In *Journal of the History of Medicine* 50 (1995): 418–20.

Rosario II, Vernon A. (ed.). *Science and Homosexualities*. New York: Routledge, 1997.

Rosario II, Vernon A. *The Erotic Imagination: French Histories of Perversity*. New York: Oxford University Press, 1997.

Rose, F. Clifford (ed.). *A Short History of Neurology: the British Contribution 1660–1910*. Oxford: Butterworth Heinemann, 1999.

Rose, F. Clifford and W. F. Bynum (eds). *Historical Aspects of the Neurosciences: a Festschrift for Macdonald Critchley*. New York: Raven Press, 1982.

Rosen, Charles. *Piano Notes: the Hidden World of the Pianist*. London: Allen Lane, 2003.

Rosen, George. 'Emotion and Sensibility in Ages of Anxiety: a Comparative Historical Review.' *American Journal of Psychiatry* 6(124) (1967): 771–83.

Rousseau, G. S. 'Doctors and Medicine in the Novels of Tobias Smollett,' PhD thesis, Princeton University, 1965.

Rousseau, G. S. 'Nerves, Spirits and Fibres: Toward the Origins of Sensibility.' In R. F. Brissenden (ed.), *Studies in the Eighteenth Century*. Canberra: The Australian National University Press, 1975, pp. 137–57.

Rousseau, G. S. 'Nervous Proliferation: the Expanding Role of the Nerves in the 1720s.' Talk delivered at Trinity College, Cambridge University, 1984.

Rousseau, G. S. (ed.). *The Languages of Psyche*. Berkeley and Los Angeles: University of California Press, 1986.

Rousseau, G. S. 'Discourses of the Nerve.' In F. Amrine (ed.), *Literature and Science as Modes of Expression*. Dordrecht: Kluwer Academic Publishers, 1989, pp. 29–60.

Rousseau, G. S. 'Cultural History in A New Key: Towards a Semiotics of the Nerve.' In Joan and Andrew Weir Pittock (eds), *Interpretation and Cultural History*. London: Macmillan, 1991, pp. 25–81.

Rousseau, G. S. 'The Perpetual Crises of Modernism and the Traditions of Enlightenment Vitalism: with a note on Mikhail Bakhtin.' In Frederick Burwick and Paul Douglass (eds), *The Crisis in Modernism: Bergson and the Vitalist Controversy*. Cambridge: Cambridge University Press, 1992, pp. 15–97.

Rousseau, G. S. 'Riddles of Interdisciplinarity: A Reply to Stanley Fish.' In Heinz Antor and Kevin Cope (eds), *Intercultural Encounters – Studies in English Literature*. Heidelberg: C. Winter Verlag, 1999, pp. 111–30.

Rousseau, G. S. 'The Inflected Voice: Attraction and Curative Properties.' In Penelope Gouk (ed.), *Musical Healing in Cultural Contexts*. Aldershot: Ashgate, 2000, pp. 92–112.

Rousseau, G. S. 'Homoplatonic, Homodepressed, Homomorbid: Some Further Genealogies of Same-Sex Attraction in Western Civilization.' In Katherine O'Donnell and Michael O'Rourke (eds), *Love, Sex, Intimacy and Friendship Between Men, 1550–1800*. Basingstoke: Palgrave, 2002, pp. 12–52.

Rousseau, G. S. 'Writing the History of the Emotions: Essay Review.' *History of Psychiatry* 15(3) (2004): 367–77.

Rouseau, G. S. 'The Decay of Scientific Theories: a Discursive Approach'. In John Heilbron (ed.), *Studies in Honor of Paolo Rossi on the Occasion of his Eightieth Birthday*. Florence: Académie internationale d'histoire des sciences, 2005.

Rousseau, G. S. '"In rapture writ": Alexander Pope and the Body of the Poet.' In David Womersley et al. (eds), *Studies on Alexander Pope in Honour of Howard Erskine-Hill*. Newark, DE: Delaware University Press, 2005.

Rousseau, G. S. 'Pope, the Body and Medicine.' In Pat Rogers (ed.), *A Companion to Alexander Pope*. Cambridge: Cambridge University Press, 2005.

Rousseau, G. S. 'Cullen's "Lady in Robes."' Forthcoming 2005.

Rousseau, G. S. and Roy Porter. *The Ferment of Knowledge*. Cambridge: Cambridge University Press, 1980.

Rousseau, G. S. et al. (eds). *Framing and Imagining Disease in Cultural History*. Basingstoke: Palgrave, 2003.

Rowe, George Robert. *Nervous Diseases, Arising from Liver and Stomach Complaints*. London, 1840.

Rowe, George Robert. *On the More Important Disorders of Females and Children*. London, 1857.

Rowley, William. *A Practical Treatise on Diseases of the Breasts of Women: Containing Directions for the Proper Management of Breasts during Lying-In.* London: F. Newbery, 1772.

Rowley, William. *A Treatise on Female, Nervous, Hysterical, Hypochondriacial, Bilious, Convulsive Disease; Apoplexy & Palsy with Thoughts on Madness & Suicide, etc.* London: C. Nourse, 1788.

Roy, P. et al. 'De l'hypochondrie.' *Archiv. Neurol.* 20 (1905): 166–83.

Rudolph, G. 'Diderot's Elemente der Physiologie.' *Gesnerus* 24 (1967): 24–45.

Rudolph, G. 'A propos de l'abrégé historique sur l'anatomie et la physiologie par Albrecht von Haller.' *Episteme* 4 (1970): 376–9.

Rush, Benjamin. *Essays, Literary, Moral & Philosophical...* Philadelphia: Thomas and Bradford, 1798.

Ruskin, John. *The Stones of Venice.* London: n. p, 1853.

Rutty, John. *Boerhaave's Men at Leyden and After.* Edinburgh: Edinburgh University Press, 1977.

Ryskamp, Charles, and Frederick Pottle. *Boswell: the Ominous Years 1774–1776.* New York: McGraw-Hill, 1963.

Sacey, Louis de. *A Discourse on Friendship in Three Books.* London, 1707.

Sacks, Oliver. *Migraine: the Evolution of a Common Disorder.* Berkeley and Los Angeles: University of California Press, 1985.

Sacks, Oliver. 'In the River of Consciousness.' 51:1 *New York Review of Books* 15 January 2004: 41–4.

Sade, Marquis de. *Oeuvres Complètes: Edition Definitive.* Paris: Cercle du livre précieux, 1966–.

Santeul, Louis de. *Des propriétés de la médecine, par rapport à la vie civile.* Paris: Briasson, 1739.

Savage, George Henry. *Insanity and Allied Neuroses.* London, 1884.

Savage-Smith, Emilie. 'Galen on Nerves, Veins and Arteries: a Critical Edition and Translation from the Arabic, with Notes...' Ann Arbor: University of Michigan, PhD thesis, 1969.

Savage-Smith, Emilie. 'Attitudes Towards Dissection in Medieval Islam.' *Journal of the History of Medicine and Allied Sciences* 50 (1995): 67–110.

Scat, John. 'An Enquiry into the Origin of the Gout.' *Critical Review* (November 1779): 347–54.

Schacht, H. O. *De melancholia hypochondriaca.* 1693.

Scheler, Max. *The Nature of Sympathy...* Trans. Peter Heath. London: Routledge and Kegan Paul, 1979.

Schenk, Hans G. *The Mind of the European Romantics.* London: Kegan Paul, Trench and Trübner, 1966.

Schofield, Alfred T. *Nerves in Disorder.* London: Funk & Wagnells, 1903.

Schofield, Alfred T. *Nervousness, a Review of the Moral Treatment of Disordered Nerves.* London: Rider, 1910.

Schreber, Daniel Paul. *Daniel Paul Schreber: Memoirs of My Nervous Illness.* London: Dawsons, 1955.

Schurig, Martin. *Spermatologia Historico-Medica, h.e. Seminis Humanis Consideratio Physico-Medico-Legalis...* Frankfurt, 1720.

Scot, John. *Remarkable cures, of gouty, bilious, and nervous cases, related by the patients themselves...* London, [1783].

Sekora, John. *Luxury: the Concept in Western Thought, Eden to Smollett.* Baltimore: Johns Hopkins University Press, 1977.

Sena, John F. *A Bibliography of Melancholy*. London: Nether Press, 1970.

Sena, John F. *The Best-Natured Man: Sir Samuel Garth, Physician and Poet*. New York: AMS Press, 1986.

Sennett, Richard. *Flesh and Stone: the Body and the City in Western Civilization*. London: Faber, 1994.

Seyffert, J. F. *De hypochondriasi*. Leipzig, 1824.

Sèze, Victor de. *Recherches physiologiques et philosophiques sur la sensibilité ou la vie animale*. Paris: Prault, 1786.

Sha, Richard C. 'Scientific Forms of Sexual Knowledge in Romanticism.' *Romanticism On the Net*. 23 August 2001, http://users.ox.ac.uk/~scat0385/23sha.html

Shamdasani, Sonu and Michael Münchow (eds). *Speculations after Freud: Psychoanalysis, Philosophy and Culture*. London: Routledge, 1994.

Shaw, Peter. *The Reflector: Representing Human Affairs as They Are; and may be Improved*. London, 1750.

Sherwood, Margaret. *Undercurrents of Influence in English Romantic Poetry*. Cambridge, MA: Harvard University Press, 1934.

Sherwood, Margaret. *Coleridge's Imaginative Conception of the Imagination*. Wellesley, MA: Hathaway, 1937.

Shorter, Edward. *A History of Women's Bodies*. London: Penguin, 1983.

Shorter, Edward. *From Paralysis to Fatigue: a History of Psychosomatic Illness in the Modern Era*. New York: Free Press, 1992.

Showalter, Elaine. *The Female Malady: Women, Madness and English Culture, 1830–1980*. London: Virago, 1987.

Shrewsbury, J. F. D. *The Plague of the Philistines, and Other Medical-Historical Essays*. London: Gollancz, 1964.

Shuttleworth, Sally. *Charlotte Bronte and Victorian Psychology*. Cambridge: Cambridge University Press, 1996.

Sickels, Eleanor Maria. *The Gloomy Egoist: Moods and Themes of Melancholy from Gray to Keats*. New York: Columbia University Press, 1932.

Siegel, Rudolph E. *Galen on Psychology, Psychopathology, and Function and Diseases of the Nervous System: an Analysis of his Doctrines, Observations and Experiments*. Basel and New York: Karger, 1973.

Siess, J. *Dissertatio sistens ideam pathematum hypochondriaco-hystericorum cum singulari huc faciente historia morbi*. Giessen, 1780.

Sill, Geoffrey M. 'Neurology and the Novel: Alexander Monro Primus and Secundus, *Robinson Crusoe*, and the Problem of Sensibility.' *Literature and Medicine* 16(2) (1997): 250–65.

Sill, Geoffrey M. *The Cure of the Passions and the Origins of the English Novel*. Cambridge: Cambridge University Press, 2001.

Silverman, Kaja. *Male Subjectivity at the Margins*. London and New York: Routledge, 1992.

Sims, James. 'Pathological Remarks upon Various Kinds of Alienation of Mind.' *Memoirs of the Medical Society of London*. 1799. V: 372–406.

Sims, James. 'Remarks on Nervous and Malignant Fevers.' *Memoirs of the Medical Society of London*. 1799. V: 222–38.

Singer, Charles. *A Short History of Anatomy and Physiology from the Greeks to Harvey*. New York: Dover, 1957.

Siraisi, Nancy. *Medieval and Early Renaissance Medicine*. Chicago: University of Chicago Press, 1990.

Slotkin, James Sydney. *Readings in Early Anthropology.* London: Routledge Library Editions, 2004. [rev. 1965]

Small, Helen. *Love's Madness: Medicine, the Novel and Female Insanity, 1800–1865.* Oxford: Clarendon Press, 1996.

Smith, Adam. *The Theory of Moral Sentiments.* Edinburgh: A Miller, 1759.

Smith, Charlotte. *The Wanderings of Warwick.* London, 1794.

Smith, Daniel. *A Dissertation upon the Nervous System to show its influence upon the Soul.* London, 1768.

Smith, Daniel. *An Apology to the Public for Commencing the Practice of Physic; Particularly in . . . Hysterical Cases.* London: Carnan & Newberry, 1770.

Smith, Daniel. *A Treatise on Melancholy and Nervous Disorders.* London, 1778.

Smith, Henry, M. D. *Nervous debility; its cause, consequences, and cure.* London: n.p., 1861.

Smith, Hugh. *An Essay on the Nerves. To which is added an essay on foreign teas.* London, 1780.

Smith, Hugh. *Essays on the Nerves illustrating their efficient, formal, material, and final causes; with a copper-plate . . . To which is added an essay on foreign teas . . . to demonstrate their pernicious consequences on the nerves . . . 2nd edn.* London: P.Norman, 1795.

Smith, Hugh. *Philosophical Inquiries into the Laws of Animal Life.* London: L. Davis, 1780.

Smith, Hugh. *A treatise on the use and abuse of mineral waters: also rules necessary to be observed by invalids who visit the chalybeate springs of the old and new Tunbridge Wells. Together with some remarks on the immoderate use of sea water.* London: G. Kearsly, [1776].

Smith, Logan Pearsall. *Words and Idioms: Studies in the English Language.* London: Constable, 1925.

Snow, C. P. *The Two Cultures and the Scientific Revolution: the Rede Lecture, 1959.* Cambridge: Cambridge University Press, 1959.

Snyder, Robert Lance. 'The Epistolary Melancholy of Thomas Gray.' *Biography* 2 (Spring 1979): 125–40.

Soemmering, Samuel Thomas von. *De basi encephali et originibus nervorum cranio egredientium libri quinque.* Gottingen, 1778.

Solmsen, Friedrich. 'Greek Philosophy and the Discovery of the Nerves.' *Museum Helvetica* 18 (1961): 150–67.

Solomon, Samuel, MD. *A Guide to Health; Or, Advice to both Sexes, In Nervous and Consumptive Complaints, Scurvy, Leprosy, and Scrofula; Also, On a Certain Disease and Sexual Debility . . . .* London: Matthews and Symonds, 1815.

Sonnenmayer, J. G. *De vero ortu mali hypochondriaci et hysterici.* 1769.

Sontag, Susan. 'Loving Dostoyevsky.' *The New Yorker* 2001: 98–105.

Spadafora, David. *The Idea of Progress in Eighteenth-Century Britain.* New Haven: Yale University Press, 1990.

Spillane, John D. *The Doctrine of the Nerves: Chapters in the History of Neurology.* Oxford, New York and Toronto: Oxford University Press, 1981.

Spurzheim, J. G. *Observations on the Deranged Manifestations of the Mind, or Insanity.* London, 1817.

Squire, Larry R. *The History of Neuroscience in Autobiography.* Washington, DC: Society for Neurosciences, 1996.

Squirrell, Robert, MD. *An Essay on Indigestion and its Consequences, Or Advice to Persons Affected with Debility of the Digestive Organs, Nervous Disorders, Gout, Dropsy, &c.* London, 1795.

Squirrell, R., MD. *Essay on Indigestion and its Consequences; or, Advice to Persons Affected by Debility of the Digestive Organs, Gout, Nervous Disorders, Dropsy, etc.* n.p., 1820. [eighth edition].

Stafford, Barbara. *Body Criticism: Imaging the Unseen.* Cambridge, MA: MIT Press, 1991.

Stafleu, Frans A. *Linnaeus and the Linnaeans: the Spreading of Their Ideas in Systematic Botany, 1735–1789.* Utrecht, 1971.

Staub, A. *Allgemeiner Leitfaden zur Bearbeitung der Hypochondrie und Hysterie.* Wurzburg, 1826.

Staum, Martin Sheldon. 'Cabanis and the Science of Man,' PhD thesis, Cornell University, 1971.

Steininger, J. A. *Centuria positionum medicarum de melancholia hypochondriaca.* Wittenberg, 1625.

Steno, Nicolaus. *Lectures on the Anatomy of the Brain.* Ed. Gustav Scherz. Copenhagen: Arnold Busk, 1960.

Stephanson, Raymond. 'Richardson's "Nerves": The Physiology of Sensibility in Clarissa.' *Journal of the History of Ideas* 49(2) (1988): 267–85.

Stephanson, Raymond. 'G. S. Rousseau as Cultural Historian.' *University of Toronto Quarterly* 62(3) (1993): 388–400.

Stephanson, Raymond. *The Yard of Wit: Male Creativity and Sexuality, 1650–1750.* Philadelphia: University of Pennsylvania Press, 2004.

Stephen, Leslie. *History of English Thought in the Eighteenth Century.* London: Smith, 1876.

Stephen, Leslie. *English Literature and Society in the Eighteenth Century.* London: Duckworth, 1904.

Stern, Bernhard J. *Social Factors in Medical Progress.* New York: Columbia University Press, 1927.

Sterne, Laurence. *A Sentimental Journey* with *The Journal to Eliza* and *A Political Romance.* Ed. Ian Jack. New York: Oxford University Press, 1968.

Sterne, Laurence. *Tristram Shandy.* Norton Critical Edition. New York: Norton, 1980.

Stevens, J. N. MD, reviewed by James Kirkpatrick, MD. 'An Essay on the Diseases of the Head and Neck to which is added a Dissertation on the Gout and Rheumatism.' *Monthly Review* (August 1758): 145–50.

Stevens, J. N. MD, reviewed by James Kirkpatrick, MD. 'A Treatise on the Medicinal Qualities of Bath-Waters.' *Monthly Review* (September 1758): 371–9.

Stevenson, John. *On the Morbid Sensibility of the Eye, commonly Called Weakness of Sight.* London, 1810.

Stevenson, William. 'A Successful Method of Treating the Gout.' *Critical Review* July, 1779: 31–6.

Sticotti, Antonio Fabio. *Garrick ou les acteurs anglois, ouvrage contenant des reflexions sur l'art dramatique, sur l'art de la representation, et le jeu des acteurs, traduit de l'anglois.* Paris: Lacombe, 1769.

Stock, R. D. *The Holy and the Daemonic from Sir Thomas Browne to William Blake.* Princeton: Princeton University Press, 1978.

Stockdale, P. *An Inquiry . . . Laws of Poetry.* London, 1778.

Stocker, Arnold. *Le traitement moral des nerveux.* Geneva: Rhone, 1945.

Swedenborg, Emmanuel. *Brain Considered Anatomically, Physiologically, and Philosophically, in Two Volumes trans. R. L. Tafel.* London: Swedenborg Society, 1882–87, rep. 1935.

Sweeney, Nicholas P. 'The Nervous Body and the Poetic Self: Poetry and Medical Literature 1660–1760.' DPhil. thesis, Oxford University, 2002.

Symonds, John Addington, MD. *Sleep and Dreams, Two Lectures.* London: 1851, rep. 1857.

Sypher, Wylie. *Enlightened England.* New York: Norton, 1947.

Szasz, Thomas. *Schizophrenia.* New York: Basic Books, 1976.

Tate, Alan (ed.). *Samuel Johnson: a Collection of Critical Essays.* New York: Random House, 1965.

Taussig, Michael. *The Nervous System.* New York: Routledge, 1992.

Taylor, J. *Montaigne and Medicine.* New York: Paul Haeber, 1921.

Taylor, John. *Le mechanisme ou le nouveau traité de l'anatomie du globe de l'oeil.* Paris, 1737.

Taylor, John. *The History of the Travels and Adventures of the Chevalier John Taylor.* London: Williams, 1761.

Teich, M. 'Circulation, Transformation, Conversation of Matter and the Balancing of the Biological World in the Eighteenth Century.' *Ambix* 29(1) (1982): 17–28.

Thomas, Keith. *Man and the Natural World: Changing Attitudes in England 1500–1800.* London: Allen Lane, 1983.

Thompson, Caroline. 'Sensibility.' *Psyche* xv (1935): 46–161.

Thompson, E. P., Hay Douglas, Linebaugh Peter, Rule John G., and Winslow, Cal. *Albion's Fatal Tree: Crime and Society in 18th Century England.* New York: Pantheon, 1975.

Thompson, Thomas, MD. *An Historical and Critical Treatise of the Gout... showing... Danger and Presumptions of all Philosophical Systems in Physick...* London, 1742.

Thomson, Alexander. *An Enquiry into the Nature, Causes, and Method of Cure, of Nervous Disorders.* London, 1781.

Thomson, James. *The Four Seasons, and Other Poems.* London, 1734.

Thomson, John, MD. *An Account of the Life, Lectures, and Writings of William Cullen, MD... 1832,* 2 vols. Edinburgh, 1859.

Thomson, William A. R. *Spas That Heal.* London: A. and C. Black, 1978.

Timbs, John. *Things not Generally Known, Familiarly Explained.* London, 1858.

Timpanaro, Sebastiano. *On Materialism, translated by Lawrence Garner.* London: Humanities Press, 1975.

Tissot, S. A. D. *Advice to the People in General.* Trans. J. Kirkpatrick. Dublin: J. Hoey, 1766.

Tissot, S. A. D., *An Essay on Diseases Incidental to Literary and Sedentary Persons. With Proper Rules for Preventing their Fatal Consequences. And Instructions for their Care. Now Translated into English.* London, 1768.

Tissot, S. A. D., *Traité des nerfs et de leurs maladies. Tome I., partie I. [-Tome II., partie II.* Lausanne, 1784.

Todd, Edwin M. *The Neuroanatomy of Leonardo da Vinci.* Park Ridge, IL: American Association of Neurological Surgeons, 1991.

Todd, Janet M. *Sensibility: an Introduction.* London: Methuen, 1986.

Tompkins, J. M. S. *The Popular Novel in England, 1770–1800*. London: Constable, 1932.

Totman, Richard. *Social Causes of Illness*. London: Souvenir Press, 1979.

Tournefort, Joeseph Pitten de. *Traité de la Matière medicale*. Paris, 1717.

Toynbee, Paget and Leonard Whibley (eds). *Correspondence of Thomas Gray*, 3 vols. Oxford: Clarendon Press, 1935.

Trahard, Pierre. *Les maîtres de la sensibilité française au XVIIIe siècle La Sensibilité révolutionnaire*. Paris: Boivin & cie, 1931–33.

Treipsac De Vergy, Pierre Henri. *The Lovers: or the Memoirs of Lady Sarah B[urhury] and the Countess P[ercy]*. London, 1769.

Trotter, Thomas. 'A Case of Gastrodynia.' *London Medical and Physical Journal* 68 and 75 (1804–5): 26–47 and 18–23.

Trotter, Thomas. *A View of the Nervous Temperament being a practical enquiry into the increasing prevalence, prevention, and treatment of those diseases commonly called nervous . . .* London: E. Walker, 1807 [2nd edn, 1807].

Trotter, Thomas. *An Essay Medical, Philosophical, and Chemical on Drunkeness and its Effects on the Human Body*. London, 1803. Reprinted Tavistock Classics in the History of Psychiatry, New York: Routledge, 1988.

Trumbach, Randolph. 'The Birth of the Queen: Sodomy and the Emergence of Gender Equality in Modern Culture, 1600–1750.' In Martin Duberman et al. (eds), *Hidden from History: Reclaiming the Gay and Lesbian Past*. New York: New American Library, 1989, pp. 129–40, 509–11.

Trusler, Reverend John. *Memoirs of the Life of the Rev. Dr. Trusler written by himself*. Bath: John Browne, 1806.

Tryon, Thomas. *A treatise of dreams & visions: wherein the causes natures and uses of nocturnal representations, and the communications both of good and evil angels, as also departed souls, to mankinde, are theosophically unfolded . . . . By Philotheos Physiologus*. London, 1689.

Tryon, Thomas. *Tryon's Letters, domestick and foreign, to several persons of quality: occasionally distributed in subjects, viz. philosophical, theological and moral*. London: Printed for G. Conyers & E. Harris, 1700.

Tucker, Susie I. *Protean Shape: a Study in Eighteenth-Century Vocabulary and Usage*. London: Athlone, 1967.

Tucker, Susie I. *Enthusiasm: a Study in Semantic Change*. Cambridge: Cambridge University Press, 1972.

Turner, W. *De morbo hypochondriaco*. Edinburgh, 1756.

Tusso, Joseph Clement. *De l'influence des passions de l'âme dans les maladies, et des moyens d'en corriger les mauvais effets . . .* Paris: Chez Amand-Koenig, 1798.

Uglow, Jenny. *The Lunar Men: the Friends who Made the Future, 1730–1810*. London: Faber, 2002.

Valéry, Paul. 'Discourse aux chirurgiens.' *Oeuvres Complètes. Volume 1*. Paris: Bibliothèque de la Pléiade, 1890.

Valéry, Paul. *Masters and Friends*. Princeton: Princeton University Press, Bollingen Series, 1968.

van Sant, Ann. *Eighteenth-Century Sensibility and the Novel: the Senses in Social Context*. New York: Cambridge University Press, 1993.

Vaucanson, Jacques de. *An Account of the Mechanism of an Automaton or Image Playing on the German Flute*. Trans. J. T. Desaguliers. London: T. Parker, 1742.

Vaughan, Meghan. *Curing Their Ills: Colonial Power and African Illness.* Stanford, CA: Stanford University Press, 1991.

Verity, Robert. *Changes Produced in the Nervous System by Civilization, Considered According to the Evidence of Physiology and the Philosophy of History.* London, 1837.

Verity, Robert. *Subject and Object; as Connected with our Double Brain, and a New Theory of Causation.* London, 1870.

Vesalius, Andreas. *The Epitome of Andreas Vesalius.* New York: Macmillan, 1949.

Vesalius, Andreas. *The Illustrations from the Works of Andreas Vesalius.* Ed. C. M. Saunders and C. D. O'Malley. New York: Dover, 1973.

Vila, Anne C. *Enlightenment and Pathology: Sensibility in the Literature and Medicine of Eighteenth-Century France.* Baltimore and London: Johns Hopkins University Press, 1998.

Vincent, Jean-Didier. *The Biology of Emotions.* Trans. John Hughes. Oxford: Blackwell, 1990.

Vincent-Buffault, Anne. *The History of Tears: Sensibility and Sentimentality in France.* London: Macmillan, 1991.

Vizzard, Michelle. 'Of Mimicry and Woman: Hysteria and Anticolonial Feminism in Tsitsi Dangaremba's *Nervous Conditions.' Journal of the South Pacific Association for Commonwealth Literature and Language Studies* 36 (1993): 10–19.

Vogli, Giovanni Giacinto. *Fluidi nerve i historia, authore J-H Vogli.* Bononiae Studiorum: Typis Julii Borzaghi, 1720.

Vrettos, Athena. *Somatic Fictions: Imagining Illness in Victorian Culture.* Stanford: Stanford University Press, 1995.

Wade, Ira. *The Clandestine Organization and Diffusion of Philosophic Ideas In France from 1700 to 1750.* Princeton and London: Princeton University Press, 1938.

Wagner, Peter (ed). *Christopher Anstey, the New Bath Guide.* Hildesheim: Georg Olms Verlag, 1989.

Wahl, L. de. *De causa hypochondriae proxima.* Berlin, 1832.

Wanley, Nathaniel. *The wonders of the little world: or, a general history of man.* London, 1678.

Warton, Joseph. *An Essay on the Genius and Writings of Pope.* London, 1762.

Wassermann, H. P. *Ethnic Pigmentation: Historical Physiological and Clinical Aspects.* Amsterdam: Excerpta Medica, 1974.

Waterhouse, B. *On the Principle of Vitality. A Discourse Delivered in the First Church in Boston, Tuesday, June 8th, 1790.* Boston, 1790.

Waters, Henry G. *The Nervous System of Jesus. By Salvaròna.* Langhorne: n.p., 1907.

Watt, Ian. *The Rise of the Novel.* Berkeley and Los Angeles: University of California Press, 1967.

Webb, Daniel. *An Inquiry into the Beauties of Painting.* London, 1760.

Webster, Charles. *The Great Instauration: Science, Medicine and Reform, 1626–1660.* London: Duckworth, 1975.

Wedlin, Michael V. *Mind and the Imagination in Aristotle.* New Haven: Yale University Press, 1989.

Weigelius, B. N. *De malo hypochondriaco.* 1745.

Weisskopf, Victor. *Knowledge and Wonder: the Natural World as Man Knows It.* Cambridge, MA: MIT Press, 1979.

Wharton, Thomas. *Thomas Wharton's Adenographia translated from the Latin by Stephen Freer, with an historical introduction by Andrew Cunningham . . . .* Oxford: Clarendon Press, 1996.

Whitehead, John. *An Essay on Liberty and Necessity.* London: Printed by R. Hawes, 1775.

Whiting, Sydney. *Memoirs of a Stomach, Written by Himself, That All Who Eat May Read. With Notes, Critical and Explanatory, by a Minister of the Interior.* London: W. E. Painter, 1853.

Whyte, Lancelot Law. *The Unconscious before Freud.* New York: Basic Books, 1960.

Whytt, Robert. *An Essay on the Vital and Other Involuntary Motions of Animals.* Edinburgh: Hamilton, 1751.

Whytt, Robert. *Observations on the nature, causes, and cure of those disorders which have been commonly called nervous, hypochondriac, or hysteric, to which are prefixed some remarks on the sympathy of the nerves,* 2nd edn. 8 vols. Edinburgh: T. Becket & P. Du Hondt, 1765.

Whytt, Robert. *Les vapeurs et maladies.* Trans. Achille Guillame Le Begue de Presle. Paris, 1767.

Wierzbicka, Anna. *Emotions Across Languages and Cultures: Diversity and Universals.* Cambridge: Cambridge University Press, 1999.

Wilkinson, Charles Henry. *Essays, Physiological and Philosophical, on the Distortion of the Spine, the Motive Power of animals, the Fallacy of the Senses, and the Properties of Matter.* London: S. Low, 1798.

Williams, A. N. and R. Sunderland, 'Thomas Willis: the First Paediatric Neurologist?' *Archives of Disease in Childhood* 85 (2001): 506–9.

Williams, Charles. *Insanity: Its Causes and Prevention.* London: Ambrose, 1908.

Williams, J. [M. D.]. 'Advice to People afflicted with the Gout.' *Critical Review* December 1773: 423–30.

Williams, J. H. *Requiem for a Great Killer: the Story of Tuberculosis.* London: Health Horizon, 1973.

Williams, Raymond. *Keywords: a Vocabulary of Culture and Society.* London: Fontana, 1976.

Willis, Thomas. *Affectionum quae dicuntur hystericae et hypochondriacae, pathologia spasmodica vindicata, contra responsionem epistolarem Nathanael Highmori.* Leiden, 1671.

Willis, Thomas. *The Anatomy of the Brain and Nerves,* tercentary edition. Ed. William Feindel et al. Montreal: McGill University Press, 1965, 2 vols. Rep. 1978.

Willis, Thomas. *The remaining medical works of that famous and renowned physician Dr Thomas Willis of Christ-Church in Oxford, and Sidley professor of Natural Philosophy in the famous University. Viz. I. Of fermentation. II. Of feavours. III. Of urines. IV. Of the accension of the bloud. V. Of musculary motion. VI. Of the anatomy of the brain. VII. Of the description and use of the nerves. VIII. Of convulsive diseases. With large alphabetical tables for the whole, and an index for the explaining all the hard and unusual words and terms of art, derived from the Latine, Greek, or other languages, for the benefit of the meer English reader* ... London: T. Dring, 1681. [Samuel Pordage translation]

Willis, Thomas. *Two discourses concerning the soul of brutes: which is that of the vital and sensitive man.* Gainesville, FLA: Scholars' Facsimiles, 1971. [Samuel Pordage translation].

Wilson, Andrew. *Human nature surveyed by philosophy and revelation, in 2 essays, by a gentleman [A. Wilson].* London, 1758.

Wilson, Andrew. *Short Remarks upon Autumnal Disorders of the Bowels.* Newcastle upon Tyne: J. White and T. Saint, 1765.

Wilson, Andrew. *The Origin of Nerves. A Lecture Delivered Before the Sunday Lecture Society on 24 Dec. 1878* ... London, 1879.

Wilson, Edward O. *On Human Nature.* Cambridge, MA: Harvard University Press, 1978.

Wilson, Elizabeth A. *Neural Geographies: Feminism and the Microstructure of Cognition.* London: Routledge, 1998.

Wilson, Frances. *Harriette Wilson: the Woman who Blackmailed the King.* London: Faber, 2003.

Wilson, M. D. 'Body and Mind from the Cartesian Point of View.' In R. W. Rieber (ed.), *Body and Mind. Past, Present and Future.* New York, 1980.

Wilson, Philip K. '"Out of Sight, out of Mind?": the Daniel Turner–James Blondel Dispute over the Power of the Maternal Imagination.' *Annals of Science* 49 (1992): 63–85.

Wiltshire, John. *Samuel Johnson and the Medical World: the Doctor and the Patient.* New York: Cambridge University Press, 1991.

Wimsatt, W. K. *Philosophic Words: a Study of Style and Meaning in the Rambler and Dictionary of Samuel Johnson.* New Haven: Yale University Press, 1948.

Winn, James A. *A Window in the Bosom: the Letters of Alexander Pope.* Hamden, CT: Archon, 1977.

Winters, W. *Memoirs of the Life and Writings of the Rev. A. M. Toplady.* London, 1778.

Wischke, C. L. *Mali hypochondriaci veri ac nervosi signa et diagnosis.* 1795.

Woof, Lawrence. 'Italian Opera and English Oratorio as Cultural Discourses within Eighteenth-Century English Literature, with Particular Reference to the Novels of Samuel Richardson and Fanny Burney.' DPhil. thesis, University of Oxford, 1994.

Woolf, Virginia, *The Common Reader, Second Series,* ed. Andrew McNeillie. London: Hogarth Press, 1986.

Worbs, Michael. *Nervenkunst: Literatur und Psychoanalyse im Wien der Jahrhundertwende.* Frankfurt: Europäische Verlagsanstalt, 1983.

Worthington, Rev. Richard. *Disquisitions on several subjects; . . . on Reason and Instinct.* London, 1787.

Wright, Walter. *Sensibility in English Prose Fiction. Illinois Studies in Language and Literature,* vol. 22. Urbana: University of Illinois Press, 1937.

Wright, Charles D. 'Melancholy Duffy and Sanguine Sinico: Humors in "A Painful Case."' *James Joyce Quarterly* III (1966): 171–81.

Wrigley, Richard and George Revill (eds). *Pathologies of Travel.* Clio Medica, vol. 56. Amsterdam: Rodopi, 2000.

Young, J. Z. *Doubt and Certainty in Science: a Biologist's Reflections on the Brain – the Reith Lectures for 1950.* Oxford: Clarendon Press, 1951.

Young, J. Z. *An Introduction to the Study of Man.* Oxford: Clarendon Press, 1971.

Young, J. Z. *The Anatomy of the Nervous System of Octopus Vulgaris.* Oxford: Clarendon Press, 1971.

Young, J. Z. *The Brains and Lives of Cephalopods.* Oxford: Oxford University Press, 2003.

Young, Robert M. *Mind, Brain, and Adaption in the 19th Century: Cerebral Localization and its Biological Context from Gall to Ferrier.* New York: Oxford University Press, 1990.

Yourcenar, Marguerite. 'Diagnostic of Europe.' *Bibliothèque Universelle et Revue de Genève* 68 (June 1929): 745–52.

Zacchia, Paolo. *De' mali hipochondriaci: libri tre.* Rome: Mascardi, 1651.

Zeviani, G. V. *Del flato a favore degli ipocondriaci.* Verona, 1794.

Zopef, J. C. *De malo hypochondriaco.* 1676.

Index

Adair, James Makkitrick, 27–8, 54–7, 66, 256, 262–81 passim
Addison, Joseph, 5
Aikin, Lucy, 31
Akenside, Mark, 34, 50
Amiel, Fréderic, 61
Animal spirits, 3–8, 11–33, 159, 166–81 passim, 255, 290
  and the Middle Ages, 18–22
  and 'juicy canals', 21
Apollo, 48
Aquinas, Thomas, 15
Aristotle, 10
Armstrong, John, 34, 50, 113, 131–2
Arne, Thomas, 51
Arnold, Matthew, 177
Arnold, Thomas, 42
  'muscular Christianity', 42
Augustine, Saint, 15
Austen, Jane, 44, 56
Avicenna, 12, 24

Bach, J. S., 48
Bachelard, Gaston, 343
Bacon, Francis, 15, 29, 175
Bayle, Pierre, 25
Beckett, Samuel, 61
Beethoven, Ludwig van, 7, 48, 52
Bell, George, 32
  The Anatomy of the Brain, 32
Benedetti, Alessandro, 13
Bergson, Henri, 43–4
Bienville, J. T. D. de, 185–210, 326
Berkeley, Bishop George, 51
Bienville, J. D. T., see nymphomania
Blackmore, Richard, 23, 90
  and nerves, 23
  as poet, 23
Blake, William, 31, 99, 270, 273, 279
  Crazy Jane, 270
Blumenbach, Johann Friedrich, 39, 55

Boerhaave, Hermann, 29, 158–9, 175, 194–8
Boyle, Robert, 104
  The Boyle Lectures, 49–50
Brain, 23–68 passim
Brocklesby, William, 46–50
Broussais, F. J. V., 55
Brontë, Charlotte and Emily, 56
Brown, John, 35
  and Brunonian medicine, 35
  and Coleridge, 35–8
Browne, Dr Richard (physician), 48–50
Browne, Sir Thomas, 5
Burette, Pierre-Jean, 47–8
Burke, Edmund, 16–17
Burney, Charles, 47
Burton, Robert, 175
  Anatomy of Melancholy, 175
Butler, Samuel, 43–4
Byron, George Gordon, 42

Calvinists, 17
Canaveri, Francesco, 64
Carlyle, Thomas, 61
da Carpa, 15
Cheyne, George, 13, 16–23, 27–38, 54–113 passim, 158–73, 193–7, 253–72
Chopin, Frederic, 7
Christ, 10–17 passim
  see also Henry Waters
Churchland, Patricia, 211
Cicero, 40–1
Coleridge, Samuel Taylor, 8, 36–58, 86, 100, 103, 159, 177, 273, 289
Collins, William, 40
corpora fabrica, 13–15, 17
Cotugno, Domenico, 13
Cousin, Victor, 55
Cowper, William (anatomist), 29–30
Cowper, William (poet), 40
Crane, R. S., 166, 179

390

Vico, 55
Victoria, Queen, 6
Vitalism, 18, 53
Volta, Alessandro, 25–6

Wagner, Richard, 52–3
  The Ring, 53
  The 'Wagner phenomenon',
    52–3
Walpole, Horace, 30
Warburton, Bishop William,
  26, 42
Waters, Henry G. ('Salvarona'), 65
Watson, Francis, 342
Whytt, Robert, 35–6, 224–36, 261

Willis, Thomas, 3–8 passim, 157–73,
  244, 265, 277, 282–3
  and animal spirits, 21–49
    passim
  and sensibility, 157–73
Wilson, Harriet, 30–1
Winslow, Jacob, 13
Wood, Dr., 27
Woolf, Virginia, 61, 346
Wordsworth, William, 8, 55–9,
  100, 177

Young, J. Z., 5
Young, Robert M., 211
Yourcenar, Marguerite, 65–6